Pharmaceutical Dosage Forms

Pharmaceutical Dosage Forms

Parenteral Medications
Third Edition

Volume 3
Regulations, Validation and the Future

Edited by

Sandeep Nema
Pfizer, Inc.
Chesterfield, Missouri, U.S.A.

John D. Ludwig
Pfizer, Inc.
Chesterfield, Missouri, U.S.A.

CRC Press
Taylor & Francis Group
Boca Raton London New York

CRC Press is an imprint of the
Taylor & Francis Group, an **informa** business

First published in paperback 2024

First published 2010 by CRC Press
2385 NW Executive Center Drive, Suite 320, Boca Raton FL 33431

and by CRC Press
4 Park Square, Milton Park, Abingdon, Oxon, OX14 4RN

CRC Press is an imprint of Taylor & Francis Group, LLC

© 2010, 2024 Taylor & Francis Group, LLC

A CIP record for this book is available from the British Library.

Library of Congress Cataloging-in-Publication Data available on application

ISBN: 978-1-4200-8653-9 (Set)
ISBN: 978-1-4200-8643-0 (hbk) (vol. 1)
ISBN: 978-1-4200-8645-4 (hbk) (vol. 2)
ISBN: 978-1-4200-8647-8 (hbk) (vol. 3)
ISBN: 978-1-03-292201-0 (pbk) (vol. 3)
ISBN: 978-0-429-14227-7 (ebk)

DOI: 10.3109/9781420086485

**Visit the Taylor & Francis Web site at
http://www.taylorandfrancis.com**

**and the CRC Press Web site at
http://www.crcpress.com**

We dedicate this work to those who have inspired us.
To my parents Walter and Ruth Ludwig and my wife Sue Ludwig
To my parents Hari and Pratibha Nema and my wife Tina Busch-Nema

Foreword

I was a faculty member at the University of Tennessee and a colleague of Dr. Kenneth Avis when he conceived, organized, and edited (along with H.A. Lieberman and L. Lachman) the first edition of this book series that was published in 1984. It was so well received by the pharmaceutical science community that an expanded three-volume second edition was published in 1992. Dr. Avis did not survive long enough to oversee a third edition, and it was questionable whether a third edition would ever be published until two of his graduate students, Drs. Nema and Ludwig, took it upon themselves to carry on Dr. Avis' tradition.

Their oversight of this third edition is work that their mentor would be highly pleased and proud of. From 29 chapters in the second edition to 43 chapters in this new edition, this three-volume series comprehensively covers both the traditional subjects in parenteral science and technology as well as new and expanded subjects. For example, separate chapter topics in this edition not found in previous editions include solubility and solubilization, depot delivery systems, biophysical and biochemical characterization of peptides and proteins, container-closure integrity testing, water systems, endotoxin testing, focused chapters on different sterilization methods, risk assessment in aseptic processing, visual inspection, advances in injection devices, RNAi delivery, regulatory considerations for excipients, techniques to evaluate pain on injection, product specifications, extractables and leachables, process analytical technology, and quality by design.

The editors have done an outstanding job of convincing so many top experts in their fields to author these 43 chapters. The excellent reputations of the authors and editors of this book will guarantee superb content of each chapter. There is no other book in the world that covers the breadth and depth of parenteral science and technology better than this one. In my opinion, the editors have achieved their primary objectives—publishing a book that contains current and emerging sterile product development and manufacturing information, and maintaining the high standard of quality that readers would expect.

Michael J. Akers
Baxter BioPharma Solutions
Bloomington, Indiana, U.S.A.

Preface

Pharmaceutical Dosage Forms: Parenteral Medications was originally published in 1984 and immediately accepted as a definitive reference in academic institutions and the pharmaceutical industry. The second edition was published in 1993. The ensuing years have produced incredible technological advancement. Classic small-molecule drugs are now complemented by complex molecules such as monoclonal antibodies, antibody fragments, aptamers, antisense, RNAi therapeutics, and DNA vaccines. There have been significant innovations in delivery devices, analytical techniques, in-silico modeling, and manufacturing and control technologies. In addition, the global regulatory environment has shifted toward greater emphasis on science-based risk assessment as evidenced by the evolving cGMPs, quality by design (QbD), process analytical technology (PAT), continuous processing, real time release, and other initiatives. The rapidly changing landscape in the parenteral field was the primary reason we undertook the challenging task of updating the three volumes. Our objectives were to (*i*) revise the text with current and emerging sterile product development and manufacturing science and (*ii*) maintain the high standard of quality the readers expect.

The third edition not only reflects enhanced content in all the chapters, but also more than half of the chapters are new underscoring the rapidly advancing technology. We have divided the volumes into logical subunits—volume 1 addresses formulation and packaging aspects; volume 2, facility design, sterilization and processing; and volume 3, regulations, validation and future directions. The authors invited to contribute chapters are established leaders with proven track records in their specialty areas. Hence, the textbook is authoritative and contains much of the collective experience gained in the (bio)pharmaceutical industry over the last two decades. *We are deeply grateful to all the authors who made this work possible.*

Volume 1 begins with a historical perspective of injectable drug therapy and common routes of administration. Formulation of small molecules and large molecules is presented in depth, including ophthalmic dosage forms. Parenteral packaging options are discussed relative to glass and plastic containers, as well as elastomeric closures. A definitive chapter is provided on container closure integrity.

Volume 2 presents chapters on facility design, cleanroom operations, and control of the environment. A chapter discussing pharmaceutical water systems is included. Key quality attributes of sterile dosage forms are discussed, including particulate matter, endotoxin, and sterility testing. The most widely used sterilization techniques as well as processing technologies are presented. Volume 2 concludes with an in-depth chapter on lyophilization.

Volume 3 focuses on regulatory requirements, risk-based process design, specifications, QbD, and extractables/leachables. In addition, we have included chapters on parenteral administration devices, siRNA delivery systems, injection site pain assessment, and control, PAT, and rapid microbiology test methods. Volume 3 concludes with a forward-looking chapter discussing the future of parenteral product manufacturing.

These three volumes differ from other textbooks in that they provide a learned review on developing parenteral dosage forms for *both* small molecules and biologics. Practical guidance is provided, in addition to theoretical aspects, for how to bring a drug candidate forward from discovery, through preclinical and clinical development, manufacturing, validation, and eventual registration.

The editors wish to thank Judy Clarkston and Lynn O'Toole-Bird (Pfizer, Inc.) for their invaluable assistance and organizational support during this project, and Sherri Niziolek and Bianca Turnbull (Informa Healthcare) for patiently leading us through the publishing process.

We also acknowledge the assistance of Pfizer, Inc. colleagues Lin Chen and Min Huang for reviewing several of the chapters.

We would like to express special gratitude to the late Kenneth E. Avis (University of Tennessee College of Pharmacy) for his dedication to teaching and sharing practical knowledge in the area of parenteral medications to so many students over the years, including us. Finally, we acknowledge the contributions of Dr Avis, Leon Lachman, and Herbert A. Lieberman who edited the earlier editions of this book series.

Sandeep Nema
John D. Ludwig

Contents

Contributors

James Agalloco Agalloco & Associates, Belle Mead, New Jersey, U.S.A.

James Akers Akers Kennedy & Associates, Kansas City, Missouri, U.S.A.

Amit Banerjee BioTherapeutics Pharmaceutical Sciences, Pfizer, Inc., Chesterfield, Missouri, U.S.A.

Karoline Bechtold-Peters Process Science Department/Pharma Development, Boehringer Ingelheim Pharma GmbH & Co. KG, Biberach, Germany

Michael Bergren JHP Pharmaceuticals LLC, Rochester, Michigan, U.S.A.

Gayle A. Brazeau Department of Pharmacy Practice and Pharmaceutical Sciences, School of Pharmacy and Pharmaceutical Sciences, State University of New York, Buffalo, New York, U.S.A.

Ronald J. Brendel Covidien, Hazelwood, Missouri, U.S.A.

James J. Collins Eli Lilly and Company, Indianapolis, Indiana, U.S.A.

Tony Cundell Schering-Plough Research Institute, Union, New Jersey, U.S.A.

Lan Feng UNC Eshelman School of Pharmacy, University of North Carolina at Chapel Hill, Chapel Hill, North Carolina, U.S.A.

Donna L. French Genentech, Inc., South San Francisco, California, U.S.A.

Thomas Garcia PharmaTherapeutics Pharmaceutical Sciences, Pfizer, Inc., Groton, Connecticut, U.S.A.

Michael Gorman St. Louis Laboratories, Pfizer, Inc., Chesterfield, Missouri, U.S.A.

Pramod Gupta Bausch & Lomb, Rochester, New York, U.S.A.

Carol F. Kirchhoff BioTherapeutics Pharmaceutical Sciences, Pfizer, Inc., Chesterfield, Missouri, U.S.A.

Jessica Klapa Department of Pharmaceutical Sciences, School of Pharmacy and Pharmaceutical Sciences, University at Buffalo, State University of New York, Buffalo, New York, U.S.A.

Russell Madsen The Williamsburg Group, LLC, Gaithersburg, Maryland, U.S.A.

Vince McCurdy PharmaTherapeutics Pharmaceutical Sciences, Pfizer, Inc., Groton, Connecticut, U.S.A.

Russell J. Mumper   UNC Eshelman School of Pharmacy, University of North Carolina at Chapel Hill, Chapel Hill, North Carolina, U.S.A.

Terry E. Munson   PAREXEL Consulting, PAREXEL International, LLC, Waltham, Massachusetts, U.S.A.

Sandeep Nema   BioTherapeutics Pharmaceutical Sciences, Pfizer, Inc., Chesterfield, Missouri, U.S.A.

Roger Nosal   PharmaTherapeutics Pharmaceutical Sciences, Pfizer, Inc., Groton, Connecticut, U.S.A.

Diane M. Paskiet   West Pharmaceutical Services, Inc., Lionville, Pennsylvania, U.S.A.

Satish K. Singh   BioTherapeutics Pharmaceutical Sciences, Pfizer, Inc., Chesterfield, Missouri, U.S.A.

Edward J. Smith   Packaging Science Resources, King of Prussia, Pennsylvania, U.S.A.

Maria Toler   BioTherapeutics Pharmaceutical Sciences, Pfizer, Inc., Chesterfield, Missouri, U.S.A.

Geert Verdonk   Schering-Plough Pharmaceuticals, Oss, The Netherlands

1 | cGMP regulations of parenteral drugs

Terry E. Munson

INTRODUCTION

This chapter presents an overview of the current Good Manufacturing Practice (cGMP) regulations for parenteral drugs. Since most of the major world regulatory authorities follow either the U.S. or the European Union (EU) model for their GMP regulations, this chapter will focus only on these two regulations.

U.S. REGULATIONS

Food, Drug, and Cosmetic Act

In the United States, the law that is violated when a parenteral drug product is not manufactured according to cGMPs is the Food, Drug, and Cosmetic Act (Act). Failure to follow GMPs is covered under Section 501, Adulterated Drugs and Devices section of the Act. Section 501(a)(2)(B) states:

> A drug or device shall be deemed to be adulterated—... if it is a drug and the methods used in, or the facilities or controls used for its manufacture, processing, packing, or holding do not conform to or are not operated or administered in conformity with current good manufacturing practice to assure that such drug meets the requirements of this Act as to safety and has the identity and strength, and meets the quality and purity characteristics, which it purports or is represented to posses....

This is the section of the law that requires manufacturers to produce drug in conformity with GMP practices. Adverse observations from manufacturing site inspections typically fall under this section of the Act.

Two other sections of the action that should be noted, although they do not directly apply to the GMP regulations, are as follows:

> Section 501(b): "A drug or device shall be deemed to be adulterated—... If it purports to be or is represented as a drug the name of which is recognized in an official compendium, and its strength, quality, or purity falls below, the standards set forth in such compendium."
>
> Section 501(c): "A drug or device shall be deemed to be adulterated—... If it is not subject to the provisions of paragraph (b) of this section and its strength differs from, or its purity or quality falls below, that which it purports or is represented to posses."

Sections 501(b) and (c) citations result from the analysis of drug samples picked up either during inspections or at end-user sites in one of the Food and Drug Administration (FDA) laboratories.

To fail the requirements in 501(b), the product must be tested exactly by the methods in the compendial monograph. One key element of this section is that the product only needs to be represented in the compendia to fall under the jurisdiction of this section, irrespective of whether or not it purports to be United States Pharmacopeia (USP)/National Formulary (NF). For example, Sodium Chloride Injection USP, Bacteriostatic Sodium Chloride Injection, and allergenic extract diluent (0.9% sodium chloride in water) are all represented in the compendia and thus subject to 501(b) of the Act.

Section 501(c) is used for drugs not meeting the requirements that are in the drug application or in-house specifications. This could also apply to specifications that are in addition to the requirements in a compendial monograph.

cGMP Regulations

To implement the provisions of the Federal Food, Drug, and Cosmetic Act, Congress delegated to the FDA, through the Secretary of Health, Education, and Welfare, broad authority to promulgate regulations for the efficient enforcement of the Act under Section 701(a). The exceptions to this authority are those provisions of the Act that are cited in Section 701(e). These include several drug provisions relating to certain types of adulteration and misbranding. Regulations issued under Section 701(e) require an opportunity for a public hearing under formal rule-making procedures, referred to as an evidentiary public hearing. Regulations promulgated under Section 701(a) on the contrary are subject only to notice and comment or informal rule-making under the provisions of the Administrative Procedures Act (1).

To implement Section 501(a)(2)(B) of the Act, the FDA issued regulations, in accordance with Section 701(a) of the Act, defining what it considered "current good manufacturing practice." The latest revisions of the cGMP regulations for human and veterinary drugs were published in the Federal Register of September 29, 1978, and became effective on March 28, 1979 (2). Because these regulations provide legal standards for controlling the quality of drugs, they should be of interest to all health professionals. They can also provide an insight into standard operating procedures that may serve those who are called upon to set up a quality *control program on the handling and administration of* parenterals in health care facilities. Unlike other regulations, regulatory controls are based primarily on inspections of establishments manufacturing, processing, packing, or holding human and veterinary drugs.

The cGMP regulations are contained in Title 21 Code of Federal Regulations parts 210 and 211.

Part 210—Current Good Manufacturing Practice in Manufacturing, Processing, Packing, or Holding of Drugs; General
Part 210 gives the status of the cGMP regulations. It indicates that failure to follow the regulations in parts 211 through 226 would render drug adulterated under Section 501(a)(2)(B) of the act. It also states that the person who is responsible for the failure to comply shall be subject to regulatory action. In the United States, typically the president or chief executive officer of the company is held responsible. In some cases the top manager of quality has also been cited in regulatory actions.

Paragraph 210.2 describes the applicability of cGMP regulations. It states that parts 210 through 226 pertain to drugs, parts 600 through 680 pertain to biological products, and part 1271 pertains to human cell, tissue, or cellular or tissue-based products subject to Section 505 of the act or Section 351 of the Public Health Service Act, shall be considered to supplement, not supersede, each other, unless the regulation explicitly provide otherwise. This means that the cGMP regulation in parts 210 and 211 also apply to biological products and biotechnology products. It also states that investigational drugs used in phase 1, 2 or 3 clinical studies must comply with the regulations in part 211. The FDA has further clarified that all cGMP provisions except labeling and process validation requirements are to be applied to clinical products.

The last section of part 210 lists definitions for some of the terms used in part 211. On September 8, 2008, the FDA published a revision to the non-fiber-releasing filter definition in part 210.3 (3). The reference to asbestos filters as fiber-releasing filters was deleted.

Part 211—Current Good Manufacturing Practice for Finished Pharmaceuticals
Subpart A: general provisions, including scope and definitions. This subpart is a repeat of the sections in 210.1 and 210.3. There is a statement that cGMPs in part 211 do not apply to Over-The-Counter (OTC) drug products that are ordinarily marketed and consumed as human foods. This is the case for vitamins and herbal products.

Subpart B: organization and personnel.

1. Responsibilities and authority of a quality control unit are to be spelled out in writing.
2. Personnel qualification for assigned functions and training in CGMP shall be conducted on a continuing basis.
3. Only authorized personnel shall enter those areas of the buildings and facilities designated as limited-access areas.
4. Consultants advising on CGMP shall be qualified, and records shall be maintained on their employment and qualifications.

Subpart C: buildings and facilities.

1. Buildings—their size, construction, and operational areas—are to be designed so that they are suited to the types of products produced or held therein to prevent contamination or mix-ups.
2. Special operations require more detailed criteria as to the adequacy of the building and facilities. Thus, the requirements for aseptic processing must include floors, walls, and ceilings of smooth, hard surfaces that are easily cleanable.

    Temperature and humidity controls.

    An air supply filtered through high-efficiency particulate filters under positive pressure, regardless of whether the flow is laminar or nonlaminar.

    A system of monitoring environmental conditions.

    A system for cleaning and disinfecting the room and equipment to produce aseptic conditions.
3. Equipment for adequate[a] control over air pressure, microorganisms, dust, humidity, and temperature shall be provided when appropriate[a] for the manufacture, processing, packing, or holding of a drug product.
4. Sanitation shall be assured by requiring written procedures for cleaning and assigning responsibility of seeing that they are followed. Rodenticides, insecticides, or fumigating agents shall not be used unless registered and used in accordance with the Federal Insecticide, Fungicide, and Rodenticide Act (FIFRA). Sanitation procedures shall apply to work performed by contractors.

Subpart D: equipment.

1. Adequacy of equipment design, size, and location should be validated.
2. Equipment cleaning and maintenance record keeping is essential.

Subpart E: control of components and drug product containers and closures.

1. Take appropriate measures to establish suitable specifications, and assure conformance with the specifications by proper records.[a]
2. Retest components, drug product containers, and closures as necessary due to conditions or passage of time that might adversely affect them.
3. Assure that drug product containers and closures are not reactive, additive, or absorptive, so as to alter the strength, quality, or purity of the drug beyond the applicable specifications.
4. Enforce standards and specifications to ensure that such hazards as pyrogens are eliminated from containers and closures for parenterals.

Subpart F: production and process controls.

1. Provide written procedures and change control with approval by the quality control unit.
2. Validate each process to demonstrate that it will consistently do what it purports to do.

---

[a]Such words as *adequate* and *appropriate* are used frequently in this and other sections of the CGMP. This puts the burden on the manufacturer of showing through data and performance records that the selections are "adequate" and "appropriate." Such flexibility is viewed by industry as a desirable attribute in the cGMP.

3. Control against microbiological contamination, including validation of the sterilization process.
4. Reprocessing must be based on procedures that *will ensure* that reprocessed batches will conform to *all* established standards, specifications, and product characteristics.

Subpart G: packaging and labeling control.

1. Provide written procedures and documentation to assure that every stage, from the design, receipt, identification, storage, and handling of labeling and packaging to their application to the drug product, is adequately controlled.
2. Use of gang-printed labeling for different drug products, or different strengths or net contents of the same drug product, is prohibited unless the labeling from gang-printing sheets is adequately differentiated by size, shape, or color.
3. Labeling reconciliation is waved for cut or roll labeling if a 100% examination for correct labeling is performed.
4. All prescription drug products and most OTC drug products shall have expiration dates on their labeling on the basis of adequate stability studies. However, the commissioner proposed in a separate *Federal Register* document published at the same time as the cGMP final rule that certain OTC drug products be exempted from expiration dates. These included those OTC drug products used without dosage limitation provided that it could be shown that they are stable for at least three years. Drug products to be reconstituted at time of dispensing shall bear expiration information for both the reconstituted and unreconstituted products.

Subpart H: holding and distribution.

1. There shall be written procedures to describe the warehousing. Where necessary to produce product, there shall be appropriate environmental controls.
2. There shall be written distribution procedures so that any recalls, if required, can be handled expeditiously.

Subpart I: laboratory controls.

1. Any specifications, standards, sampling plans, test procedures, or other laboratory controls, such as stability testing, are to be approved by the quality control unit.
2. The laboratory controls required are specified.
3. Testing and release procedures are specified for the usual drug products and exceptions in the case of short-lived radiopharmaceutical parenterals where batches may be released prior to completion of sterility and/or pyrogen testing. Appropriate laboratory testing is provided for, as necessary, of each batch of drug product required to be free of objectionable microorganisms.
4. A written stability program on the basis of studies conducted in the same container-closure system in which it will be marketed is required.
5. Products purporting to be sterile and/or pyrogen-free must be batch tested prior to release.
6. Reserve samples must have at least twice the quantity necessary for all tests except for sterility and pyrogen testing.
7. Animals used in testing components, in-process materials, or drug products for compliance with established specifications shall be maintained and controlled in a manner that assures their suitability for their intended use.
8. If a reasonable possibility exists that a non-penicillin drug product has been exposed to cross-contamination with penicillin, the non-penicillin drug product shall be tested for the presence of penicillin.

Subpart J: records and reports.

1. Documentation through written procedures and records is now required for practically all operations. The items to be reported in a laboratory assay report are spelled out for the first time in the cGMP.
2. Complaints must be documented, and procedures must be followed in the investigation of complaints by quality control.
3. The quality control unit is responsible for review of all production and control records.

Guidelines

Guidelines are a tool used by the FDA to explain its interpretation of what is required to meet the regulations. Whenever the FDA perceives that the pharmaceutical industry does not understand the regulations or their intent, a group of in-house experts is assembled to generate the proposed guideline. Once the proposed document is submitted for public review and comment, it is finalized. Below is a list of guidelines that have been written and can be applied to parenteral drug products.

GMP Guidelines

Guideline on General Principles of Process Validation, May, 1987, and November 2008 Draft

Guideline on Sterile Drug Products Produced by Aseptic Processing, September, 2004

Guideline on Validation of the Limulus Amebocyte Lysate Test as an End-Product Endotoxin Test for Human and Animal Parenteral Drugs, Biological Products, and Medical Devices, December, 1987

Guideline on the Preparation of Investigational New Drug Products, February, 1990

Guideline on Submission of Documentation in Applications for Parametric Release of Human and Veterinary Drug Products Terminally Sterilized by Moist Heat Processing, August 2008, Draft

The most significant guideline for parenteral drug manufacture is the guideline for aseptic processing.

The first aseptic guideline was issued in 1987. In an aseptic process, the drug product, container, and closure are first subjected to sterilization methods separately, as appropriate, and then brought together. It described the facilities, equipment, environmental conditions, and process validation requirements for products produced using aseptic processing. In addition, it described environmental monitoring and laboratory testing requirements. The only specific acceptance criteria given was nonviable particulate limits for clean rooms and adjacent areas and the acceptance criteria for the media fills used to validate the aseptic processes. Particulate limits are based on Federal Standard 209 (withdrawn). Media fill acceptance criteria was listed at 1 positive unit per 1000 units filled with media. This represents a contamination rate of 0.1%. A minimum of 3000 units was required to be able to detect the contamination rate with 95% confidence.

In 2004, the FDA issued a revised guideline (4). The revision was necessary to incorporate improvement made by the pharmaceutical industry and new concepts being promoted by the FDA. During the development of the guideline, a great deal of effort was made to try to harmonize as much as possible with the EU requirements. In addition, the FDA assembled an expert panel made up of industry experts to assist in answering specific questions. This is the first time the industry has had any influence on guidelines during the development phase. Most of the recommendations made by the expert panel were adopted by the FDA.

The most significant changes in the 2004 guideline are as follows:

- Microbial limits were added to guideline.
- The nonviable and viable airborne particulate limits that are stated in the EU guide were adopted. The major exception is that settling plates are not required or expected by the FDA. The FDA also adopted the International Organization for Standardization (ISO) designations for clean room classification instead of the A to D designations used by the EU.

Clean area classification (0.5 μm particles/ft^3)	ISO designation	>0.5 μm particles/m^3	Microbiological active air action levels (cfu/m^3)	Microbiological settling plates action levels (diam. 90 mm; cfu/4 hr)
100	5	3,520	1	1
1,000	6	35,200	7	3
10,000	7	352,000	10	5
100,000	8	3,520,000	100	50

- Differential pressure is now expressed in pascals and as a range instead of a single value, that is, 10 to 15 Pa versus 0.05 in. of water. This range is the same as that in the EU.
- The requirement that the velocity in a unidirectional flow area should be 90 ± 20 ft/min has been deleted. Instead each site must justify and validate that the velocities used are appropriate. Typically, airflow studies are used to demonstrate that at the measured velocity the airflow sweeps particles away from the product without generating turbulence. It also prevents extrinsic particulate matter from getting into the product.
- The FDA expects environmental monitoring data to be trended and analyzed for any trends that could indicate a potential risk to the products.
- A recommendation that the quality unit view all process simulations (media fill) was added. While this is a recommendation, it appears that the FDA investigators expect to see evidence that QA does observe and take notes on activities occurring during the process simulations.
- The FDA recommends that the media fill program address applicable issues such as
  - factors associated with the longest permitted run on the processing line that can pose contamination risk (e.g., operator fatigue);
  - representative number, type, and complexity of normal interventions that occur with each run, as well as nonroutine interventions and events (e.g., maintenance, stoppages, equipment adjustments);
  - lyophilization, when applicable;
  - aseptic assembly of equipment (e.g., at start-up, during processing);
  - number of personnel and their activities;
  - representative number of aseptic additions (e.g., charging containers and closures as well as sterile ingredients) or transfers;
  - shift changes, breaks, and gown changes (when applicable);
  - type of aseptic equipment disconnections/connections;
  - aseptic sample collections;
  - line speed and configuration;
  - weight checks;
  - container closure systems (e.g., sizes, type, compatibility with equipment); and
  - specific provisions in written procedures relating to aseptic processing (e.g., conditions permitted before line clearance is mandated).
     This is the first time that the FDA has given such details concerning what should be covered by the process simulation procedure/protocols.

- The process simulation acceptance criteria was changed as follows:
  When filling fewer than 5000 units, no contaminated units should be detected.
  - One contaminated unit is considered cause for revalidation, following an investigation.
  When filling from 5000 to 10,000 units
  - One contaminated unit should result in an investigation, including consideration of a repeat media fill.
  - Two contaminated units are considered cause for revalidation, following investigation.
  When filling more than 10,000 units
  - One contaminated unit should result in an investigation.
  - Two contaminated units are considered cause for revalidation, following investigation.

It should be noted that it does not matter if 10,000 units or 150,000 units are filled, the acceptance criteria is the same. This is a significant change from the previous contamination rate concept where the more units filled with media, the more positive units allowed. Both the pharmaceutical industry and the FDA agreed that this approach was inappropriate. The goal of aseptic process is to produce a sterile product, that is, zero units contaminated. On the basis of industry input it was determined that the above criteria represented the current industry capabilities.

Three new sections were added to the aseptic guideline to address the following:

- **Aseptic processing isolators**
  The main concerns raised were as follows:
  - Glove integrity.
  - Proper isolator design.
  - Pressure differentials—promotes the need for ISO 5 protection at opening in the isolator.
  - Isolators must be in classified rooms. ISO 8 is recommended. The isolator is prohibited from being in an unclassified room.
  - There is an extensive discussion of decontamination of isolators. For example, biological indicators must be used, if decontamination is used for product contact parts, a six-logarithm reduction must be proven and the frequency of decontamination must be justified and have supporting data.
- **Blow/fill/seal technology**
  Blow/fill/seal (BFS) machines must be designed to prevent extraneous contamination. The room environment can be ISO 8. Another major concern is container/closure integrity. The FDA requires that reliable and sensitive inspection processes must be established to make sure every unit is intact.
- **Processing prior to filling and sealing operations**
  Process simulations must cover all aseptic manipulations that occur to the product prior to the manufacturing process, including the holding times for sterile bulks.

Compliance Policy Guides

Compliance policy guides (CPGs) were developed as a mechanism of disseminating the FDA policy to the district offices. CPGs were developed by centers or other headquarters units to explain how the FDA will enforce various aspects of the regulations or various situations that the field investigator may find. They are usually developed in response to questions on how to interpret a specific regulation or what is the agency policy concerning a specific subject. CPGs can be obtained through the National Technical Information Service or at http://www.fda.gov.

One CPG that is particularly applicable to parenteral drugs is guide 7132a.13, "Parametric Release—Terminally Heat Sterilized Drug Products." In this guide the FDA defines parametric release as "a sterility release procedure based upon effective control, monitoring, and documentation of a validated sterilization process cycle in lieu of release based upon end-product sterility testing." The FDA will only accept parametric release for terminally heat sterilized parenteral drug products. Parametric release of drug products sterilized by filtration or ethylene oxide will not be allowed. For those products that are the subject of a new drug application, the manufacturer must submit a supplement and obtain approval prior to initiation of parametric release. Parametric release of drug products that do not require new drug applications cannot be used until the above requirements have been met. The firm should have the data to support parametric release at the manufacturing site. Firms planning on parametric release of non-New Drug Application (NDA) drug products should contact the local FDA district office prior to initiation so that the FDA can determine that they have met all the required criteria. There are four requirements listed in the guide that must be met before parametric release can be considered by the FDA.

1. The sterilization process cycle has been validated to achieve microbial bioburden reduction to 10^0 with a minimum safety factor of an additional six-logarithm reduction. All cycle parameters must be identified by the manufacturer as critical, for example, time, temperature, and pressure, or noncritical, for example, cooling time and heat-up time. Failure of one of the critical parameters must result in automatic rejection of the sterilizer load. Biological indicators can be used to evaluate cycle lethality where equipment malfunction prevents measurement of one critical cycle parameter. If more than one critical parameter is not met, the batch is considered nonsterile despite biological indicator sterility.
2. Integrity for each container/closure system has been validated to prevent in-process and postprocess contamination over the product's intended shelf life.
3. Bioburden testing, covering total aerobic and spore counts, is conducted on each batch of presterilized drug product. Resistance of any spore-forming organism found must be compared to that of the organism used to validate the sterilization cycle.
4. Chemical or biological indicators are included in each truck, tray, or pallet of each sterilizer load. Both chemical and biological indicators must be fully characterized and documented. Chemical indicators cannot be used to evaluate cycle lethality due to lack of time/temperature accuracy.

The FDA issued a draft guideline on the documentation that must be submitted in applications to support parametric release of human, biological, and veterinary drugs (4). The definition of parametric release has been revised to conform to the new focus of FDA. It is defined as "a sterility assurance release program where demonstrated control of the sterilization process enables a firm to use defined critical process controls, in lieu of the sterility test." The release program should be based on the results of a risk assessment of the terminal sterilization cycle, demonstration of process understanding, and prior knowledge of the production and sterilization process.

EU REGULATIONS
EU Directives
Directives in the EU are the same laws as in the United States. The directive that requires the EU member states to ensure that pharmaceutical manufacturers comply with GMP is in Chapter IV of Directive 75/319/EEC for human products and Chapter V of Directive 81/851/EEC for veterinary products. Another Directive, 92/25/EEC, requires all wholesale distributors to be authorized and comply with guidelines on Good Distribution Practice (GDP).

The principles and guidelines of GMP are stated in two directives. Directive 91/356/EEC and 2003/94/2003 are for human medicinal products and 91/412/EEC is for veterinary products (5). It must be noted that while these are termed guidelines they should be treated the same as the GMP regulations in the United States. They are enforceable under the member state laws.

GMP Regulations
The GMP regulations are organized into nine general chapters.

Chapter 1 Quality Management
This chapter covers the requirements for quality assurance and quality control.

The system of quality assurance appropriate for the manufacture of medicinal products should ensure that

1. medicinal products are designed and developed in a way that takes account of the requirements of GMP and Good Laboratory Practice;
2. production and control operations are clearly specified and GMP adopted;
3. managerial responsibilities are clearly specified;
4. arrangements are made for the manufacture, supply, and use of the correct starting and packaging materials;
5. all necessary controls on intermediate products and any other in-process controls and validations are carried out;
6. the finished product is correctly processed and checked, according to the defined procedures;
7. medicinal products are not sold or supplied before a qualified person has certified that each production batch has been produced and controlled in accordance with the requirements of the marketing authorization and any other regulations relevant to the production, control, and release of medicinal products;
8. satisfactory arrangements exist to ensure, as far as possible, that the medicinal products are stored, distributed, and subsequently handled so that quality is maintained throughout their shelf life; and
9. there is a procedure for self-inspection and/or quality audit that regularly appraises the effectiveness and applicability of the quality assurance system.

It should be noted that the legal responsibility for the quality of products rests with the qualified person. This is very different from the United States, where the CEO/president is ultimately held responsible for the quality of the products manufactured by the company. A qualified person must be noted in each Market Authorization Application.

The basic requirements of quality control are that

1. adequate facilities, trained personnel, and approved procedures are available for sampling, inspecting and testing starting materials, packaging materials, intermediate, bulk, and finished products, and, where appropriate, for monitoring environmental conditions for GMP purposes;
2. samples of starting materials, packaging materials, intermediate products, bulk products, and finished products are taken by personnel and by methods approved by quality control;
3. test methods are validated;
4. records are made, manually and/or by recording instruments, which demonstrate that all the required sampling, inspecting, and testing procedures were actually carried out. Any deviations are fully recorded and investigated;
5. the finished products contain active ingredients complying with the qualitative and quantitative composition of the marketing authorization, are of the purity required, and are enclosed within their proper containers and correctly labeled;
6. records are made of the results of inspection and that testing of materials, intermediate, bulk, and finished products is formally assessed against specification. Product assessment includes a review and evaluation of relevant production documentation and an assessment of deviations from specified procedures;
7. no batch of product is released for sale or supply prior to certification by a qualified person that it is in accordance with the requirements of the marketing authorization; and
8. sufficient reference samples of starting materials and products are retained to permit future examination of the product if necessary and that the product is retained in its final pack unless exceptionally large packs are produced.

In addition to describing the function of the quality assurance and control, this chapter also describes the elements of the annual review of medicinal products. They are as follows:

1. A review of raw materials used in the product, especially those from new sources
2. A review of critical in-process controls and finished product results
3. A review of all batches that failed to meet established specification(s)
4. A review of all critical deviations or nonconformances and related investigations
5. A review of all changes carried out to the processes or analytical methods
6. A review of marketing authorization variations submitted/granted/refused, including those for third country dossiers
7. A review of the results of the stability monitoring program
8. A review of all quality-related returns, complaints and recalls, including export only medicinal products
9. A review of adequacy of previous corrective actions
10. For new marketing authorizations, a review of postmarketing commitments
11. A list of validated procedures and their revalidation dates
12. A list of qualified equipment and their requalification dates

Chapter 2 Personnel

This chapter describes the duties of key personnel, including the qualified person. Since the qualified person is unique to the EU regulatory system, we will look at the duties of the qualified person. The duties are described in Article 51 of Directive 2001/83/EC and are summarized below.

1. For medicinal products manufactured within the European Community, a qualified person must ensure that each batch has been produced and tested/checked in accordance with the directives and the marketing authorization.
2. For medicinal products manufactured outside the European Community, a qualified person must ensure that each imported batch has undergone, in the importing country, the testing specified in paragraph 1 (b) of Article 51.
3. A qualified person must certify in a register or equivalent document, as operations are carried out and before any release, that each production batch satisfies the provisions of Article 51.

As can be seen, the qualified person is responsible for ensuring the quality of all drug products introduced into the EU market. The person must obtain training and pass a test to become a qualified person. This training and test is to ensure that the person understands the requirements of the job. In some of the member states there are other basic requirements, such as the person must be a pharmacist or have a minimum number of years of experience. If the qualified person releases a product that lacks the required quality, they can be fined or sent to jail for exposing the public to a potential risk of injury.

Chapter 3 Premises and Equipment

The basic requirement is that premises and equipment must be located, designed, constructed, adapted, and maintained to suit the operations to be carried out.

There is a very specific indication that separate facilities are required for highly sensitizing materials or biological preparations. In addition, certain antibiotics, certain hormones, certain cytotoxics, and highly active drugs should not be produced in the same facilities. The quality control laboratory should be separated from production areas, especially microbiology laboratories. U.S. regulations only mention penicillin as requiring separate facilities.

Basically the requirements for the EU and the United States are essentially the same. The main difference is that the EU regulations give more detail as to what is expected.

Chapter 4 Documentation
This chapter describes the types of documents that are required. Documents required are as follows:

- Specifications for starting and packaging materials, intermediate and bulk products, and finished products
- Manufacturing formula and processing instructions
- Packaging instructions
- Batch processing records
- Batch packaging records
- Procedures
  - Receipt of materials
  - Sampling
  - Testing
  - Release and rejection
- Other documents
  - Validation protocol and reports
  - Equipment assembly and calibration
  - Maintenance, cleaning, and sanitation
  - Personnel training, clothing, and hygiene
  - Environmental monitoring
  - Pest control
  - Complaints
  - Recalls
  - Returns

As can be seen there is no difference between the type of procedures that are required by both the EU and the FDA. Again the most striking difference is that the EU guide gives more details.

Chapter 5 Production
The basic requirements are as follows:

- Production operations must follow clearly defined procedure
- Production operations must prevent cross-contamination
- Production rooms and equipment should be identified with the product or material being processed
- All drug production processes and equipment should be validated and revalidated after significant changes to the equipment or process
- Starting materials must come from approved vendors and should be sampled and tested for compliance with applicable specifications
- Printed packaging materials should be controlled and issued by authorized personnel
- There should be physical separation between packaging operations so that mix-ups are prevented. In addition there should be procedure to inspect packaging area prior to use to ensure that all previous product and labels have been cleared from the area
- After packaging, all containers should be inspected to ensure that packaging is complete and contains all required information, especially lot identification and expiration date
- Rejected materials must be stored in separate restricted areas
- Reprocessing of rejected products should be a rare occurrence
- Returned product should be destroyed, but could be used in subsequent batches after testing by the quality control department

You will note that the above requirements are not that different from the U.S. GMP requirements.

Chapter 6 Quality Control

This chapter discusses the elements of Good Quality Control Laboratory Practice. Some of these elements are as follows:

- Documentation—specifications, test procedures, sampling procedures, validation records, and out-of-specification/out-of-trend investigation procedures.
- Sampling—methods, sampling equipment, storage conditions, etc.
- Testing—method validation, analytical results review process, training of analysts, reagent preparation and documentation, glassware cleaning and use, and reference standard handling.
- Special attention is given to on-going stability testing programs. The program should be applied to both bulk active pharmaceutical ingredients and finished dosage forms. Protocols should be developed that describe all testing that is to be performed on products. All on-going stability testing should extend to the end of the shelf life of the product. The results of the stability testing should be made available to key personnel and the qualified person(s).

Chapter 7 Contract Manufacture and Analysis

This chapter outlines the responsibilities of the company that hold the marketing authorization when they contract out the manufacturing and/or analysis of the product. It states that there should be a contract between the parties so that there is a clear understanding of the responsibilities of all the parties involved. The contract must clearly state how the qualified person releasing each batch will exercise their legal responsibilities.

The U.S. regulations are silent concerning this topic. While the FDA encourages quality agreements that state who is responsible for which aspects of the GMP regulations and drug application commitments, there is no regulation that requires the agreement.

Chapter 8 Complaints and Product Recall

All complaints and other information concerning potentially defective products must be reviewed carefully according to written procedures. During review, if a product defect is discovered or suspected in a batch, consideration should be given to check other batches. There should be a periodic review of complaints for trends.

There should be established written procedure to organize any recall activities. Recall operations should be capable of being initiated promptly and at any time. If recall is initiated, all regulatory authorities in all countries where the product was distributed must be notified. All recalled material must be properly identified and stored separately in a secure area. The effectiveness of the recall should be assessed during the recall.

Chapter 9 Self-Inspection

This chapter covers the requirements that manufacturers should conduct self-audits of their operations. Audits can be conducted by independent personnel within the company or by outside experts. All audits must be documented and corrective actions developed for any adverse observations during the audit. While the FDA does not have an equivalent section in their GMP regulations, they do consider it a duty of the quality organization to perform self-audits.

In addition to the above general chapters, there are 20 annexes that give more detailed requirements for specific product types or dosage forms. The annexes are as follows:

Annex 1   Manufacture of sterile medicinal products (revision November 2008). The revised annex should be implemented by March 1, 2009, except for the provisions on capping of vials, which should be implemented by March 1, 2010.

Annex 2   Manufacture of biological medicinal products for human use

Annex 3   Manufacture of radiopharmaceuticals

Annex 4 Manufacture of veterinary medicinal products other than immunological veterinary medicinal products
Annex 5 Manufacture of immunological veterinary medicinal products
Annex 6 Manufacture of medicinal gases
Annex 7 Manufacture of herbal medicinal products
Annex 8 Sampling of starting and packaging materials
Annex 9 Manufacture of liquids, creams, and ointments
Annex 10 Manufacture of pressurized metered dose aerosol preparations for inhalation
Annex 11 Computerized systems
Annex 12 Use of ionizing radiation in the manufacture of medicinal products
Annex 13 Manufacture of investigational medicinal products
Annex 14 Manufacture of products derived from human blood or human plasma
Annex 15 Qualification and validation (July 2001)
Annex 16 Certification by a qualified person and batch release (July 2001)
Annex 17 Parametric release (July 2001)
Annex 18 Good manufacturing practice for active pharmaceutical ingredients requirements for active substances used as starting materials from October 2005 covered under part II
Annex 19 Reference and retention samples (December 2005)
Annex 20 Quality risk management (February 2008)

Annex 1—Manufacturer of Sterile Medicinal Products
Of the annexes, the one that has received the most attention is Annex 1 concerning the manufacture of sterile medicinal products. This annex gives very specific guidance as to what is required to produce sterile drug products.

While the FDA guidelines, in general, do not give specific values, the FDA does require manufacturers to have documentation to justify that what they are doing is appropriate. In Annex 1, the EU gives very specific requirements and limits. The current version became official on March 1, 2009, except for the provision on capping of freeze-dried vials, which becomes official on March 1, 2010.

Probably the most confusing requirement in the annex is the testing of nonviable airborne particulates in the "at-rest" and "in-operation" states. Limits are given for both states. These non-viable particulate limits are now closer to the limits in ISO14644 than in the previous version of annex 1. Except for grade A there is a different limit for the two states. Normally the at-rest state is only tested during commissioning of a new clean room or after significant changes are made to an existing clean room. But it appears that the EU investigators are requesting that at-rest testing be performed periodically on a routine basis. Currently, the FDA only considers the in-operation state for all of its recommendations. They are concerned with the condition of the environment when product is being exposed to it.

The following table gives the particulate limits for the four clean room grades.

	Maximum permitted number of particles per m^3 equal to or greater than the tabulated size			
	At rest		In operation	
Grade	0.5 μm	5.0 μm	0.5 μm	5.0 μm
A	3,520	20	3,520	20
B	3,520	29	352,000	2,900
C	352,000	2,900	3,520,000	29,000
D	3,520,000	29,000	Not defined	Not defined

As can be seen, the other major difference between the EU and the FDA is the requirement to measure the particles at 5.0 μm. This has been a very controversial difference. The industry has argued that the 5 μm particles do not have to be measured. If you are going to market a product in the EU, you will have to monitor for both 0.5 and 5.0 μm particles.

In addition, there is a requirement that after operations have been completed, the at-rest limits should be attained after a 15 to 20 minutes "clean-up" period. During the at-rest stage no activities or personnel are present.

The annex also gives limits for microbiological monitoring of clean room, as shown in the following table.

| | Recommended limits for microbial contamination | | | |
Grade	Air sample cfu/m^3	Settle plates (diameter 90 mm) cfu/4 hr	Contact plates (diameter 55 mm) cfu/plate	Glove print 5 fingers cfu/glove
A	<1	<1	<1	<1
B	10	5	5	5
C	100	50	25	–
D	200	100	50	–

It should be noted that all of the values in the above table are average values. There is no explanation of what is to be averaged. From discussions with EU inspectors, it appears that all of the plates in a grade A zone can be averaged and then compared to the given limits. This is a dangerous concept since a problem area could be averaged out by the other areas in the same zone. For example, there could be a high count at the area where sterile stopper are placed in a hopper but very low values at other locations. The average would indicate no problem but the individual value would indicate a potential contamination risk at the stopper hopper area.

The annex gives grades under which operations should be performed. For instance filling of high-risk terminally sterilized products should be done in a grade A area with a grade C background. Formulation is to be performed in a grade C area. For aseptically filled products, filling must be conducted in grade A with a grade B background. For isolators, the EU allows the background to be grade D at rest, while the FDA requires grade C in operation. But for a BFS machine, the background must be grade C with personnel in grade A/B clothing.

One other area of difference between the EU and the FDA is the issue of capping of vials. The EU requires that vials with missing or displaced stoppers be rejected prior to capping and that the capper be either in the clean room or in a restricted access barrier system (RABS) supplied with grade A air. The major problem with this concept is that caps are very dirty from a particulate matter viewpoint and should never be in a clean room. In addition, if the capper is in the clean room then all of the caps have to be sterilized. Since the caps take up a lot of space, this could have a significant impact on the sterilization capacity of the facility. While the FDA requires that the vials be protected with grade A air prior to capping, they have not required that the enclosure be a RABS design. Capping machines typically take a lot of adjustments in the beginning and sometimes during production. Making the adjustments through glove ports would be a problem, not only from an equipment design viewpoint but also from a personnel safety viewpoint. Currently when one of the doors is opened, the capper automatically shuts off. With glove ports, a new detector system would have to be developed and installed that would shut off the capper when someone inserts their hand into a glove. The requirements in this section of the EU guidance will have very little impact on the sterility assurance of parenteral products. Missing stoppers are already detected during the 100% inspection that takes place after capping. Slightly raised stoppers are only a concern for lyophilized products where the stopper has slits in the sides. Even here the stopper would have to be significantly displace to present a high risk and should be able to be removed during lyophilizer unloading operations.

The other section of the EU GMP guide gives the same information as required by the other regulatory authorities and has caused little discussion over the years.

This chapter would be remiss if it did not discuss one of the guidelines in volume 3 of "The rules governing medicinal products in the European Union." The EU has published a series of scientific guidelines to harmonize the manner in which the EU member states and the EMEA interpret and apply the detailed requirements for the demonstration of quality, safety, and efficacy container in the Community Directives. In addition, these guidelines are also intended to help companies make sure that their marketing authorization applications will be recognized as valid by the EMEA. This means the EMEA investigators will audit against these guidelines.

In guideline CPMP/QWP/486/95 concerning the manufacture of the finished dosage form, there is a statement that

> "For sterilisation by filtration the maximum acceptable bioburden prior to the filtration must be stated in the application (6). In most situations NMT 10 CFU's/100 mL will be acceptable, depending on the volume to be filtered in relation to the diameter of the filter. If this requirement is not met, it is necessary to use a pre-filtration through a bacteria-retaining filter to obtain a sufficiently low bioburden."

Over the years this has become a mandatory expectation by the EMEA and has forced manufacturers to dual filter parenteral preparations that are filter sterilized. This author has never heard any justification for the 10 cfu/100 mL limit. EU's requirement for dual filtration has caused a lot of discussions and misinterpretations. Many companies have interpreted this to mean that to sterilize a product you need two 0.2 μm filters, and both must be integrity tested and pass to declare the sterilization process successful. But the statement indicates that the first filter is a bioburden reduction filter that will reduce the bioburden to NMT 10 cfu/100 mL. The second filter is the sterilizing filter. So, only the final filter should be integrity tested. If the second filter fails and the first filter passes, you still have a sterility failure from a regulatory viewpoint because the sterilizing filter did not pass integrity. This is another example where the EU has enforceable guidelines, whereas the FDA guidelines are not enforceable.

In this chapter we have looked at the GMP regulations in the United States and Europe. These regulations form the basis for most of the regulations in the other markets of the world. The good news is that the regulatory authorities have been working hard to harmonize their requirements to lessen the burden on pharmaceutical companies, while at the same time protecting the users of the medicinal products from harm.

REFERENCES

1. Administrative Procedures Act of 1946, 5 U.S.C. 551 et seq. For a detailed description of this Act, see A Guide to Federal Agency Rule-making, GPO, 1984. Stock #436-0661-796.
2. Federal Register (FR), Friday, September 29, 1978, Part II, p. 45021–45026.
3. Federal Register (FR), Monday, September 9, 2008, p. 51919.
4. U.S. Food and Drug Administration. Available at: www.fda.gov/Drugs/GuidanceComplianceRegulatoryInformation/Guidances/default.htm.
5. European Commission. Enterprise and Industry Pharmaceuticals. Available at: http://ec.europa.eu/enterprise/sectors/pharmaceuticals/documents/eudralex/index_en.htm.
6. CPMP/QWP/486/95, Manufacture of the Finished Dosage Form, issued May 1996, obtained from EU. Available at: http://ec.europa.eu/enterprise/sectors/pharmaceuticals/documents/eudralex/index_en.htm. Located in Volume 3 Guidelines, Section Quality Guidelines.

2 | Risk assessment and mitigation in aseptic processing

James Agalloco and James Akers

INTRODUCTION

Sterile products are frequently administered to patients through the dermal layer to attain rapid therapeutic response and accurate dosing. Delivery in this manner intentionally circumvents the body's protective mechanisms, and mandates that the product be largely free of infectious microorganisms and endotoxin. These concerns are heightened when the drug is delivered to patients whose health is already compromised as is common in clinical settings. Awareness of the patient has prompted regulatory preference for the use of terminal sterilization (1,2). While the use of lethal processes on finished formulations in their final container is favored because of their lethality, material considerations have limited their application such that an estimated 85% of all sterile products are manufactured by aseptic processing that are less abusive of essential material and container properties.[a]

Aseptic processing customarily use a variety of sterilization procedures for the individual components of the formulated product, container, and product contact parts, enabling the sterilizing process to be chosen for preservation of the key quality attributes of the materials. The core aseptic process assembles the sterilized items into the final dosage form in an environment specifically designed for that purpose. Because product containers are closed after the individual sterilization processes are carried out, the potential for contamination ingress is ever present during aseptic processing. In the belief that knowledge of the conditions under which the aseptic process is carried out would be valuable in determining the acceptability of the resulting product, environmental monitoring has historically been considered essential. Microbial sampling of air and surfaces as well as personnel gloves and gown within the aseptic environment were instituted as a means of environmental monitoring, which ultimately evolved into a program thought to provide critical information regarding sterility assurance. When monitoring was first instituted, the environmental conditions and gowning systems were markedly less capable than those presently in use. As a consequence, performance expectations and demonstrated performance were understood to be nonabsolute. Nevertheless, it was certainly understood that improvement in contamination control performance was both desirable and attainable. The gradual refinements in aseptic processing technology and performance expectations took place over a period of some 50 years.

THE MYTH OF STERILITY

The manufacture of sterile products is closely associated with expectations for sterility of the finished dosage form. This is customarily defined by the probability of a nonsterile unit (PNSU) or sterility assurance level (SAL).[b] The minimum expectation for PNSU in terminally sterilized products is that it be no greater than 1 nonsterile unit in one million units or 1×10^{-6}. The origins of this target lie in the food industry as it was initially developed for canned foods where the concern was the avoidance of *Clostridium botulinum*, an anaerobic spore former. The goal was not sterility of the canned foods but an acceptable level of safety for the consumer. In actuality, it defines a maximum level of risk that a consumer might be exposed to in the consumption of the sterilized material. This approach is essentially the same as that employed for the terminal sterilization of pharmaceutical products, which while stated as a PNSU, it is really a statement of the level of material safety (or risk minimization). Aseptic processing relies on the component and material sterilization methods for success, but differs in that

---

[a]This is estimated as a percentage of products and not as a percentage of the number of containers.
[b]The current preference is for the use of PNSU rather that SAL, because PNSU is a far easier concept for the novice to interpret.

calculation of a PNSU (or SAL) is impossible there being no directly lethal element of the aseptic manufacturing process. Aseptic processing performance is evaluated using process simulation studies in which the maximum contamination rate in the exercise demonstrates the capability of the overall aseptic process during that exercise and that exercise alone. Suggestions that the success in a process simulation defines the sterility assurance capability of an aseptic process are entirely fallacious. The process simulation is a singular event comprised of a number of individual sterilization, manual decontamination, and manipulations that cannot support the ability of those practices under different circumstance. The simulation demonstrates potential capability in a limited manner, but there are no means to extend the results to the same aseptic procedures in a separate event.

There is a common belief that the environmental monitoring performed in conjunction with every aseptic process provides a means for extension of the simulation performance to production operations. The limitations of microbial recovery from environmental samples in present-day manned clean rooms are such that these claims are certainly spurious. Extension of this thinking to advanced aseptic processing technologies is similarly inappropriate. What has been demonstrated for every aseptic processing is that it can be most realistically described as safety. Aseptic processes are essentially considered safe because the patient outcomes are successful and contamination in aseptically filled products has only infrequently been linked to known product contamination derived from the aseptic process. Our industry's ability to detect contamination in aseptic processes through any form of environment monitoring is extremely limited both in terms of analytical limit of detection and sampling statistics. The monitoring sample sizes are too small to afford any meaningful evaluation of the conditions, and the cultural methods employed do not have a limit of detection approaching zero. The sterility test is of such limited value in assessing process efficacy that it could be more aptly termed "the test for gross microbial contamination." The FDA's recalls of aseptically produced products are rarely the result of demonstrated contamination in the finished product, but rather an absence of appropriate conditions or inadequate documentation during the production operations. What has been attained with aseptic processing is more properly described as "safety." Sterility of aseptically filled products is completely unprovable, as it would require the evaluation of an infinite sample size with analytical method capable of detecting *any* contamination present. This is simply impossible, so realistically proving sterility in aseptic processing is not simply a matter of being willing to make a greater effort in terms of sampling intensity.

The improvements in aseptic processing were instituted to effect greater control over the environment, as influenced by its basic design, decontamination method, and operator involvement with a singular goal of reducing the contamination potential. The true objective has always been reduction of risk to the patient receiving the aseptically produced product. Aseptic processing systems in their most evolved forms have reached the point where the means to establish their acceptability are no longer adequate to provide any meaningful indication of performance.

As the processing capability evolved, closely followed (or at times preceded) by regulatory expectations, a critical component of the monitoring system remained unchanged. With each refinement of aseptic processing technology, the microbial sampling methods were increasingly taxed. In today's advanced aseptic processing systems, the environmental monitoring is being asked to prove an absolute negative—that no microorganisms are present anywhere in either the processing environment or the product. While particularly true of isolators, Restricted Access Barrier Systems (RABS) and many newer conventional aseptic facilities suffer the same limitations. This presents industry with a substantial conundrum of some magnitude with respect to the evaluation, selection, and ongoing control of aseptic processing technologies.

RISK ASSESSMENT

We first noted the limitations of monitoring programs nearly 20 years ago when new facilities began to exhibit environmental control capabilities that challenged the sensitivity and resolution of monitoring methods then available. When contamination was detected in these environments, it was increasingly associated with personnel. This was consistent with the long-standing understanding across the industry and regulators that personnel are responsible for the majority of contamination in an aseptic process (3,4). Deceased former FDA inspector

Hank Avallone had expressed this exact belief in direct manner during the 1980s, "It is useful to assume that the operator is always contaminated while operating in the aseptic area. If the procedures are viewed from this perspective, those practices which are exposing the product to contamination are more easily identified" (5). Actually, from experience and published research, we need not merely assume that an operator is a source of contamination, rather we can take it as an absolute certainty that clean room personnel function as mobile contamination generators. The idea that it is possible to have "sterile" clean rooms or sterile gowned operators has in fact been completely debunked.

In late 2004, when we began the development of our risk analysis method, we drew heavily on the concept that the release of contamination by operators was not merely possible but rather inevitable. With this simple truth in mind we focused our method on the human actions that are central to any aseptic process (6,7). The Agalloco-Akers (A-A) methods attribute risk almost exclusively to human activity within the aseptic process. The closer, longer, and more invasive the personnel intervention, the greater the risk is for contamination introduction. In taking this tact we discounted the more traditional approaches to risk assessment that endeavored to associate risk with contamination transfer to open containers from the air or surfaces (8,9). While we agree with the basic premises of these methods, the calculations required to calculate a risk value include values for which there is no reliable input data. It is our belief that because these methods utilize microbial recovery determination as a fundamental factor in the determination of the contamination ingress potential (and thus risk), they are inherently limited where the background microbial levels are largely devoid of recoverable microorganisms. Also, it is clear, given the variability of microbial analysis and the extremely limited sample size, that it is not possible for monitoring to give us much insight regarding patient risk. It therefore follows that it is not possible to determine through monitoring that an appropriate level of "sterility assurance" has been attained, or to assess anything but truly gross changes in environmental control.

Since the publication of the A-A method, it has been successfully utilized by several firms to evaluate and improve their aseptic processing operations.[c] Katayama and his colleagues compared its application to other risk methods and concluded that the A-A method offered the closest correlation to the historical performance at several aseptic sites when compared with other aseptic risk methodologies (10). A more general means for risk assessment related to sterile products has been developed by Parenteral Drug Association (PDA) (11). Regardless of the risk assessment methodology employed, it is essential that firms consider how their designs, practices, and expectations impact the contamination potential. Assumptions about outcomes must be evaluated in a rigorous manner to provide the greater confidence in the eventual design. Not only is this scientifically sound but is also expected by regulators (12,13).

Discussion of aseptic processing risk, or truly any risk assessment, should not end with completion of the assessment. It must be acknowledged that while risk assessment is an important task, it is not an end onto itself. It must be followed by a far more important activity, which is risk mitigation. Consider the driver of an automobile who notices that it is beginning to rain. This is the risk assessment, and although necessary it does not effect any improvement in the driver's safety. Until the driver mitigates the risk in an effective manner, there is no real benefit. Unless the driver adjusts the speed and turns on the wipers and the lights, there is no reduction in its risk potential. The assessment of risk is only the first step that must be accomplished, and aseptic processing is no different.

RISK MITIGATION

Risk assessment alone however is not enough; if the fundamental concepts adopted are inadequate, the resulting risk might be lowest for a specific design, but not the lowest possible. It would be far preferable to define and utilize design principles that ultimately result in the best aseptic processing design for a specific application. In considering what criteria to utilize, we believe adherence to the core principles of advanced aseptic processing is most appropriate: "An advanced aseptic process is one in which direct intervention with open product containers or exposed product contact surfaces by operators wearing conventional cleanroom garments is not required and never permitted" (14).

---

[c]Akers J and Agalloco J, personal communications, 2005–2009.

Full consideration of this expectation can be utilized to define the elements of facility design, equipment selection, container/closure selection, product delivery, personnel, procedure definition, and environmental monitoring.

Facilities

The selection of an appropriate advanced aseptic technology is central to nearly all of the subsequent design choices. The choice is often between closed RABS and isolators; however, other designs and technologies should be given due consideration. Once that basic choice has been made, there are options within those alternatives that should be considered as well to further define the technology to be implemented. The design process for an aseptic facility is a lengthy process: proceeding from conceptualization to preliminary and detailed design with a myriad of choices and decisions to be made throughout. Considering the core objective, the following preferences can be defined:

- Design for ease of execution through the choice of construction materials, design for ease of access, and detail elements that facilitate both cleaning and decontamination of the core environment.
- The material, personnel, and equipment flows should be defined to minimize mix-ups and contamination potential.
- The heating, ventilating, and air conditioning (HVAC) system should provide adequate air quality and pressurization to prevent the ingress of contamination.
- Air flow should be sufficient to provide high dilution rates, particularly within the most risk intensive locations within the environment.[d]
- Air systems should be supplied with high efficiency particle air (HEPA) filters that are periodically integrity tested.
- Differential pressures for the system should be controlled, monitored, and alarmed to support continuous integrity of the critical core area.
- Temperature and humidity should be controlled to maximize personnel comfort during operations consistent with product stability/safety requirements.
- Materials and personnel airlocks should be utilized to increase separation between environments of different classification
- Facility and enclosure surfaces must be resistant to the potential corrosive action of sanitizing and decontamination agents, especially sporicidal agents because of their generally greater chemical activity.
- The core aseptic environment should be maintained in an "aseptic" condition when in an operational state and periodically sanitized or decontaminated. Isolators and closed RABS should be decontaminated with sporicidal agents on a periodic basis.
- Only a minimum of materials should be retained in the aseptic portion of the facility through the utilization of just-in-time delivery to the aseptic area.
- Subjective regulatory tenants of aseptic processing such as smoke studies, air velocity measurements, unidirectional air flow, absence of eddy's should be considered but not overly weighted in the definition of HVAC design details. The absence of turbulence in any aseptic production environment is not physically achievable, and there are no objective metrics to define acceptable or unacceptable conditions.
- The completion of operation of RABS should be possible in a "closed" mode. Open door interventions during aseptic processing are never acceptable

Equipment/Utensils

In aseptic processing, the processing equipment located within the enclosure is always critical to success. The reliability of the equipment and the sophistication of its design are paramount in minimizing the need for interventions within the enclosure.

---

[d]The use of high air dilution rates in isolators has not been demonstrated to be of any meaningful benefit as it is with other aseptic processing designs.

- All product contact surfaces should be sterilized using validated methods (vibratory feed systems may be exempted from this requirement) provided they are high-level decontaminated with a sporicidal agent in situ. Their installation following sterilization often entails substantial and lengthy interventions that can result in contamination risk. Even in separative technologies, the need to curtail interventions persists.
- Sterilization-in-place and clean-in-place should be utilized wherever possible for product and gas delivery lines and filters. At the current state of technology, sterilization-in-place is possible for all types of aseptic filling processes including powder fill.
- Equipment and utensils should be sterilized in hermetically sealed containers/wrapping. The container design should be supported by scientific proof of their integrity. In separative technologies decontaminating utensils in situ may be the best alternative.
- Equipment and utensils should remain within sterile containers/wraps until entry into the critical zone just prior to use to avoid contamination that would occur if they were exposed in the adjacent less-clean environment.
- Equipment and utensils should be sterilized/depyrogenated using a just-in-time approach.
- Processing equipment should be selected for high reliability, ease of changeover, and remote adjustment. Wherever possible they should be self-clearing to eliminate the need for personnel intervention in the event of a miss-feed, jam, or other fault.
- Equipment change over from one format to another should be possible with a minimum of manual intervention.
- The process equipment should use Process Analytical Technologies (PAT) and other feedback systems for ease of control, operation, and documentation. This can result in fewer interventions in both the critical and background environments.
- Non–product contact portions of the equipment should be easily decontaminated and noninvasive of the critical zone.
- Equipment and enclosure surfaces should be resistant to the potential corrosive action of sanitizing and decontamination agents.
- Equipment surfaces within closed RABS should be easily accessible for high-level decontamination, automatic decontamination systems in RABS should be favored over manual decontamination activities as they are inherently lower risk since they can be accomplished with the system in a fully closed configuration and without human contamination.

Containers/Closures/Components

The containers and closures necessary are perhaps the most important items in an aseptic process. The ease of their introduction, transfer, movement, placement, and closure must all be successfully accomplished by the equipment with a minimum of human intervention. It should be immediately evident that they need to consistently process throughout the system, and thus high-quality components with extremely tight dimensional tolerances may be a required when compared to what might be customary in a less advanced (and thus markedly less capable) processing system part. As more complex and multifaceted combination products and medical devices are aseptically produced, the ability to sterilize, introduce, and feed components with complex shapes and special fitments have become an absolute requirement. Robotics, which can now withstand frequent exposure to agents such as vapor phase hydrogen peroxide (VPHP), can often handle complex parts and by utilizing vision systems and laser guidance achieve levels of flexibility and precision that would be impossible by more conventional means.

- Containers/closures/components must be prepared and sterilized/depyrogenated using validated processes.
- Containers/closures/components should be introduced in a manner that retains at least one layer of sterilized container or hermetic wrapping until entry into the critical zone. It is important to remember that in advanced aseptic processing systems such as

isolators, the entire enclosure must be considered the critical zone. The container should have a defined level of integrity. An important rule in isolator systems or closed RABS is that nothing should ever be transferred into the enclosure that is not equal to or lower than that environment. This necessitates the use of VPHP pass boxes, E-beam tunnels, or pass through systems that can be validated using biological indicators or in the case of radiation dosimetry. With proper design, execution, and procedures, Rapid Transfer Ports (RTPs) can also serve as transfer devices that ensure that the objective of taking in objects of equal or better contamination control quality than the environment is met.

- Containers/closures/components should be selected for reliability of handling in the processing equipment to avoid the need for corrective interventions. Higher Acceptable Quality Levels (AQLs) for defects can result in a reduction in the need for interventions.
- Containers/closures/components should be sterilized/depyrogenated using a just-in-time approach. Inventories of materials within the aseptic environment (especially the critical environment) should be minimized. In a typical separative technology–based aseptic processing space does not allow for substantial accumulation of parts and they are therefore typically transferred in on an as needed basis. However, if the criterion of transferring only materials of equal or better contamination controls quality than the enclosure environment, there is no reason to be concerned that such objects might become contaminated within the validated use or campaign period of an enclosure. Materials do not become less microbiologically clean over time in a well-controlled, separative, and unmanned environment.

Product
Delivery of sterile product to the critical zone is easily accomplished with minimal risk using either directly piped connections or RTP connection systems.

- Production materials must be prepared and sterilized using validated methods.
- Liquid product delivery piping should be cleaned and sterilized in place. Gas delivery piping should be sterilized in place.
- Any product delivery and other aseptic connections (e.g., inert gas) should be made within the enclosure.
- Where product is supplied to the critical zone in sterile container (e.g., sterile powders), it should be introduced in a manner that retains at least one layer of sterilized protective covering or wrap until entry into the critical zone.

Personnel
The operating personnel must be diligent in the operation of the equipment and adherence to the core principles of aseptic processing technique at all times. The permanent use of thicker gloves on an enclosure must not be misinterpreted as permission to operate in violation of defined aseptic procedures.

- Personnel must receive initial and periodic formal training in current Good Manufacturing Practices, aseptic processing, microbiology, aseptic gowning, and job specific tasks they must perform.
- Where appropriate personnel should be initially and periodically thereafter assessed for their proficiency in aseptic gowning; of course in many advanced aseptic processing systems gowning is limited and nonaseptic and will require no real training since it is not a critical risk mitigation factor in isolators and potentially in closed RABS as well.
- Personnel should be initially and periodically thereafter assessed for their proficiency in aseptic technique. Specific training should also be provided for those individuals performing the initial set-up of the equipment prior to the aseptic process. Obviously, in highly automated systems that do not rely in personnel or gowned operators, conventional clean room practices and traditions are not necessary.

- Personnel shall conform to the highest standards of aseptic technique at all times even when working with a closed RABS or isolator.
- Personnel should be periodically monitoring when exiting from the aseptic core. Isolator or closed RABS gloves, however, need only be tested at the end of a production run or campaign. It is not desirable to leave media residues on gloves and sleeve assemblies. Also, in separative technologies, glove integrity is the key to risk mitigation. That which does not leak cannot pass microorganisms, therefore physical testing is generally a better solution.
- Gown materials should be cleaned and sterilized using validated methods. It is not necessary to use sterile gowns in rooms surrounding isolators. Also, it is worth remembering that sterility is always a trade-off between microbial "kill" and damage to materials. Thus, extreme overkill is unwarranted where gowning materials are concerned; damage to the gown's integrity is a far greater concern than achieving extreme sterilization lethality levels that are meaningless anyway.
- Gloves must never contact product contact surfaces within an enclosure. Also, the gloves when used to make adjustments must never be put at risk from punctures, tears, or pinching. The operator should also avoid stretching glove/sleeve assemblies in an attempt to reach something within an enclosure. Stretching beyond the initial point of resistance can lead to wear at the glove/sleeve junction and perhaps even a full-blown separation. We cannot overemphasize the need for careful ergonomic design, and should flaws in ergonomics be found in operations they should be corrected immediately. It is possible in many enclosures to relocate gloves to better access positions.

Procedures

Interventions always increase the risk of contamination in an aseptic process even those using advanced technologies (however, the superior environmental control inherent in advanced aseptic technologies makes personal risk a far lower risk factor than in conventional clean rooms). The design of the facility, equipment, component, and product supply should serve to reduce the complexity, duration, and number of interventions. The "perfect" intervention is the one that is not necessary (15).

- Procedures should be critically reviewed to eliminate and/or simplify interventions throughout the aseptic processes.
- All interventions should be designed for minimal risk of contaminating sterile materials.
- Interventions performed during aseptic processing must be recognized as increasing the risk of contamination dissemination.
- All interventions should be performed using sterilized tools whenever possible.
- Defined procedures should be established in detail for all inherent interventions and more broadly for expected corrective interventions (where some flexibility in execution is necessary due to their greater diversity).

Monitoring

The monitoring of an advanced aseptic processing system plays a substantially less important role than it does in ordinary manned aseptic clean rooms, and it is important to recognize that even in standard clean rooms monitoring has a point of diminished return. Its eventual elimination as an anachronism in these extremely clean environments can be expected at some future time. In the interim, any monitoring performed must be accomplished in as minimally invasive and disruptive manner as possible.

- Monitoring of any type must not subject the product to increased risk of contamination. No monitoring is preferable to monitoring that increases the risk of contamination for sterile materials.

- Environmental monitoring activities must be recognized as interventional activities and subject to the similar constraints and expectations (including detailed procedures) as any other intervention.
- Monitoring must be recognized as subject to adventitious contamination pre- and postsampling that is unrelated to the environment, material, or surface being sampled. Methods to minimize that potential beyond what is incorporated into monitoring of conventional manned clean rooms may be necessary.
- Viable monitoring should not be considered an "in-process sterility test" regardless of whether the sample is taken in the enclosure or of a so-called "critical" product contact surface.
- Environmental monitoring results should not be considered as "proof" of either sterility or nonsterility.
- It must be recognized that microbial monitoring can never recover all microorganisms present in an environment nor on a surface.
- The absence or presence of microorganisms in an environmental sample is not confirmation of asepsis nor is it uniformly indicative of process inadequacy.
- Significant excursions from the routine microbial prolife within the enclosure and background environments should be investigated.
- Detection of low numbers of microorganisms in manned clean rooms should be considered a rare, but not unusual event.
- Investigations into recoveries of low numbers of microorganisms in manned clean rooms should be recognized as predominantly make work exercises.
- Process simulations are indicators of process capability but cannot definitely establish the sterility of material produced at another time.
- Process simulations in excess of 5 to 10,000 units are of relatively limited value; their greatest utility is in the evaluation of aseptic set-up practices.

CONCLUSION

What has been presented above represents a major departure from the established doctrine for aseptic processing control. The ever increasing capabilities of aseptic processing technologies have dramatically reduced the utility of the classical monitoring tools that this industry has used for decades. If some future technology enables effective monitoring at the extremely sensitive levels that advanced aseptic processing systems presently provide, there may be justification in their use. We might postulate that by the time those systems become available, future aseptic processing technologies demonstrating superior capabilities to those presently available might make those new monitoring tools moot as well. In the interim, it is clearly time to shift the paradigm for advanced aseptic processing systems away from monitoring and toward their design. Where monitoring is used, total particulate monitoring has significant advantages over microbiological monitoring. Total particulate monitoring provides a real-time indication of a major change in the physical performance of HEPA filters or a significant increase in particles produced by processing equipment, something that conventional microbiological monitoring cannot do.

Our industry has always sought to improve the sterility of aseptically produced products. For many years this was accomplished through measures that were more or less instinctive rather than reflective of real science and engineering. The adoption of risk-based approaches is a relatively new concept, but it is essential that the practitioner take the next step. Mitigation is of far greater importance in the overall effort and provides a greater measure of safety to aseptic operations.

REFERENCES

1. FDA. Guideline on Sterile Drug Products Produced by Aseptic Processing, 2004.
2. Decision Trees for the Selection of Sterilization Methods (CPMP/QWP/155/96), 1999.
3. Agalloco J, Gordon B. Current practices in the use of media fills in the validation of aseptic processing. J Parenter Sci Technol 1987; 41(4):128–141.

4. PDA TR# 36. Current practices in the validation of aseptic processing—2001. PDA J Pharm Sci Technol 2002; 56(3).
5. Avallone H. FDA Field Investigator Training Curriculum, Circa 1985.
6. Agalloco J, Akers J. Risk analysis for aseptic processing: the Akers-Agalloco method. Pharm Technol 2005; 29(11):74–88.
7. Agalloco J, Akers J. Simplified risk analysis for aseptic processing: the Akers-Agalloco method. Pharm Technol 2006; 30(7):60–76.
8. Whyte W, Eaton T. Microbiological contamination models for use in risk assessment during pharmaceutical production. Eur J Parenter Pharm Sci 2004; 9(1).
9. Tidswell E, McGarvey B. Quantitative risk modeling in aseptic manufacture. PDA J Pharm Sci Technol 2006; 60(5):267–283.
10. Katayama H, Toda A, Tokunaga Y, et al. Proposal for a new categorization of aseptic processing facilities based on risk assessment scores. PDA J Pharm Sci Technol 2008; 62(4):235–243.
11. PDA, TR #44. Quality risk management for aseptic processes. PDA J Pharm Sci Technol 2008; 62(suppl 1).
12. FDA. Pharmaceutical cGMPs for the 21st CENTURY: A Risk-Based Approach, 2002.
13. ICH, Q9. Quality Risk Management, 2005.
14. Akers J, Agalloco J, Madsen R. What is advanced aseptic processing? Pharm Manuf 2006; 4(2):25–27.
15. Agalloco J, Akers J. The truth about interventions in aseptic processing. Pharm Technol 2007; 31(5): S8–S11.

3 | Development challenges and validation of fill and finish processes for biotherapeutics

Karoline Bechtold-Peters

INTRODUCTION

The therapeutic antibodies presently sold on the international markets are all administered via parenteral route, that is, by subcutaneous or intramuscular injection or intravenous infusion, with one product injected into the vitreous humor of the eye (Table 1). Approximately, two-thirds of these preparations are liquid-stable preparations with 1 to 100 mL nominal volume; one-third of the preparations are marketed as lyophilized powder for reconstitution. With this selection of routes of application and dosage forms, it becomes quickly clear that the manufacturing procedure of choice for therapeutic antibody products is a liquid filling process under aseptic conditions, optionally with an additional freeze-drying process directly in the vial. Sterilization in the final container is practically excluded because of the thermolability of proteins.

Monoclonal antibodies, as well as other therapeutic proteins, are sensitive toward various stresses such as heat, shear, interfaces, and foaming (Fig. 1), and since aseptic processes in general are high-level risk processes, suitable attention must be dedicated to the development of the manufacturing procedure, process parameters, as well as their validation.

In the first section of this chapter, manufacturing steps up to the final product will be described, including challenges in the development of the fill and finish process based on various case studies. The second part will go into details as regards to the validation of the pharmaceutical manufacturing process taking into account current authority views like the "risk-based approach" of the FDA.

COMPONENTS AND PROCESS STEPS IN FILL AND FINISH

Typically the process chain for the finished drug product starts with the final formulation step. Often times this is the last step in the downstream drug substance process (Fig. 2). The concentrated bulk is diluted, if necessary, after thawing, preferentially under controlled conditions with a formulation buffer containing functional excipients to obtain the required protein concentration. The bulk solution is generally isotonic at a physiologically acceptable pH range and contains various stabilizers, for example, nonreducing sugars, amino acids, complexing agents, antioxidants, or cake formers.

The tanks containing the formulated solution are placed in proximity to the grade A area. For aseptic processes in isolators, the tank resides in grade D (or better), while for classical clean room operation the tank resides in grade C area. One or two sterile filters in series are placed as a part of transfer assembly to sterile filter the bulk solution into the grade A room for filling. The sterilizing grade filters are checked for integrity immediately before the sterile filtration and are again checked for membrane integrity following completion of the filtration step. The sterile filtration can occur off-line into a second hold container (Fig. 3). After finalization of filtration this hold container is connected to the filling machine and the filling starts. For additional sterility assurance, an additional sterile filter may be placed very close to the filling needle (EU GMP Guide, Annex 1, 2008).

If large volume batches are to be filled, an in-line filtration (Fig. 3) directly from the bulk tank, via one or two sterile filters, into a smaller intermediate container (surge vessel) does make sense. In this case, the bulk tank can be placed outside of grade A area avoiding extra manipulations and equipment in the clean area. Reduction in the process time occurs as the sterile filtration and filling proceed in parallel and there is no need to clean and sterilize at least one or more stainless steel tanks. Before the start of the sterile filtration, a sample for bioburden is collected for every batch of drug product (EU Guidelines to GMP, Annex 1, 2008). To avoid false positive results for the bioburden, closed, presterilized disposable systems are useful (1).

Table 1 Therapeutic[a] Monoclonal Antibodies on the Market (December 2009)

INN name	Trade name	Type of antibody	Dosage form	Application route
Oncology				
Alemtuzumab	MabCampath®	Humanized	Liquid, vial	IV infusion
Bevacizumab	Avastin®	Humanized	Liquid, vial	IV infusion
Cetuximab	Erbitux®	Chimeric	Liquid, vial	IV infusion
Panitumumab	Vectibix®	Human	Liquid, vial	IV infusion
Rituximab	MabThera®, Rituxan®	Chimeric	Liquid, vial	IV infusion
Gemtuzumab	Mylotarg®	Humanized, conjugated to calicheamicin	Lyophilized powder, amber vial	IV infusion
Ibritumomab	Zevalin®	Murine, ^{90}Y conjugated	Liquid, vial, radiolabeled	IV infusion
Tositumomab	Bexxar®	Murine, ^{131}I-conjugated	Liquid, vial, radiolabeled	IV infusion
Trastuzumab	Herceptin®	Humanized	Lyophilized powder, vial	IV infusion
Ofatumumab	Arzerra	Human	Liquid, vial	IV infusion
Catumaxomab	Removab	Rat-murine	Liquid, prefilled syringe	IV infusion
Autoimmune diseases, transplantation therapy				
Adalimumab	Humira®	Human	Liquid, prefilled syringe	SC injection
Basiliximab	Simulect®	Chimeric	Lyophilized powder, vial	IV infusion or injection (bolus)
Daclizumab	Zenapax®	Humanized	Liquid, vial	IV infusion
Infliximab	Remicade®	Chimeric	Lyophilized powder, vial	IV infusion
Muromonab	Orthoclon OKT3®	Murine	Liquid, ampoule	IV injection (bolus)
Natalizumab	Tysabri®	Humanized	Liquid, vial	IV infusion
Efalizumab	Raptiva®	Humanized	Lyophilized powder, vial	SC Injection
Certolizumab pegol	Cimzia®	Humanized	Lyophilized powder, vial AND Liquid, prefilled syringe	SC Injection
Golimumab	Simponi®	Human	Prefilled syringe with needle guard or in autoinjector	SC injection
Tocilizumab	Actemra®, RoActemra®	Humanized	Liquid, vial	IV infusion
Canakinumab	Ilaris®	Human	Lyophilized powder, vial	SC injection
Other indications				
Abciximab	ReoPro®	Chimeric (Fab$_2$ fragment)	Liquid, vial	IV injection followed by infusion
Palivizumab	Synagis®	Humanized	Liquid, vial (new dosage form)	IM Injection
Ranibizumab	Lucentis®	Humanized (Fab fragment)	Liquid, vial	Intravitreal injection
Omalizumab	Xolair®	Humanized	Lyophilized powder, vial	SC injection
Eculizumab	Soliris™	Humanized	Liquid, vial	IV infusion

[a]In addition, monoclonal antibody-based imaging agents are available; e.g., CEA-Scan (Arcitumomab Tc-99), LeukoScan (Sulesomab Tc-99), ProstaScint (Capromab pendetide Indium-111), and Verluma (Nofetumomab merpentan Tc-99).

Dispensing into the vials, cartridges, or syringes occurs by pumps, that is, peristaltic pumps, piston pumps, or rolling diaphragm pumps, or by time-pressure or mass flow filling systems. Formulation solution is placed into sterile containers via 1 to 16 needles. The containers are closed immediately by sterilized stoppers or sealing discs (with cap). Filling and stoppering process steps are highly critical as they are ran while the containers are open. Particles and microbes can reach the open product. With therapeutic proteins such as monoclonal antibodies, a heat sterilization in the final container is not feasible. The foreign particulate matter specification is particularly stringent for large-volume solutions for infusion to avoid embolism in the patient (Table 2).

Biological activity dependent on primary sequence, and on secondary and tertiary (or quarternary) structure, i.e. the **3D conformation**

Reversible or irreversible **denaturation, if unfolded**: chemical degradation, aggregation, dissociation of oligomer to monomers, adsorption to surfaces, incorrect folding

Hydrophilic to hydrophobic dependant on structure and glycosylation pattern, but mostly **hardly soluble in organic solvents**

Sensitive to various agents like high pressure, interfaces/surfaces, pH changes, high temperature, freezing, oxidation, dehydration, chemicals (e.g. chaotropic salts)

Figure 1   Special properties of biologics as drugs.

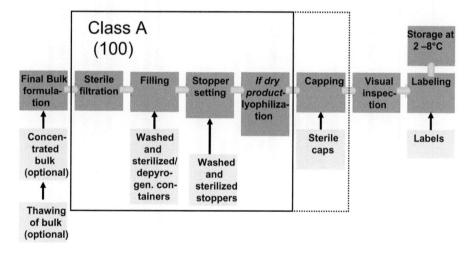

Figure 2   Typical fill and finish process into vials.

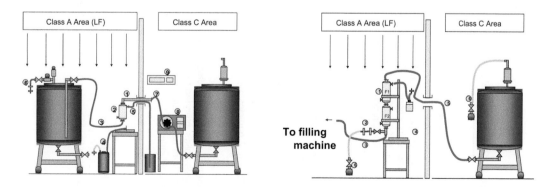

Figure 3   Examples for off-line and in-line sterile filtration setups.

Table 2 Pharmacopoeial Requirements as Regards to Particles in Injectables

USP 30-NF25 <788>	Ph Eur. 6.8	JP XIV (foreign insoluble matter test and insoluble particulate matter test)
Subvisible particles		
Limits for Light Obscuration Test Particle Count[a]	Limits for Light Obscuration Test Particle Count[b]	Criteria
Small volume injections (≤100 mL volume)	≤100 mL volume	≥10 μm: 25 particles/mL
≥10 μm: 6000 particles/container	≥10 μm: 6000 particles/container	≥25 μm: 3 particles/mL
≥25 μm: 600 particles/container	≥25 μm: 600 particles/container	
Large volume injections (>100 mL volume)	>100 mL volume:	
≥10 μm: 25 particles/mL	≥10 μm: 25 particles/mL	
≥25 μm: 3 particles/mL	≥25 μm: 3 particles/mL	
Limits for Microscopic Method Particle Count[a]	Limits for Microscopic Method Particle Count[b]	
Small volume injections	≤100 mL volume	
≥10 μm: 3000 particles/container	≥10 μm: 3000 particles/container	
≥25 μm: 300 particles/container	≥25 μm: 300 particles/container	
Large volume injections	>100 mL volume:	
≥10 μm: 12 particles/mL	≥10 μm: 12 particles/mL	
≥25 μm: 2 particles/mL	≥25 μm: 2 particles/mL	
Visible particles	*Parenteral preparations*: solutions for injection/infusion examined under suitable conditions of visibility (2,000–3,750 lux) are clear and practically free from particles	Injections must be clear and free from readily detectable foreign insoluble matter (8,000–10,000 lux)
Injectable solutions, including solutions constituted from sterile solids intended for parenteral use, are essentially free from particulate matter that can be observed on visual inspection.	*Monoclonal antibodies for human use*: liquid or reconstituted freeze-dried preparations are clear or slightly opalescent and colourless or slightly yellow, without visible particles	

[a]Unless solely intended for IM and SC administration or the label states that the product is to be used with a final filter are exempt from these requirements.
[b]In the case of preparations for SC or IM injection, higher limits may be appropriate; radiopharmaceuticals and preparations for which the label states that the product is to be used with a final filter are exempt from these requirements.

In the last step of the manufacturing process, the stopper is fixed, in case of vials, by a cap. Because the vial is sealed using a suitable stopper/vial combination and the fact that the stopper is pushed-in (2), the vial capping theoretically can be performed in grade C area—although under laminar air flow. However, due to current EU regulation (EU GMP Guide, Annex 1), many manufacturers have upgraded the capping process to occur under grade A conditions.

The manufactured vials, syringes, or cartridges are 100% visually inspected by trained examiners for various defects, either manually or using automatic machines. The quality of 100% visual inspection is verified by taking random product samples according to Acceptable Quality Level (AQL) tables that are based on Military Standard 105D.

In the case of lyophilizates, the freeze-drying stopper is pressed only partially into the container and the half-closed containers are then loaded into the freeze-dryer. At the end of the lyophilization process, while the vials are still in the freeze-dryer, the stoppers are pushed into the containers as the shelves are pressed together. Following unloading, the capping is performed in grade A or C areas under laminar flow according to the procedure specified by the company.

The filling process is preceded by the preparation of the packaging materials that are washed and if necessary siliconized and sterilized after unpacking. However, the use of RTS (ready-to-sterilize) or RTF (ready-to-fill) quality stoppers or caps are widespread throughout the industry. A special "Closed-Vial" technology (3) is based on prewashed, irradiated, already stoppered cyclic olefin copolymer (COC) vials that are filled through the stopper and then resealed by laser light and equipped with a flip-off cap. For lyophilization process, closed stopper systems that allow moisture to escape but prevent bacteria or particulates from entering are offered by at least two companies.

SELECTED CASE STUDIES EXEMPLIFYING DEVELOPMENT CHALLENGES DURING FILL AND FINISH

Problems can occur with each of the mentioned process steps under production conditions, if the procedures, process parameters, and materials in contact with the product were not examined carefully for possible influence on the product (Table 3). The whole manufacturing process must be broken down, during development, into unit operation steps and the criticality of every step be examined.

Detailed descriptions of this process are exemplified in the following case studies.

Filtration

Because of the thermolability of the biologically active substance, the sterile filtration constitutes the only possible sterilization procedure for the protein or antibody-containing formulated bulk. The aspects that are to be followed and examined for definition of the sterile

Table 3 Instability of Therapeutic Proteins and Stress Factors Occurring During Final Drug Product Manufacture

Protein instability		
Type of instability	Stress factors	Time point of occurrence during manufacture of final drug product
Physical	pH	Mixing
Denaturation	Ionic strength	Transport/transfer
Aggregation (reversible,	Denaturants	Filtration
irreversible)	Metal ions	Filling
Precipitation	Oxygen	Freezing
Adsorption	Light	Drying
Chemical	Temperature	Visual inspection
	Shear	
Hydrolysis/deamidation/fragmentation	Interfaces, surfaces (air/liquid,	
Oxidation	liquid/glass, or metal or plastic)	
Isomerization		
Disulfide exchange		

Table 4   Considerations for Sterile Filter Selection

Nature and type of membrane material
- Cellulose esters
- Polyvinylidenfluoride
- Polyethersulfone
- Nylon
- Polycarbonate

As capillary (with cylindrical straight-through pores) or noncapillary filters (8)

Membrane and solution factors	Filtration parameter
<ul><li>Adsorptives behavior</li><li>Expected bioburden/particle burden</li><li>Total filtration volume</li><li>Logistics/supplier relation</li></ul>	<ul><li>Product contact time, filtration time</li><li>Filtration pressure</li><li>Flow rates per unit area</li><li>Filtration temperature</li><li>Area-specific bioload per unit area (and specific bioburden microorganisms if relevant)</li></ul>

Types of filter studies
- Adsorption of formulation components
- Flow rate over time/after interruption
- Membrane integrity (bubble point, diffusive flow, forward flow), with WFI/purified water and product
- Bacterial challenge/bacterial retention performance including viability of test bacteria in formulation
- Correlation between bacterial retention and integrity test method
- Extractables (chemicals, particles)
- Sterilizability (sterility with/without manifolds attached, effect of sterilization on filter integrity)
- Chemical compatibility

filter are summarized in Table 4. The PDA Technical Report No. 26 (4) gives a good overview of the key considerations during sterile filtration of solutions.

As a rule, 0.2- or 0.22-μm filters are used for the sterile filtration. This filter rating refers to the smallest size of microorganisms and particles that are removed rather than to the actual size or form of the filter pores (5). Usually pressure applied during sterile-filtration process is ≤ 2 bars. Because the composition of the bulk formulation can influence the ability of the filter to hold back bacterial loads, bacteria-retaining filters must be validated specifically for a product. Bracketing can be considered with very similar products (6). During the bacterial retention study, the filter is challenged with >10^7 cfu of *Brevundimonas diminuta* (ATCC 19146) per cm^2 under worst-case process conditions expected during filtration of the drug product, and the ability to retain challenge microorganism (and hence ability to achieve a sterile product) is examined (4,7). Unfortunately, the bulk volumes required for these studies are substantial (several liters), even after usual scale-down to 47-mm diameter disks, which renders the study an expensive investigation, especially for highly concentrated formulations. Although complete filter validation package is only available at the time of process validation, it is recommended that at least a risk assessment or an abbreviated bacterial retention should be performed during clinical development for novel formulations, because filtration is the only sterilization step for the heat-unstable protein preparations.

Coarser, foreign particles and large aggregates are removed during sterile filtration using a 0.2-μm filter. Consideration must be given to the desired removal of contaminants like bioburden and the undesirable removal of active substance or excipients by adsorption to the large surface of the filter membranes. The typical area of a 0.22-μm filter is >100 m^2/g (8). Furthermore, the sterile filter must be compatible with the bulk protein solution under filtration conditions. For most protein formulations, this is uncritical given the aqueous nature, low-processing temperature between 0 and 30°C, and a pH range between 5 and 8. However, during the process development stage, studies must be conducted to confirm the compatibility. Studies by Pitt (9) show that different filter materials bind protein weakly or very strongly depending on the polymer material. Pitt found that mixed cellulose ester and nylon have stronger affinity to proteins than poly(vinylidene fluoride), or PVDF, and polysulfone.

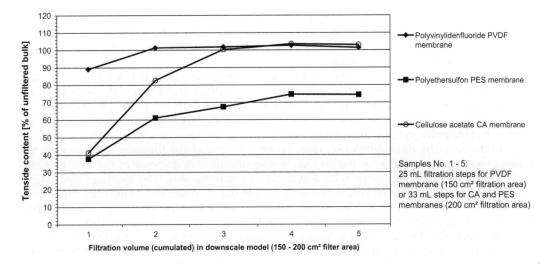

Figure 4   Adsorption of surfactant (Tenside) from a monoclonal antibody containing formulated bulk (0.01% of polysorbate 80, pH 5.5) during filtration through three filter membranes, filter area 150 to 200 cm^2.

Excipients can also bind to filters with varying avidity. Surfactants, for example, are preferentially distributed at surfaces because of their interfacial properties, their amounts can be reduced significantly if the formulation has low concentration of surfactant and the process involves multiple filtrations. In an unpublished study with antibody-containing bulk (0.01% polysorbate 80 and 5 mg/mL of an antibody), three sterile filters, composed of PVDF, polyether sulfone (PES), and cellulose acetate (CA), were examined (Fig. 4). The following order of loss of polysorbate 80 was observed: PVDF < CA <<PES.

Similar investigations must also be undertaken for preservatives added to multiple-dose preparations, for example, growth hormone or insulin, as they can also be removed in significant amounts by filtration.

In the case of undesirable adsorption of formulation components or protein, saturating the membrane by prerinsing with formulation buffer can be a mitigation strategy. In addition, choosing a suitable, small surface area with low absorption filter material will be important. With in-line filtration, filtrate might be discarded before the beginning of the main filling process to guarantee the homogeneity of the drug product over the whole batch.

During process development, migration of potential extractables from filter membrane, filter housing, supporting fabrics, sealing disks, O-rings, and tubing into the formulation should be examined. These extractable studies are conducted following the sterilization of the filters by steam or by irradiation. Potential extractable sources from sterilizing filters may include surfactants and wetting agents, additives used in the plastic component manufacture, manufacturing debris, monomers of materials of construction, etc. Unfortunately, neither the pharmacopoeias nor official guidances give concrete information regarding the acceptable amount of extractable substance from the filters. Therefore, the toxicological assessment and check of compatibility is left eventually with the user. Current filter manufacturers offer services where they can conduct studies on extractables or provide information to the user on the basis of previous experience.

As a part of filter extractable study, filters from several lots are eluted following worst-case pretreatment (several sterilization cycles, higher sterilization temperature, and sterilization time) and under more aggressive model conditions (as regards to time, temperature, solvent/power of elution, pH). The eluates are analyzed gravimetrically (NVR, nonvolatile residues), by TOC (total organic carbon), liquid chromatography (RP-HPLC), capillary electrophoresis (HPCE), gas chromatography (GC and GC-MS), or Fourier transform infrared (FTIR) spectroscopy. Because protein and formulation components interfere with the analyses, water-based solvent system is mostly used as representative extraction media. The found extractables must be classified afterward as toxic or nontoxic. Additional testing like biological reactivity in vitro and in vivo according to USP <87> and <88> may have to be performed to provide added safety assurance. In the end, a recommendation for a preflush volume (e.g., 2 L for a 4 in. capsule) is made, so that possible extractables are already removed to a great extent before the main filtration commences.

With protein preparations, the testing for membrane integrity is not trivial. The filter must be rinsed with large amounts of water to remove surface active materials, often at raised temperature, before the established value for bubble point, forward flow/diffusive flow, or pressure decay is reached. For the postuse membrane integrity determination, a product-specific bubble-point value is recommended.

In the end, the filter dimension must be chosen so that the filtration rate fits with the filling process, in particular for the on-line/in-line filtration. A decrease of the filtration rate over time or after an interruption with downtime due to increasing filter blockage resulting in protein fouling can be mostly avoided by determining the adequate size of the filter membrane surface. Nevertheless, an oversizing should also not occur because of the already described adsorption losses and large dead volumes in the filter of expensive drug product.

Maa and Hsu (8) compared the fouling behavior of different proteins [recombinant human growth hormone (rhGH), recombinant human deoxyribonuclease (rhDNAse), recombinant tissue plasminogen activator (rt-PA), an anti–immunoglobulin E (anti-IgE) antibody] during sterile filtration and examined some possible fouling mechanisms: pore narrowing, adsorption because of nonspecific binding between membrane and protein, shear-induced adsorption and aggregation, and adsorption due to hydrophobic surfaces. The filtration flow over the time of 0.09 molar solutions was noted. All solutions, including anti-IgE solutions, showed a decline of filtration flux with increasing filtration volume due to membrane fouling, although this was relatively slightly pronounced for the anti-IgE solution (slope, -0.004). A clear correlation between initial flow and molecular weight was noted with the monoclonal antibody solution filtering being very slow due to the high molecular weight (~ 2.7 mL/min/cm^2 at about 13 mg/mL concentration, Millipore filter from mixed cellulose ester, 9.4 psi filtration pressure). Fouling at filtration membranes, tubing, or other surfaces can be affected by the respective protein formulation (pH, ionic strength) and is often reduced by the addition of surfactants such as polysorbates (8–10).

Filling

Protein-containing solutions tend to foam so in some cases they not only compromise filling process but also cause a reduction of the protein integrity. In downscale models, which shake the formulated bulk or press it with increasing velocity through narrow cannulas, the foam behavior can be explored a priori as a part of the formulation development. Because excessive foaming leads to inaccuracy in dosing, the filling speed must be lowered drastically or filling of larger volumes performed in two steps by means of two filling stations in series. In the end, the filling mode also plays an important role. Most filling machines at production scale are equipped with a movable filling needle that follows during the filling process the upward moving liquid meniscus. Hereby, the filling needle can be led below the meniscus, at the meniscus, or above the meniscus.

In a filling test with an antibody-containing bulk solution (protein concentration 5 mg/mL), the filling "under the meniscus" turned out to be advantageous compared with the filling "above the meniscus" (unpublished results) (Fig. 5). For a further model describing foaming behavior see Maa and Hsu (11).

A decisive influence on the filling process is the choice of filling or dosing system (principle). The following systems are available:

- Peristaltic pumps
- Piston pumps (180° or 360°)
- Rolling diaphragm pumps
- Time-pressure filling systems
- Mass flow filling systems

These dosing systems differ both in dosing precision and in shear stress applied to the protein solution. While peristaltic pumps enable very smooth filling, piston pumps may be

Filling ABOVE meniscus

Filling BELOW meniscus

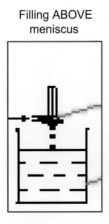
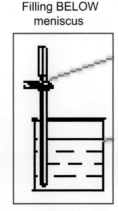

Figure 5 Filling modes into vials, cartridges, and syringes.

Table 5 The Excess Volumes Recommended by USP Usually Sufficient to Permit Withdrawal and Administration of the Labeled Volumes (USP <1151>)

Labeled size	Recommended excess volume	
	For mobile liquids	For viscous liquids
0.5 mL	0.10 mL	0.12 mL
1.0 mL	0.10 mL	0.15 mL
2.0 mL	0.15 mL	0.25 mL
5.0 mL	0.30 mL	0.50 mL
10.0 mL	0.50 mL	0.70 mL
20.0 mL	0.60 mL	0.90 mL
30.0 mL	0.80 mL	1.20 mL
50.0 mL or more	2%	3%

more accurate dosing devices. If feasible, at least in development, various dosing systems should be tested to meet the needs of the product.

At the beginning of the filling process, a target in-process specification with alert and action limits is defined by determination of the minimal fill volume, which will assure that specified extractable volume can be withdrawn. USP Chapter <1151> provides recommendation for injectable excess fill volumes for low- and high-viscosity products (Table 5). Proteinaceous solutions are typically between the two classes. Because of the high molecular weight of proteins and antibodies (5–150 kDa), the colloidal solutions formed may be rather viscous at high concentrations. To cope with the volume restrictions as regards subcutaneous application (max. 2–3 mL volume), concentrations of 100 to 200 mg/mL may be the goal with viscosities far beyond water-like liquids (Fig. 6A). Also, since viscosity varies with temperature as depicted in Figure 6B, a dosing system such as the time-pressure filler is specifically challenged and requires an experimentally obtained temperature compensation algorithm into the process control system.

Effect of Contact Surfaces
During final formulation, storage of bulk and filling of the antibody preparations glass, steel, and plastics are typical materials in contact with product (e.g., silicone tubes, polypropylene filter housings, polyvinylidene difluoride filter membranes).

Although the pharmaceutical industry uses high-quality stainless steels, 1.4404/1.4435 (316L) and 1.4539 (904L), the potential to release small quantities of metal ions must not be neglected. Lam et al. (13) described considerable reduction in the oxidation during the storage, particularly at Met255, by replacing the stainless steel filler by a filler from an alternative material (52% oxidized Fc compared with 18% after 2 weeks at 40°C). This oxidation was ascribed to the corrosion of steel by chloride ions after contact with NaCl containing formulation buffer at low pH (pH 5.0) and extraction of iron ions. Metal ions such as iron ions can react with peroxide

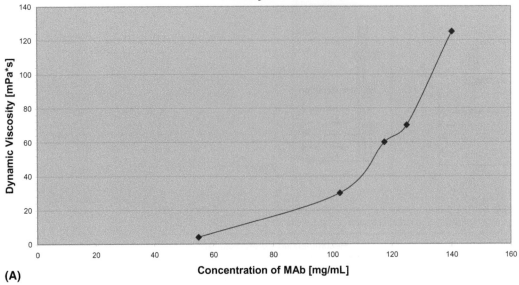

(A)

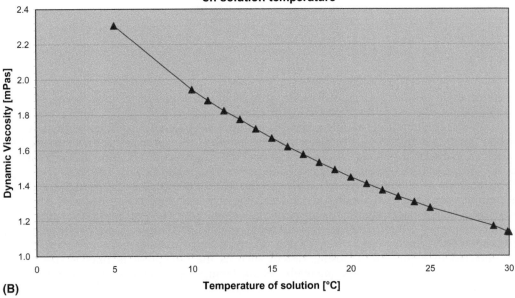

(B)

Figure 6 **(A,B)** Dynamic viscosities of monoclonal antibody containing bulks. Influence of temperature and antibody concentration on viscosity. *Source*: From Ref. 12.

impurities (e.g., from polysorbates) or the protein itself in the formulation to form free radicals that initiate oxidative degradation. In the same study, Lam et. al. could generate 26% increase of the anti-HER2 oxidation after storage only by manufacture of the NaCl containing diluent buffer in a steel tank. After production of the buffer in a glass container no oxidation occurred (13). Furthermore after three months of storage of the NaCl containing anti-HER2 formulation in a 30 mL stainless steel container, up to 3 ppm of extracted iron ions were determined. However, such data can be transferred only to a limited extent to tanks in the pilot and production scale because of the unfavorable surfaces to volume ratio in such minitanks.

Metallurgically it is important to distinguish reactions that lead to either rouging or leaching of metal ions from steel surfaces. The latter is favored in acidic solution and in the presence of chloride ions. The rouging progress on the other hand represents an inversion of the formerly chromoxide-rich passive layer to an iron oxide–dominated porous surface layer due to the environmental influences (medium, temperature). These environmental factors are responsible for disturbing the dynamic balance of the former passive layer (14). Rouging on stainless-steel surface is often detected during preventive maintenance or by presence of reddish to brownish coloring, for example, in water for injection conduit systems, reddish iron oxide particles are then found on particle filters and sieve inserts. By chemical cleansing cycles, the corrosion layer can be stripped off (e.g., citric acid, EDTA, H_3PO_4, HF + HNO_3) and the surface repassivated (oxidizing materials such as HNO_3), so that the corrosion resistance of the stainless steel is restored (15). The cleanliness and absence of particles from stainless steel components is necessary for the production of parenterals. Substantially higher quality stainless steel can be obtained after electropolishing than after mechanical polishing procedures (removal of Beilby layer) (16).

Air-Water Interface

Antibodies and other therapeutic proteins can be damaged by shear or stressing at interfaces in various steps of the process. These stresses occur not only in upstream and downstream processes such as during aeration and agitation, recirculation, centrifugation and filtering for cell separation, purification, buffer exchange, and concentration, but also during production of the pharmaceutical final product. Interface stress occurs during final mixing of drug product, sterile filtration, filling, conveying, transporting, and shaking during visual control. A summary of stress factors during production of the final product is represented in Table 3.

Proteins diffuse to and subsequently orient to interfaces formed. This process may be followed by unfolding of globular configurations to the denatured state. Ultimately aggregation and precipitation may occur dependent on the conformational stability of flexible segments of the protein molecule. During many of the processes mentioned in the previous paragraph, the interfacial film is continually renewed with progressively more protein exposed to the interface resulting in further loss in activity (17). Harrison et al. examined the effect of high shearing (20,000/sec) on scFv fragments in an underfilled cylindrical minireactor with impeller. In the absence of protective additives, 80% irreversible loss of activity was measured compared with the unsheared scFv fragment (17). Other reports suggest that shear, without generation of new interface, has minimal effect on protein activity.

Listed in Table 6 are the shear rates during pharmaceutical process steps under usual conditions. For a noncompressible Newtonian liquid, the shear rate occurring at the capillary walls γ_{wall} in sec^{-1} is calculated according to equation (1) (18,19):

$$\dot{\gamma}_{wall} = \frac{4Q}{\pi R^3} = \frac{8V}{D} = \frac{4V}{R} = \frac{32Q}{\pi D^3} \qquad (1)$$

where Q is the volumetric flow rate in cm^3/sec; V, the average flow velocity in cm/sec; D, the inner diameter of the capillary in cm; and R, the inner radius of the capillary in cm.

Table 6 Estimates of Shear Stress Applied During Fill and Finish of Antibody Solutions (20 mL Fill Volume, 10 mg/mL Antibody Concentration, Newtonian Behavior, One Filling Station)

Process step	Conditions and assumptions for calculation	Shear rate per orifice (sec^{-1})
0.2 μm filtration	130 L filtered within 30 min (off-line) or 6 hr (on-line), filter with 10^8 pores/cm^2 at a diameter of 0.2 μm/pore, 1900 cm^2 filtration area (Opticap XL4)	4.8×10^5 (off-line) 4.0×10^4 (on-line)
Rotary pump	130 L filled within 6 hr, 25% of time flow through outlet, ID outlet 5 mm	2.0×10^3
Time-pressure filler or peristaltic filler	130 L filled within 6 hr, 80% of time valves opened, ID 7 mm	2.2×10^2
Needle size A	130 L filled within 6 hr, 50% of time flow though needle, ID 7 mm	3.6×10^2
Needle size B	130 L filled within 6 hr, 50% of time flow though needle, ID 5 mm	9.8×10^2

Abbreviations: ID, inner diameter.

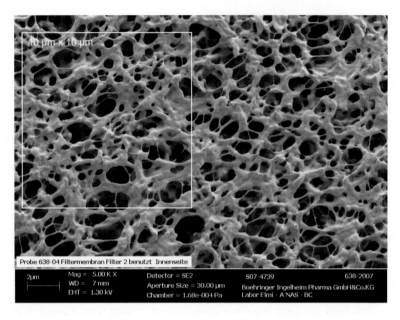

Figure 7   SEM photograph of sterile filter membrane (Durapore) 0.2 μm pore structure after use.

It becomes obvious that filtration through narrow pores under pressure (Fig. 7) imposes the comparatively highest stress effects on the antibody, with the filter structure and the pore size having a considerable influence. Shear stress in piston pumps or time-pressure pumps as well as during dosing by needles is considerably lower. Nevertheless, the calculations show that with the reduction of the inner diameter of the filling needle the shear rate increases and that peristaltic pumps as well as time-pressure pumps are markedly more gently dosing than piston pumps.

For the development of protein products stress studies during the process development are important. The following systems may be used as downscale models: vial and bottle shakers, pumps, impeller systems, Couette flow systems, ultrafiltration modules, capillaries, static and dynamic concentric cylinder systems, and rotating disk reactors. These differ clearly in the applied shear rates in the range of a few 100/sec up to 26,000/sec (20). Examples of such shear stress test systems are represented in Figure 8.

Light, Oxygen, and Temperature

The influence of light must be considered on the stability of protein. Under the influence of light (2 weeks, 20,000 lux, 27°C), the recombinant antibody anti-HER2 in liquid formulation showed oxidation at Met255 in the heavy chain of the Fc region. The light-induced oxidation of recombinant monoclonal antibody anti-HER2 occurs through the singlet oxygen pathway and could be controlled effectively by antioxidants like 3.5 mM methionine or 6.3 mM sodium thiosulfate (13). With very photosensitive proteins, precautionary controls to reduce photodegradation during manufacturing should be in place, for example, use of stainless steel tanks, in place of glass vessels, since they are impervious to light, use of opaque tubings, colored plastic carton boxes to minimize light exposition during intermediate storage and visual inspection, etc. If necessary manufacturing and filling under red or yellow light is an option for the parenteral product based on the knowledge of the molecule sensitivity to the various wavelengths of light. In general, the room light mapping is also performed, and work processes are designed that avoid unnecessary light exposure and define the maximum total light exposure to the product over the entire manufacturing, packaging, and labeling operations.

The effect of molecular oxygen on oxidation of recombinant monoclonal antibody anti-HER2 was examined by Lam et al. by replacing headspace oxygen with nitrogen before stress

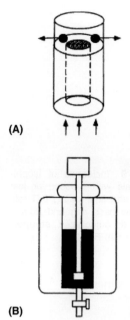

(A)

(B)

Figure 8 Schematic representation of a rotor/stator assembly (**A**) and of a homogenizer system (**B**) used by Maa and Hsu in shear stress studies for proteins. *Source*: From Ref. 20.

storage at 40°C for two weeks (13). Whereas the control sample developed 52% oxidized Fc after two weeks at 40°C, the removal of oxygen in the sample vials after repeated pulling of vacuum and replacement by nitrogen was as effective as the addition of antioxidants.

Packaging System and Its Preparation

For the parenteral products containing proteinaceous solutions vials, prefilled syringes and cartridges are the most relevant packaging systems. The vial, syringe, and cartridge bodies are mostly made of type I tubing glass. Plastic bags or vials (e.g., blow-fill-seal bottles) play a minor role except for plasma products and the storage of intermediate or final bulk. Rubber is used as sealing discs and stoppers.

The selected packaging system has to be checked for the following main attributes:

- Container closure integrity
- Potential interactions between protein solution and packaging material

Examples are given below to illustrate that packaging selection is a vital task during drug product development of protein products.

Package Integrity

The primary sealing zones of the vial and stopper combinations are located at the flange and neck of the vial; hence, appropriate stoppers should preferably be uncoated in this area.

If packaging manufacturers provide material with too broad dimension tolerances, the stoppered vials may not maintain integrity prior to capping (Fig. 9). Adequate investigations of stopper sealing performance, supplier and incoming controls of the packaging materials, for example, conformance to predefined dimensional tolerances, are mandatory.

Protein—Packaging Material Interaction

The interaction can have two directions: components may leach from the packaging material into the proteinaceous solution, but also components from the proteinaceous formulation may adsorb to the packaging material.

In regards to the former phenomenon, essentially silicone detaching from syringe or cartridge bodies or siliconized stoppers are of concern because silicone can cause aggregation

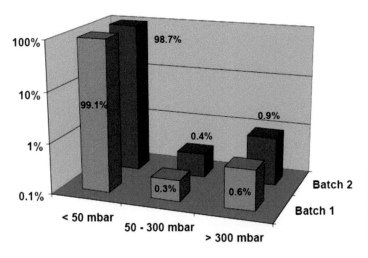

Figure 9 Incidence of lyophilized vials with reduced or lost vacuum after final sealing, 6 R vials with blow back ring and 20-mm uncoated, siliconized stopper. *Source*: From Ref. 21.

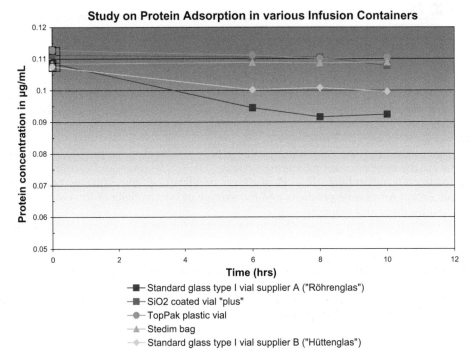

Figure 10 Comparison of various container materials as regards to adsorption of protein from a low concentration protein solution. *Source*: From Ref. 22.

of sensitive proteins. Tungsten, in trace amounts, from the syringe forming process has been described to cause aggregation of therapeutic proteins; however, now tungsten-free syringes are on the market.

Concerning adsorption of the protein to the packaging walls, less thoughts must be spent in case of high-concentration liquid antibody formulations than in case of low-protein concentrations. A study could nicely show that for a small chemokine protein, at a concentration of as low as 0.1 μg/mL, the packaging material indeed mattered. Glass revealed massive adsorption of the protein as opposed to EVAM bags, COC vials, and SiO_2 plasma-coated glass vials (Fig. 10).

Lyophilization

In case of problems with the stability during storage of liquid protein preparations, which cannot be repaired by formulation measures such as pH optimization, buffer exchange, or addition of surfactants, the product will be mostly freeze-dried. Other measures such as freezing are usually not an attractive option for commercial preparations, except for intermediate products before conjugation. Lyophilization belongs to a category of rather difficult pharmaceutical processes, since many parameters must be adjusted. Although smart freeze-drier technology and predictive modeling have advanced significantly, often experiments conducted at small scale are of limited value.

After filling the product into vials or other containers (e.g., double chamber cartridges), special lyo-stoppers are at first only partially pushed into the containers and the lot is loaded manually using a loading cart or automatically onto the lyophilizer shelves that may be precooled. The subsequent freezing process may be in some instances interrupted by annealing to drive partially amorphous sugars to completely crystallize (predominantly mannitol containing formulations). Following primary drying process, the shelf temperature is typically increased stepwise, for example, from -50 to $0°C$. The solvent is collected by the condenser. At a vacuum of typically ≤ 100 mbar energy must be introduced into the frozen product via the shelves to sublimate the water, since vacuum is a bad heat conductor. Care must be taken not to exceed the glass transition temperature T_g', because this would lead to a collapse of the cake and hence to unacceptable appearance of the lyophilizate, and in some cases even lead to damage of the protein. In other cases, to the contrary, collapse of the lyo cake had no negative effect on protein activity, monomer content, or even on the secondary structure (infra-red spectra) of an IgG and of lactate dehydrogenase (LDH) initially and following storage (23,24). As ice sublimes during primary drying, collapse is prevented by maintaining the structural integrity of the maximally freeze-concentrated amorphous phase that surrounds the ice crystals. Below its glass transitions temperature T_g', this amorphous phase exists as a "glass," which is hard and brittle and has negligible mobility. The final residual moisture, which is also very much dependent on the formulation (25), is reached during secondary drying by the removal of adsorbed water. Here the shelf temperature is again raised, for example, to $+25°C$. A low residual moisture is prerequisite for storage stability since residual water lowers the glass transition temperature T_g of the lyophilizate. According to a general rule, T_g should be at least $10°C$ (preferably $50°C$) above the targeted storage temperature. Under vacuum or after partial or complete break of vacuum with sterile air or nitrogen, the partially pushed-in stoppers are moved to their final, closed, position by collapsing the shelves. After unloading of the freeze-dryer, the caps or seal caps are fixed to the container by crimping. There exist a large number of books that deal exclusively with drying and lyophilization (26–30). These books along with a chapter on freeze-drying in this book series cover this topic and hence the rest of the chapter will not focus on lyophilization process.

Residual Moisture of the Stopper

The moisture content of lyophilized product, during storage, is influenced by insufficiently dry stoppers that may lead to instabilities. The ratio of mass of stopper to mass of lyophilizate is decisive: the smaller the mass of lyophilizate, the more significant the effect (31). In Figure 11, the storage-related moisture contents of lyophilizates upon usage of stoppers with a residual moisture of 0.3% and 0.05% are compared. The permeation of water through the stopper, to the contrary, is of subordinate relevance.

Change of Native Structure of the Monoclonal Antibody by Lyophilization

In an aqueous environment, the driving forces for the protein conformation are the hydrophobic effect. Nonpolar amino acid residues are pushed into the inner of the protein core and thus removed from the solvent; hydrogen bonds form between the polar amino acid residues at the protein surface and the surrounding water. For this reason, the removal of tightly bound water can change the protein structure and lead to insoluble or soluble protein aggregates. Described in various publications, the partial exchange of the binding partner water by water substitutes like sugars can physically stabilize the protein. Andya et al.

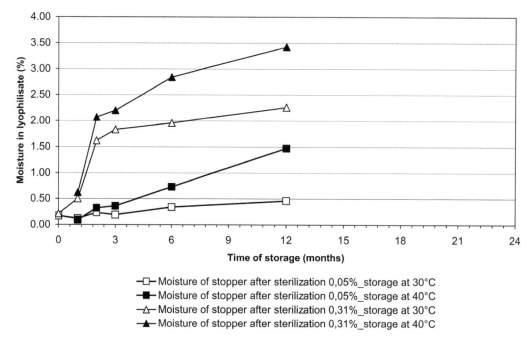

Figure 11 Dependence of moisture content of lyophilized cake during storage on stopper moisture after stopper sterilization.

compared, by FTIR spectrometry and circular dichroism (CD), the secondary structure of a purified recombinant human monoclonal antibody after lyophilization without excipients and together with sucrose or trehalose in ratios of 260:1 up to 2000:1 (25). Indeed in the presence of sugars, a structure very similar to the native one was found even in the solid state. After one year of storage at 30°C, covalent aggregates through free thiol groups had formed without the sugars. A comparison of the measured and calculated water monolayer by use of the Brunauer–Emmett–Teller equation found that the residual moisture values of the excipient-free lyophilizate showed approximately 25% less water was present in the cake than needed for the complete saturation of all surface-accessible hydrophilic groups (25). The authors attributed the observed differences in the solid state and the storage-related instabilities of the sugar-free preparation to the uncovered charged sites due to dehydration.

QUALITY PROCESSES BY DESIGN
(Refer to chap. 13, "Application of Quality by Design in CMC development.")

Philosophy Changes Due to Risk-Based Approach and QbD
In the context of the Quality-by-Design (QbD) philosophy, as outlined in the ICH Guideline Q8 Pharmaceutical Development (2006, followed by revisions), during development profound process understanding must be *systematically* attained and be developed further in the product life cycle. By definition, a well defined design space and convincing justification in the regulatory submission documents should allow future flexibility as far as the proposed change is within the established design space. The prerequisites for allowing regulatory flexibility are good design of product and process, good risk management strategies (ICH Q9), and good quality systems (ICH Q10).

The introduction of new technologies and continuous process improvements have been purposely omitted so far by (bio)pharmaceutical manufacturers, since regulatory requirements made implementation of optimized production almost impossible. Very detailed batch records, which defined numerous non-key parameters, led to deviations and laborious investigations. Primarily, compliance led to rising costs, reduced yields with minimal to no improvement in the product quality. The FDA and EMEA have acknowledged these problems and embraced the Process Analytical Technology (PAT) and developed the 21st century Good Manufacturing Practice (GMP) initiative. The goals are to allow industry to implement the latest technology to produce high-quality products with optimized processes. Assurance of regulatory relief, if a QbD concept was implemented, is also a potential benefit.

However, this benefit for the submitter is yet to be shown in practice broadly. Large skepticism exists around the concept of allowing more freedom during change control, if additional information in the submission documents and the development studies are provided to establish design space. Little concrete, executable instructions are available from the regulatory authorities in how to request and implement flexibility.

A multivariate experimental study should be conducted to define the design space. Traditionally, proven acceptable ranges (PARs) were defined for each parameter on the basis of development studies that were often univariate in design. Useful, but not a must, is to know the edge of failure, the range in which the process no longer works as desired. For validation and the dossier, the question arises whether a target range is sufficient, or whether the PARs are to be indicated. The concept of PAR and NOR (normal operating range) may be in conflict with the design space concept, since they are typically derived from one-factor-at-a-time (OFAT) experimentation or univariate analysis where the interaction terms are not taken into account (32).

QbD—Systemic Approach and Use of Risk Analysis

Key to QbD is based on good science and performing risk assessment. Critical step for assuring the product quality is the identification and control of parameters within the manufacturing process, active substance, excipient, components, and packaging materials.

Practice of performing a *systematic risk analyses* has proven to be a very helpful tool for the process development, process verification, and preparation of the process validation. Different systems for the risk analysis are shown in Figures 12 to 14.

- Failure mode and effect analysis (FMEA)
- Ishikawa's cause-and-effect diagram (commonly referred to as fishbone diagram)
- Tree analysis

The result of the risk analyses can serve as the basis for a systematic and detailed understanding, for example, by design of experiment (DoE). According to ICH Q8, it is essential to establish to what extent the variation of the process parameter settings affect product quality. It is also important to know where the variability of the process parameters and the product quality attributes stem from. Fluctuations in the measuring systems should be accounted for prior to drawing any conclusions. Different experimental designs are available including full factorial design, fractional factorial design, central composite design, or Box-Behnken design. These designs can be provided by off the shelf DoE software such as JMP (SAS), Statistica (StatSoft), Modde (Umetrics), etc.

A systematic representation of the approach is provided in Figure 15. As an example parameter, the pressure during sterile filtration of the proteinaceous bulk was selected. Depending on sensitivity of the protein, filtration pressure can cause damage to the active substance (input = parameter filtration pressure, output = critical quality attribute, for example, aggregate content).

PROCESS VALIDATION
Legal Basis for Process Validation
United States
U.S. guidances that give details on how to validate (aseptic) pharmaceutical processes include the following:

- 21 CFR 211.100 and 110
- Guidance for Industry: Process Validation: General Principles and Practices (Draft published Nov 2008 replacing the Guidance in place since 1987)

Criticality	Characteristic (main branch)	Potential failures	Potential effects	Potential causes	O	S	D	RPN	Action required	To be actioned by	Review / Documentation
	FMEA - Risk analysis for product XXX mg **Purpose of this FMEA: process robustness testing in Pilot Plant and transfer to Production Plant**										
	Bulk										
1	Bulk container (type - stainless steel)	Vessel leaks	High bioburden, endotoxins in finished product	Defective valve / flange / seal	3	8	1	24			
4		Bag leaks	High bioburden endotoxins in finished product	Material defects excessive mechanical stress, embrittlement of material	8	10	1	80			
5	Bulk container (type - plastic bag)	Leachables	Product quality inadequate (foreign substances)	Bag material not suitable for formulation	3	5	10	150			
6		Gas permeation during long/short term storage	Product quality inadequate (oxidatation of drug or excipients, loss of solvent)	Bag material does not inhibit gas permeation such as oxygen uptake and water vapour permeation	8	10	8	640			
7	Holding of bulk after thawing	Hold time/temp too long/high	Reduction in product quality (OOS result)	Wrong storage temp. Wrong storage time	1	10	1	10			
	Packaging material Rubber stoppers										
12	Machinability (in Pilot Plant and in Production Plant)	Poor machinability	Increased aseptic risk, increased process time	Stopper dimensions Incorrect degree of siliconisation Incorrect sterilisation / drying process Rubber formulation (stoppers too tacky) Change parts unsuitable or unavailable	3	8	1	24			
15	Sterilisation (in Pilot Plant and in Production Plant)	Incorrect sterilisation process	Non-sterile product	Sterilisation process not valid	1	10	8	80			
17	Quality	Stoppers contain endotoxins and/or particles, degree of siliconisation not as per order, non-compliance with dimensional tolerances, stopper defects	OOS result for stoppers	Manufacturing / cleaning process at stopper manufacturer's not valid	3	5	1	15			
18	Delivery capability Delivery lead times	Stoppers cannot be delivered on time	Filling of batches delayed	Order placed too late Supply problems on manufacturer's part	3	5	1	15			
19	Sensitivity of product to silicone oil (from stoppers)	Sensitivity to silicone oil	Reduction in product quality (OOS result)	Siliconisation of stoppers	5	10	5	250			

Figure 12　Example of an FMEA (failure mode and effect analysis).

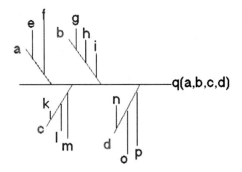

Figure 13　Ishikawa's cause-and-effect diagram ("fishbone").

- Compliance Policy Guide (CPG), Process Validation Requirements for Drug Products and Active Pharmaceutical Ingredients Subject to Pre-Market Approval, rev. Dec. 2004
- FDA Guidance for Industry: Sterile Drug Products Produced by Aseptic Processing— Current Good Manufacturing Practice, Sep 2004
- Guidance for Industry for the Submission of Documentation for Sterilization Process Validation in Applications for Human and Veterinary Drug Products, Nov. 1994

	CTQ$_1$	CTQ$_2$	CTQ$_3$	CTQ$_4$	CTQ$_5$	CTQ$_6$	CTQ$_7$	CTQ$_8$	CTQ$_9$
Processsstep 1	O	O	●	X	O	O	O	O	O
Processsstep 2	O	XX	O	O	●	O		O	O
Processsstep 3	O	X	O	O	O	●	O	O	X
Processsstep 4	X	●	●	O	XX		O	●	X

O No influence
● Small influence
X Big influence

CTQ = critical to quality attribute

Figure 14 Fault tree analysis.

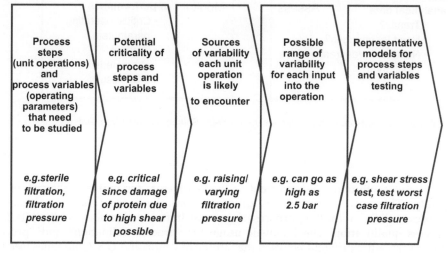

Process steps (unit operations) and process variables (operating parameters) that need to be studied	Potential criticality of process steps and variables	Sources of variability each unit operation is likely to encounter	Possible range of variability for each input into the operation	Representative models for process steps and variables testing
e.g. sterile filtration, filtration pressure	*e.g. critical since damage of protein due to high shear possible*	*e.g. raising/ varying filtration pressure*	*e.g. can go as high as 2.5 bar*	*e.g. shear stress test, test worst case filtration pressure*

➡ a "robust process" is able to tolerate input variability and still produce consistent acceptable output

Figure 15 Systematic analysis of critical process variables and example from aseptic fill and finish manufacture of a protein product.

- FDA Guidance for Industry, PAT—A Framework for Innovative Pharmaceutical Development, Manufacturing, and Quality Assurance, Sep 2004
- Pharmaceutical cGMPs for the 21st Century—A Risk-Based Approach, FDA Final Report, Sep 2004

In the guidance "Process Validation Requirements for Drug Products and Active Pharmaceutical Ingredients Subject to Pre-Market Approval" of 2004 (which however is actually not directly applicable to recombinant proteins) it is stated that—"the proof of validation is obtained through rational experimental design and the evaluation of data, preferably beginning from the process development phase and continuing through the commercial production phase."

The process validation guideline from 1987 has been revised by the FDA and a draft has been published for public comment in November 2008. The draft revision picks up the ideas of the CPG and addresses the relationship between modern quality systems and manufacturing science advances to the conduct of process validation. The revision focuses on a process validation life-cycle approach including four key phases—design, confirm, verify (three of four listed) (Fig. 16). The current focus on the commercial process "validation" will be shifted toward the design phase in agreement with the new FDA science-based approach, the application of QbD by gathering complete product/process knowledge, a "continuous quality verification system," and an effective monitoring/assessment program to address effective

Process Validation Life Cycle

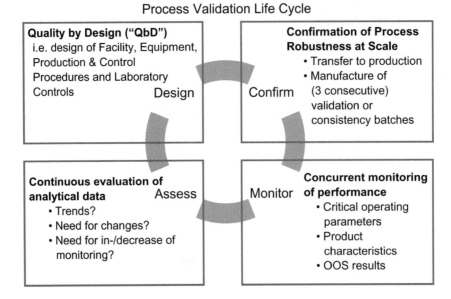

Figure 16   Validation concept according to three-stage concept. *Source*: Adapted from Ref. 33.

process control and continuous improvement as the key factors for reducing the risk to the product quality. This new perspective is different from the current process validation approach. Process validation for the purposes of the new draft directive does not limit itself to the pure process qualification. The linguistic usage for "process validation" and "process qualification" is somewhat different between the European Union and U.S. Guidelines (see EU GMP Guide, Annex 15).

Other important basic ideas and elements of the validation in the new guideline are as follows:

- Integrated team approach to process validation that embraces expertise from a variety of disciplines including process engineering and statistics.
- Project plans, along with the full support of senior management "... all studies should be planned and conducted according to sound scientific principles, appropriately documented, and should be approved in accordance with the established procedure appropriate for the stage of the lifecycle ...."
- Modeling of procedures, ideally at small scale, and transfer by appropriate simulations or virtual programs from the pilot to production scale; at least the process understanding can be improved significantly by modeling.
- Importance of process controls that can reduce input variation, adjust for input variation during manufacturing, or combine both approaches. PAT for in-process controls is essential in those cases where quality is not readily measurable later in the product, for example, microbial contamination.
- Statistician or person with adequate training in statistical process control techniques to develop the data collection plan and statistical methods and procedures used in measuring and evaluating process stability and process capability.
- Representation of the processes by process flow diagrams.

The terms conformance batches and validation batches are equivalent. The approval of the product can take place still before the successful manufacture of the conformance batches (for small molecule parenterals). However, the validation should be completed prior to the start of marketing. For biologics, validation batches need to be manufactured prior to filing the

BLA. For biotherapeutics, typically drug product process validation is already described in the dossier and can thus be reviewed during preapproval inspection (PAI).

Exemptions to process validation requirements are made for orphan drugs for health-economic reasons, drugs with very limited shelf life, and drugs with limited use like radiopharmaceuticals. In these cases, release for the market and validation can take place in parallel (concurrent validation).

EU and International
EU and international guidance that address validation of pharmaceutical processes include the following:

- CPMP Note for Guidance on Process Validation + Annex I Process Validation Scheme, Sep 2001
- CPMP Note for Guidance on Development Pharmaceutics, Jan 1998
- Annex II, Note for Guidance on Process Validation—Non Standard Processes, Jan 2005
- EU Guide to GMP, Annex 15, Qualification and Validation, Sep 2001
- ICH Q8 and ICH Q9

In the *Note for Guidance (NfG) on Process Validation* with its Annex I, validation is described as the verification of the process at scale. Usually on the basis of three full-scale batches, and development data, small to pilot scale that is 10% of production scale, proof of process validation is established (*critical steps, critical parameters*). It is expected that in the process validation, additional tests are accomplished beyond the spectrum of the release tests.

Also in the EU, revalidation is understood as periodic continuum. As already mentioned in the previous section, definitions of PV (process validation), new processes/products and PQ (process qualification) validation using product or simulating product, are somewhat different from the linguistic usage in the United States, in particular from the FDA draft on process validation.

An accomplished process validation is not necessary at the time of the submission, but the protocol of the planned studies for the production batches is part of the dossier and/or the Pharmaceutical Expert Report. This is not valid for nonstandard methods of production (*Annex II to NfG on Process Validation*), nonstandard sterilization procedures, aseptic manufacturing processes, certain lyophilization procedures, microencapsulation procedures, and sustained release products. In these cases, before approval, three consecutive batches at production scale are demanded. Comparable to the FDA regulations, the *EU GMP Guide, Annex 15*, defines that even though process validation should normally be completed prior to the distribution and sale of the medicinal product (prospective validation), "in exceptional circumstances, where this is not possible, it may be necessary to validate processes during routine production (concurrent validation)." There is also a retrospective validation, which is applicable only to well-established processes without changes.

Significant changes in the manufacturing process can initiate variations in the market authorization that need approval by the authorities prior to implementation (type II variations). In case of products with biological active substances, even small changes of process are nearly always type II variations in accordance with *guideline on the details of the various categories of variations to the terms of marketing authorizations for medicinal products for human use and veterinary medicinal products*. Very detailed descriptions of process and equipment design are expressly not demanded for the dossier in the *NfG on Process Validation*.

In the EU, the basic ideas of parametric release have been introduced. It however covers only terminally sterilized products (*NfG on Parametric Release, 2001; Annex 17 to the EU Guide to GMP*).

In the *Annex 13 of the EU GMP Guide*, it is stated that manufacturing processes for clinical supplies do not have to be validated, with the exception of buildings and equipment. For products labeled as sterile, sterilization processes must be validated according to the same standard as market products. The same holds true for virus inactivation or removal processes. Aseptic processes must be validated at this stage, whereby the smaller batch sizes and the semimanual steps during the production are valid.

Cleaning validation for similar products and processes can be accomplished with a representative product and on the basis of one worst-case consideration.

Validation of an Aseptic Fill and Finish Process for a Monoclonal Antibody or Therapeutic Protein Product

Before the product-specific validation can be started, it is understood that any process validation facilities and equipment must be qualified for the purpose by DQ/IQ/OQ. The focus of the drug product validation runs should be to monitor and control critical operational parameters (COPs).

The validation of the pharmaceutical process will be described referencing the example of a classical monoclonal antibody product. The following steps must be considered in the context of the validation, and the list is not meant to be exhaustive:

- Thawing and pooling of the frozen bulk, potential additions and dilutions
- Cleaning and sterilization of multiprocess equipment [e.g., needles, pumps, tanks, pipes, disposable tubings, and bottles (unless provided in ready-to-use quality); also the hold time between end of use and cleaning as well as between cleaning and reuse are part of the cleaning validation]
- Bioburden reduction filtration and sterile filtration
- Cleaning and sterilization of packaging materials (e.g., vials, stoppers, and crimp caps)
- Filling and sealing procedures, to include freeze-drying if necessary
- Aseptic procedures, interventions, and the facility/personnel involved (media fills)
- Visual inspection
- Transportation of the filled and controlled vials (e.g., to the final packer, warehouse, etc.)
- Hold times of the bulk after thawing, the final product after the filling at various temperatures, and the hold times during process stops
- Decontamination of equipment in vaporized hydrogen peroxide (VHP) material locks

Naturally the analytical and in-process control procedures must be validated. Starting from phase 2 and 3, methods complying to cGMP are expected in accordance with the Draft FDA Guidance on Process Validation.

Because of fact that aseptic processes for therapeutic protein preparations belong to high-risk processes, some process steps are already validated during the clinical phase. For aseptic manipulations, sterilization and decontamination procedures guarantee the absence of contamination. Process steps, that determine other quality parameters, are usually validated prospectively and fully only with the conformance batches and are covered during the clinical phase by concurrent validation.

Design of Process

As shown in sections "Selected Case Studies Exemplifying Development Challenges During Fill and Finish" and "Quality Processes by Design," prior to validation, the process must be developed and investigated. Hereby "design" means to identify critical operation parameters and acceptable operating ranges by development studies such as

- Design of experiments
- Laboratory or scale-up experimental batches to gain process understanding

Furthermore mechanisms to limit or control variability, based on experimental data, must be established. A "robust process" is able to tolerate input variability and still produce consistent, acceptable output.

What exactly are COPs? The definitions given in the PDA Technical Report No. 42 (1998) are very useful even though the report does not explicitly target the aseptic processes:

- Operational parameter = *input variable* or condition of manufacturing process that can be directly controlled in the process.
- Critical operational parameter (COP): input process parameter that should be controlled within a *narrow* operating range to ensure *quality* attributes meet specifications.

- Non–critical operational parameter (Non-COP): input process parameters that fall outside definition for COPs. Divided into:
    - Key operational parameters (KOP): input parameter that should be carefully controlled within a *narrow* range and is essential for *process performance* (does *not* affect quality).
    - Non–key operational parameters (Non-KOP): input parameter that has been demonstrated to be easily controlled and has a *wide acceptable limit* (quality or process performance impacted if acceptable limits exceeded).

Other sources differentiate between COP or non-COP and KOP and non-KOP depending on whether (critical) *quality* attributes (CQA) or *performance* attributes (= *output variables*) of the product or the process are affected. Non-COP or non-KOP are parameters that have a wide tolerance and need not be narrowly controlled. This is depicted in Figure 17. Table 7 lists some of the operational parameters during drug product manufacturing and their hypothetical classification. The table should be considered an example since the classification of parameters as COP or KOP will change depending on the product and process used. In ICH guideline, the terminology used is "process" rather than "operational," for example, instead of COP or KOP, the terms CPP and KPP are used.

Conformance Batches
The conformance batches, that are exactly consistent with the classical validation batches, are manufactured after the transfer of the process into the production facility. The EU GMP Guide,

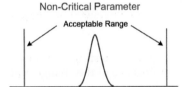

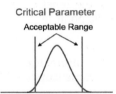

Critical process Parameter: a process parameter that must be controlled within predetermined criteria to ensure the bulk drug substance or drug product meets its specification/quality attributes. A process parameter is critical if the target range (TR) is near the acceptable range (AR) or as determined by a technical expert

Figure 17 Definition of operational parameters.

Table 7 Example for a List of Operational Parameters and Their Classification

Controllable operational parameter	Classification[a]	Acceptable range (as of development)	Normal range (conformance batches)
Hold time of bulk outside cold room (hours)	COP	3 mo at 25°C within specification	Up to 2 wk at room temperature
Filtration pressure or rate (bar or mL/min)	Non-KOP	Up to 2 bar	1.0–1.5 bar
Fill volume and precision (mL)	COP	At 10.5 mL fill with 10.1 mL extractable volume	10.7 mL ± 0.2 mL
Bulk temperature (°C)	Non-KOP	No density change between 0 and 30°C	5–25°C
Fill rate (vials/min/needle)	Non-KOP	Up to 30 vials/needle	25 vials/needle
Hold time of product outside cold room (hours)	COP	3 mo at 25°C within specs (drug product or drug substance)	Up to 2 wk at room temperature
Hold time of product exposed to light (hours)	Non-KOP	No impact after 1.2 Million lux hours (ICH)	Up to 24 hr exposed to room light
Capping force or conditions (N or mm)	COP	± . . . mm capping height	± . . . mm capping height

[a]This classification is hypothetical and is for illustration purpose only.
Abbreviations: COP, critical operational parameter; KOP, key operational parameter.

Annex 15 states that "It is generally considered acceptable that *three* consecutive batches/runs within the finally agreed parameters, would constitute a validation of the process." According to the current Compliance Policy Guides Manual of the FDA, however, no concrete number of conformance batches is required (CPG 7132c.08 Sec. 490.100).

Hereby, "transfer" means the transfer of developmental knowledge to production (technology transfer), that is, after transfer:

- Batch records and standard operating procedures (SOPs) are written and equipment and facilities equivalency is established.
- All raw materials and the suppliers are qualified.
- Measurement systems are qualified to include QC lab as well as production floor test instrumentation.
- Personnel training is completed.

Conformance stands for the following:

- Execution of conformance batches (usually 3 per dose strength) with appropriate sampling points and sampling level, that is,
  - Evidence that process can function at commercial scale by production personnel.
  - Demonstrate reproducibility.
  - New process and packaging components are considered in the media fill concept (validation of aseptic procedures).
- Full sample and data analysis of the consistency batches
  - Data may confirm process as is, point to major process design change(s) or suggest process improvement(s).
  - Changes are implemented via change control procedures
    - Assess need for additional conformance batch(es) or limited testing. Amount/ degree of additional work commensurate with the significance of the change and its impact on product quality.

All activities and the underlying validation policy must be established prospectively in a Validation Master Plan (VMP) as well as in individual validation protocols (EU GMP Guide, Annex 15, 2001). The review process must be adhered to. The validation protocols define sampling, analysis, and acceptance criteria for judging whether validation can be classified as successful or not. Furthermore, operating parameters, processing limits, and component (raw material) inputs should be described in the validation protocol. It is very helpful to have tested these acceptance criteria and also the statistical approach during process transfer. The acceptance criteria, however, should have been derived from development studies and clinical material manufactured previously, except for those tests that are predefined by pharmaco-poeias or other guidelines.

In our example of a monoclonal antibody product, it would be reasonable to fix acceptance criteria during validation of the fill and finish process for the following parameters (as appropriate):

- Homogeneity and quality of the bulk after thawing and pooling, for example, determined by protein concentration, aggregates, monomers, particle number, and turbidity.
- Bioburden of the bulk prior to sterile filtration.
- Homogeneity and quality of the bulk after sterile filtration, for example, surfactant concentration, protein concentration, aggregates, monomers, number of particles, turbidity, further biophysical or chemical properties depending on the product.
- Quality of the final product at the beginning, middle, and end (where appropriate). This is important for suspension products.
- Sterility of the bulk formulation.
- Additional IPC (in-process control) specifications, for example, fill volume, stopper seating, quality of capping.

- Final product properties, for example, extractable volume, residual moisture, and aggregate or monomer content.
- Yield after filling and capping.
- Yield after visual inspection
- Maximal filtration time, filling time, and capping time.
- Maximal hold time of bulk and final product at room temperature or at other relevant temperatures.

After performance and analysis, validation or conformance drug product manufacturing report is written and must be reviewed against the protocol. Deviations from the protocol must be explained and their impact (if any) on the validation should be evaluated.

Validation During Product Life Cycle
After successful manufacture of the conformance batches, PAI and market authorization, validation is to be continued, according to the current FDA philosophy, during the entire product lifetime. This means:

- Monitor
  - Routine commercial manufacturing
    - Monitor critical operating and performance parameters
      - Utilize appropriate tools, for example, Statistical Process Control (SPC)
      - Monitor product characteristics, for example, stability, product specifications.
    - Monitor state of personnel training and material, facility/equipment, and SOP changes.
    - Investigate out of specifications (OOS) for root cause and implement corrective and preventive actions (CAPA).

- Assess
  - Analyze monitoring data
    - Trend data upon regular review.
    - Evaluate need to increase/decrease level of monitoring/sampling on the basis of accumulated data.
  - Periodic evaluation
    - To determine the need for changes, for example, manufacturing procedures, control procedures, drug product specifications.
  - Study OOS and OOT (out-of-trend) data.
  - Assess impact of process and product changes made over time.
  - Feedback into design stage for significant process shifts or changes.

The CFR [section 211.180(e)] requires "that information and data about product performance and manufacturing experience be periodically reviewed to determine whether any changes to the established process are warranted. Ongoing feedback about product performance is an essential feature of process maintenance." EU GMP Guide, Annex 15, also demands periodic examination whether the process is still in the validated condition or whether changes make revalidation necessary. It is not explicit whether or not changes originate from the feedback of the product performance.

SPECIFICATIONS AND CONTROL LIMITS
The definition of specifications and control ranges, both during the development phases and prospective validation are complex and not clearly agreed upon across the (bio)pharmaceutical industry or regulated at this time.

According to the QbD concept, there should be differentiation between critical quality attributes (CQAs) and parameters used to monitor process consistency. "Specifications and the corresponding limits as applied to CQAs serve to ensure that the product is fit for use, whereas control limits are a manufacturer's tool to monitor shifts and trends in the manufacturing process. In the current paradigm, inappropriate use of specifications creates a disincentive for continuous process understanding …" (34). The vision of QbD and design space cannot be achieved when control limits are used as specifications. The design space may be viewed as the region of process settings that yields acceptable product (*i.e., product that meets specifications*). When control limits are used as specifications, the design space reverts to the control space for the process, leaving no opportunity for process improvement (35).

This risk is likewise borne in validation studies. Samples are taken from multiple locations of the production process, or at multiple levels of a process parameter, and subject to specifications. As with stability testing, in which multiple samples are taken over time, validation samples are subject to excess risk of OOS due to multiplicity. This acts as a disincentive to collecting data for better process understanding (35). It is likely that as industry and regulators gain experience, some of these difficulties could be overcome so that process can be improved over time without undue regulatory burden.

SUMMARY
In this chapter, key factors in developing and validating aseptic drug product process for biologics has been covered. Readers are advised that a sound scientific practice should always be used in conjunction with a knowledge of current regulatory environment. Although, this chapter provides a high-level overview of QbD as relates to drug product validation, a more thorough discussion is included in chapter "Application of Quality by Design in CMC Development" in this volume.

REFERENCES
1. Bechtold-Peters K. Aseptic fill & finish processes employing solely disposable materials—focus on pharmaceutical pilot plant. Presentation at the Annual Conference of the AAPS, Atlanta, GA, November 16–19, 2008.
2. Hemminger M. Presentation at the National Biotech Meeting NBC of the AAPS, San Diego, CA, June 25–27, 2007.
3. Verjans B, Thilly J, Vandecasserie C. A new concept in aseptic filling: closed-vial technology. Pharm Technol Europe 2006; 18:45–48.
4. PDA Technical Report #26. Sterilizing filtration of liquids. J Pharm Sci Technol 1998; 52(suppl):1–31.
5. Technical Sheet Millipore TB1025EN00. Effect of Membrane Filter Pore Size on Microbial Recovery and Clonony Morphology.
6. McBurnie L, Bardo B. Validation of sterile filtration. Pharm Technol 2004:13–23.
7. Technical Sheet Millipore SDS1009E00. Bacterial Retention Testing.
8. Maa YF, Hsu CC. Investigation on fouling mechanism for recombinant human growth hormone sterile filtration. J Pharm Sci 1998; 87:808–812.
9. Pitt AM. The non-specific protein binding of polymeric microporous membranes. J Parenter Sci Technol 1987; 41:110–113.
10. Bowen WR, Gan Q. Properties of microfiltration membranes: flux loss during constant pressure permeation of bovine serum albumin. Biotechnol Bioeng 1991; 38:688–696.
11. Maa YF, Hsu CC. Protein denaturation by combined effect of shear and air-liquid interface. Biotechnol Bioeng 1997; 54:503–512.
12. Harris RJ, Shire SJ, Winter C. Commercial manufacturing scale formulation and analytical characterization of therapeutic recombinant antibodies. Drug Dev Res 2004; 61:137–154.
13. Lam XM, Yang JY, Cleland JL. Antioxidants for prevention of methionine oxidation in recombinant monoclonal antibody HER2. J Pharm Sci 1997; 86:1250—1255.
14. Henkel G. Praxisrelevante Untersuchungen zum Rougeproblem in WFI-Systemen. apv-news 2005:12–15.
15. Henkel G. Die Passivierung von austenitischen Edelstahloberflächen in der pharmazeutischen Industrie unter Berücksichtigung von Rougingphänomenen. apv-news 2004.
16. Henkel G. Der Reinheitszustand einer Edelstahloberfläche und diesbezügliche Messmethoden. apv-news 2005.
17. Harrison JS, Gill A, Hoare M. Stability of a single-chain Fv antibody fragment when exposed to a high shear environment combined with air-liquid interfaces. Biotechnol Bioeng 1998; 59:517–519.
18. Mezger T. Das Rheologie-Handbuch. 2nd ed. Hannover, Germany: Vincentz Network GmbH, 2007. ISBN 978-3-87870-175-0.
19. Vichroy B, Lorenz K, Kelly W. Modeling shear damage to suspended CHO cells during cross-flow filtration. Biotechnol Prog 2007; 23:194–199.
20. Maa YF, Hsu CC. Effect of high shear on proteins. Biotechnol Bioeng 1996; 51:458–465.

21. Sigg J, Schwinger G. Investigations on container/closure integrity of vials sealed under vacuum. Presentation at the 2004 PDA International Congress, Basel, February 16, 2004.
22. Bechtold-Peters K. Challenges to overcome with regard to packaging of biopharmaceuticals. Presentation at the SCHOTT Forma Vitrum Pharma Symposium in Mainz, Germany, October 2004.
23. Schersch K, Betz O, Garidel P, et al. Effect of collapse on protein lyophilizates using L-lactic dehydrogenase as sensitive model protein. AAPS Annual Meeting, San Antonio, November 2006.
24. Schersch K, Betz O, Garidel P, et al. Scrutinizing the effect of collaps during lyophilization: stability of L-Lactic dehydrogenase lyophilizates during storage at elevated temperatures. AAPS Annual Meeting, San Diego, November 2007.
25. Andya J, Hsu CC, Shire SJ. Mechanisms of aggregate formation and carbohydrate excipient stabilization of lyophilized humanized monoclonal antibody formulations. AAPS Pharm Sci 2003; 5: Article 10.
26. Costatino HR, Pikal MJ, eds. Lyophilization of Biopharmaceuticals. Biotechnology: Pharmaceutical Aspects. Arlington, VA: AAPS Press, 2004. ISBN 0-9711767-6-0.
27. Jennings TA. Lyophilization: Introduction and Basic Principles. Denver, Colorado: Interpharm/CRC, 1999. ISBN 1-57491-081-7.
28. Rey L, May JC. Freeze-Drying/Lyophilization of Pharmaceutical and Biological Products. Series: Drugs and the Pharmaceutical Sciences. Vol. 96, 2nd ed. New York: Marcel Dekker, 1999:385–424. ISBN 0-08247-1983-2.
29. Oetjen GW. Freeze-Drying. Weinheim, Germany: Wiley-VCH, 1999. ISBN 3-527-29571-2.
30. Mujumdar AS, ed. Handbook of Industrial Drying. New York: Marcel Dekker Inc, 1995. ISBN 0-8247-9644-6.
31. Earle JP, Bennett PS, Larson KA, et al. The effect of stopper drying on moisture levels of haemophilus influenzae conjugate vaccine. International symposium on biological product freeze-drying and formulation, Bethesda, 1990. Develop Biol Stand 1991; 74:203–210.
32. Berridge JC. Quality by design—a modern system approach. An industry perspective. Presentation at the AAPS & ISPE Workshop US Food and Drug Administration's Pharmaceutical Quality Initiatives. Implementation of a Modern Risk-based Approach. Bethesda North Marriott, North Bethesda, MD, February 28–March 2, 2007.
33. McNally G. Lifecycle Approach to Process Validation. Presentation at the 41st Annual Meeting of the Drug Information Association, Washington DC, June 26–30, 2005.
34. Apostol I, Schofield T, Koeller G, et al. A Rational Approach for Setting and Maintaining Specifications for Biological and Biotechnology–Derived Products—Part 1. BioPharm Int 2008.
35. Schofield T, Apostol I, Koeller G, et al. A Rational approach for setting and maintaining specifications for biological and biotechnology–derived products—Part 2. BioPharm Int 2008.

4 | Visual inspection

Maria Toler and Sandeep Nema

OVERVIEW

The inspection of parenteral products is driven by the need to minimize the introduction of unintended particulate matter to patients during the administration of injectable medications. Visual inspection also allows for the opportunity to detect and reject other categories of nonconforming units, such as those with cracks and or incomplete seals, which can affect the integrity and sterility of the product. In most cases, these defects will occur randomly and at low frequency. This has led to the current expectation that each finished unit will be individually inspected 100%.

Particulate matter is defined by the United States Pharmacopeia (USP) as "mobile undissolved particles, other than gas bubbles, unintentionally present in the solutions." It is the expectation of the USP that "each final container of injection be subjected individually to a physical inspection, whenever the nature of the container permits, and that every container whose contents show evidence of contamination with visible foreign material be rejected." Visible particulate matter can be defined by the size of the particles. It is generally accepted that the human eye can detect particles once the size approaches 50 μm. The detection of a particle is based on the probability of being able to see it within a container, with the probability increasing with increasing particle size. Analysis of visual inspection results from several studies involving different groups of inspectors showed that the probability of detecting a single 50-μm particle, in a clear solution within a 10-mL vial with diffuse illumination between 2000 and 3000 lux, is just slightly above 0%. However, this probability increases to approximately 40% for a 100-μm particle and becomes >95% for particles ≥200 μm (1). This and similar studies show the dependence of visual detection and particle size. Other factors, such as the refractive index and luster of the particle will also affect the ability to detect.

Why inspect for visible particulates? There is no clear consensus on the safety of having a small number of visible particles in an injectable drug product. The primary evidence for safety can be found in the literature on drug abuse. There is some evidence that addicts who injected drugs had manifestations of pulmonary foreign body emboli and granulomas, along with abnormal pulmonary function (2–4). It has been observed that granulomas are generally associated with fibers and silicosis with glass particulates, while fungal particles have been associated with pyretic issues (5). Protein particles, both subvisible and visible are being investigated for their effect on immune responses. There is a lack of controlled studies in humans to better understand the effect of small amounts of visible particles. Rather, it is generally accepted that injectables should be clear and essentially free of particles that can be seen by the unaided eye. This primarily applies to drugs being infused via the peripheral veins (IV). The presence of particulate matter in intramuscular or subcutaneous injections is not of great concern, especially since small volumes are usually injected and tissue phagocytosis as well as local immobilization of the particles would make them almost harmless.

The major effects and pathological conditions that have been linked in the literature to the injection of particulate matter include the following (6):

- Direct blockage of a blood vessel by foreign particulate matter
- Platelet agglutination, leading to the formation of emboli
- Local inflammatory reactions caused by the impaction of particles in the tissues
- Antigenic reactions with subsequent allergenic consequences
- The distribution of injected particles will depend on size and to lesser extent on particle composition
  - Large particles (≥50 μm) on the basis of circulation (venous infusion → right heart → lung) will be retained in the lung
  - Particles that are ≥10 μm pass the pulmonary vasculature slowly

○ Particles <10 μm in size are retained in the liver and spleen for long periods
○ Particles <10 μm in size are significantly cleared by phagocytosis by cells of the reticuloendothelial system

In addition to safety issues, the presence of particulate matter in the product can be an indication of formulation unsuitability, improper container closure system, degradation or lack of process cleanliness. Because it can be considered a product quality attribute, particulate matter should be controlled in intramuscular and subcutaneous dosage forms as well as intravenously injected products.

REGULATORY ASPECTS

The purpose of a visual inspection is to satisfy the regulatory agencies and ensure the safety and quality of the drug product. A survey was presented at a Parenteral Drug Association (PDA) forum in 2008 describing regulatory observations that had been reported over the last 12 years (7). Fifty percent of the firms surveyed were challenged on their inspection programs. Having an appropriate inspection program can aid an organization in avoiding a Form 483 or Warning Letter from the U.S. Food and Drug Administration (FDA). The Form 483 is referred to as a "Notice of Inspectional Observations." It is issued by an FDA field investigator after an on-site inspection and will list areas of noncompliance with current Good Manufacturing Practices (cGMPs) or other deficiencies in the quality system. The organization must respond to the Form 483 and identify a course of action to correct the findings, along with a timeframe that issues will be addressed.

Some findings from Form 483s issued over the last 10 years include the areas of documentation, quality limits, and process. Documentation findings included lack of training procedures, standard operating procedures (SOPs) for visual inspection, and inspector retesting schedules. An example quoted from a Form 483 was "Observation #3 from the FDA-483 states that there was no documentation that your firm performed a visual, unit-by-unit examination of containers, vials, and ampoules for defects. You also did not visually inspect each component, diluent, or product for visible contamination." Another Form 483 example, relating to inspector training stated that:

> "There is no written procedure to describe the training required for employees performing visual inspection of containers from either media fill operations or of final products. Additionally, the control standards used to train individuals who perform visual inspections are incomplete in that there are no standards that describe the criteria for sizing and characterizing particulate matter, examples of over or under filled containers or bottles containing glass, metal or rubber contaminants."

Poor documentation can indicate a lack of proper control over the inspection program. Findings related to quality limits include no definition of critical, major, and minor defects that would trigger an investigation. Defining the acceptance criteria and level of defect is a good manufacturing practice and allows for an unbiased approach to rejection of defective product. Examples quoted from Form 483s include "SOP BV1019, entitled 'Visual Inspection of BoTox Product,' does not specify limits for critical, major and minor defects which, when exceeded, would trigger an investigation, and does not instruct operators to place rejected vials in the specific bins"; "there is no specified action level or limit for the filled product container visual inspection performed by the filling department."

Findings related to process include the lack of a separate labeled container/area for rejected samples and a failure of a machine inspection station to completely clear a lot of product before running a new lot of product. There is a greater chance of mixing lots or introducing rejects into an acceptable lot of product, as described in this Form 483:

> A Seidenader™ inspection machine was being used for visual inspections of a parenteral. The machine had an exit arm in which several vials would remain in the machine and were not pushed out. These vials could not be clearly seen without bending down to look at the exit arm. As a result of this, an incident occurred where the line was not cleared of these vials and they became mixed with the beginning of the next lot of product. The next lot of product looked the same as the previous product and as a result mislabeled product was distributed.

VISIBLE PARTICULATE MATTER INSPECTION REQUIREMENTS

In the early part of the 1900s, the USP recognized the need for injectable compounds to be true solutions. In 1936, a requirement for the "clarity" of injectable solutions was specified.

Currently the USP states that all "inspection processes shall be designed and qualified to ensure that every lot of all parenteral preparations is essentially free from visible particulate" with no inspection method specified. The pharmacopeias from other countries that participate in the International Conference on Harmonization (ICH) are similar in requirements, having some differences in the amount of detail provided in the description of the inspection methods.

The USP (Chapter <1> Injections) states that all

> articles intended for parenteral administration shall be prepared in a manner designed to exclude particulate matter as defined in Particulate Matter in Injections <788> and other foreign matter ... The inspection process shall be designed and qualified to ensure that every lot of parenteral preparations is essentially free from visible particulates. Qualification of the inspection process shall be performed with reference to particulates in the visible range of a type that might emanate from the manufacturing or filling process. Every container that shows evidence of visible particulates shall be rejected.

The Japanese Pharmacopoeia (JP) states that unless otherwise stated, injections should meet the requirements of the Foreign Insoluble Matter Test for Injections <6.06>. There are two inspection methods described.

Method 1 "is applied to injections either in solutions or in solution constituted from sterile drug solids" and uses the following procedure: "Clean the exterior of containers, and inspect with the unaided eyes at a position of light intensity at approximately 8000 to 10,000 lux, with an incandescent lamp at appropriate distances above and below the container."

Method 2 "is applied to injections with constituted solution" and uses the following procedure: "Clean the exterior of the containers, and dissolve the contents with constituted solution or with water for injection carefully, avoiding any contamination with extraneous foreign substances. The solution thus constituted must be clear and free from foreign insoluble matter that is clearly detectable when inspected with the unaided eyes at a position of light intensity of approximately 1000 lux, right under an incandescent lamp."

The requirements for the freedom of parenteral solutions from the presence of particulate matter are very strict in Japan. The inspection process for individual containers is more rigorous than the USP (8).

The European Pharmacopoeia (EP) states that "Solutions for injection, examined under suitable conditions of visibility, are clear and practically free from particles." The inspection procedure states that "Gently swirl or invert the container and observe for about 5 seconds in front of the white panel. Repeat the procedure in front of the black panel. Record the presence of any particles." The method includes a description of the viewing station and lighting requirements (2000–3750 lux at the viewing point). The presence of particles is recorded but no sample quantity or acceptance criteria are provided.

The various compendia have similar statements for the requirements for visible particulate matter in injectables:

> USP: "essentially free from particles that can be observed on visual inspection"
> EP: "clear and practically free from particles"
> JP: "clear and free from readily detectable foreign insoluble matter"

There is currently a proposal to the USP to revise the General Chapter—Injections <1> sampling requirements (9). The proposal is based on the General Inspection Level II sampling plan as described in ANSI/ASQ Z1.4 with an Acceptable Quality Limit (AQL) of 0.65%. This AQL was chosen based on the median value obtained from a recent benchmarking survey of industry practice that was conducted by the PDA (7). This inspection procedure would apply to retesting of product in distribution (having undergone 100% inspection) or when a limited subset of the batch is available for inspection (e.g., from retained or returned samples). A sampling of 60 units would be inspected. The batch would be considered to meet the requirement "essentially free" when no more than one (1) unit with one or more particles is observed. This sampling plan has an AQL of 0.60%, which is acceptably close to the ANSI/ ASQ Z1.4 AQL of 0.65%.

It does, however, need to be noted that several parenteral products have recently been approved that contain visible particulate matter. Vectibix (Amgen, Thousand Oaks, California, U.S.) is a marketed product that specifies in the product insert that visible particles may be observed. Upon examination, it is observed that this product contains a great deal of visible particles in solution. It is also specified that the drug is filtered right before administration. Stelara (ustekinumab, Centocor, Radnor, Pennsylvania, U.S.) is described as a colorless to light yellow product that may contain a few small translucent or white particles. Arzerra (GlaxoSmithKline, Brentford, Middlesex, U.K.) is a product stating that it is a colorless solution that may contain a small amount of visible translucent-to-white amorphous drug product particles. The product is supplied with an in-line filter. Similarly, Erbitux (Imclone/Bristol-Myers Squibb, New York, New York, U.S.) is a product that is also instructed to be used with an in-line filter due to the presence of protein particles. All four cases (Vectibix, Stelara, Arzerra, and Erbitux) are a result of the inherent nature of protein formulations, which can result in the formation of intrinsic protein particles. These visible, translucent, white particles are removed via in-line filtration during IV administration without noticeable loss of assay/potency since only a small fraction of protein is in the form of visible particulates. It could be debated if a better formulation could be developed to eliminate such protein particles is or is not possible. However, these types of products present a significant challenge during visual inspection. A reference to an acceptable lot must be made and the inspection assay should be validated to ensure that the product meets this reference. It would be a good practice to provide a quantitative or semiquantitative determination of these visible particles. In addition, some attempt should be made to understand the composition of the visible particles and to ensure that it remains consistent during storage under preferred conditions.

The Inspection Process

The levels of subvisible and visible particulates in a sample are a useful measure of the product quality. Monitoring of visible particles is an important product attribute and a regulatory requirement. There are two very different approaches to detection of visible particulate matter in parenterals. One method utilizes people and the other utilizes machines for detection.

Human Visual Inspection

The inspection apparatus that is normally used is comprised of an inspection station containing a lamp at a specified intensity. The EP provides a figure of the type of apparatus to be used in visual inspections. The lighting may be fluorescent, incandescent, spot and/or polarized, with fluorescent being the most common. The light source may be positioned above, below, or behind the units being inspected, with a range of intensity from 100- to 350-ft candles (note that the JP requires 740- to 930-ft candle light intensity for inspection of plastic containers) (19). The inspection station has both white and black backgrounds. Some inspection stations also include a magnification lens. The variables that are of concern during human inspection, such as fatigue and visual acuity, are addressed when appropriate training procedures are in place. The parameters that should be adhered to during visual inspection are referred to in the compendia. Figure 1 shows a manual inspection room. The inspectors in this

Figure 1   A manual inspection room. Inspectors are situated at inspection stations.

facility sit comfortably at stations with a gray/black background (black and white backgrounds are most commonly used). In this example, they are inspecting prefilled syringe products. The inspectors will view the syringes at several positions (Figs. 2–4), they are not only looking for particulate matter within the syringe but also looking at the condition of the container, inspecting all dimensions of the units for any defects. Figure 5 shows an inspector looking at vials. Vials are placed into a clear holder to allow for more than one unit to be inspected at a time for defects.

Some of the factors that are of concern during human visual inspection are listed below. The factors are categorized as either inspection process variables (these variables are controllable) or product characteristics (may not be controllable).

Inspection process variables:

Visual acuity (close vision capability)
Proper motion of the container to suspend the particles

Figure 2   An inspector begins by observing the prefilled syringes in their packaging trays.

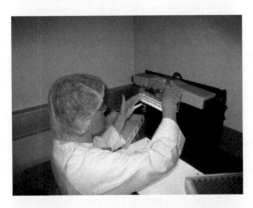

Figure 3   Prefilled syringes are viewed at the light source.

Figure 4   Prefilled syringes are inspected, ensuring that all dimensions of the product are monitored. Here the inspector is viewing the top of the syringe for any defects.

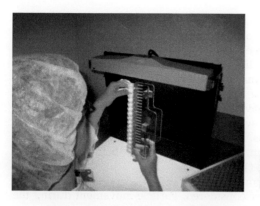

Figure 5   An inspector is monitoring vials. Here vials are placed in a clear holder so more than one unit can be inspected at a time.

Manual dexterity of the inspector
Type and intensity of lighting
Background to particle contrast
Time of inspection
Total background illumination
Accurate illumination at the point of inspection
Use of magnification
Initial position of the particle at the commencement of inspection
The presence of interfering signals such as cavitation bubbles or glare
Inspector fatigue

Product characteristics:

Total volume in container
Container clarity
Particle density and optical properties of particles and solution
Optical defects in the wall of the container and markings on container
Foaming properties and viscosity of the product

In general, a manual inspection procedure should include the following steps:

1. Any labels, if present, must be removed from the container. Container should be cleaned using low particle shedding wipes.
2. Hold the container by the neck and swirl container to set particles in motion. Care should be taken to avoid creating air bubbles. Air bubbles will rise to the surface, which will help differentiate them from particulate matter.
3. The container is inspected while being held at a 45° to 60° angle from the vertical, about 10 in. below the light source [some sources state 4 in. (8)]. The container should be inspected in front of both a black and white background. Light should be directed away from the inspector's eyes and the container should not be placed directly under lighting to avoid glare.
4. If no particles are seen so far, the container can be slowly inverted and inspected for any heavy particles that may not have been suspended during swirling.
5. Containers with visible particles should be set aside for further investigation/ rejection.
6. Small volume vials and ampoules (<50 mL) should be viewed for approximately 5 seconds (10 seconds per the JP). Containers that are 50 to 100 mL should be viewed for 10 to 20 seconds. Containers that are 250 to 1000 mL should be viewed for 30 to 60 seconds.

The manual inspection is considered the benchmark for all other particle inspection methods and devices, therefore no validation studies are required. However, it is important to note that human inspectors must undergo appropriate training and testing to become qualified for inspection. The qualification is then maintained through good supervision, proper procedures, and continued retraining. It has been shown that none of the commercially available inspection systems can, with a single reading, show performance that is equivalent to a single manual inspection (10). According to Knapp, even two serial machine inspections cannot achieve the security and discrimination of a trained inspector. With the advances in automated inspection machines, this statement may not be as accurate today, but it does show the power of the visual inspection procedure. However, it is imperative that the operators be properly trained for performing visual inspections. The accept/reject decision by the inspector has been shown to be a probabilistic determination, with the probability of being able to detect a particle increasing with increasing particle size (11).

The initial position of the particle at the commencement of inspection is an interesting aspect of a good inspection and should be further considered. Particle movement during the inspection is necessary to differentiate a contaminating particle from the container markings or other optical distortions. The inspector should be trained to provide maximum velocity to the particle while avoiding an excess of energy so that bubbles are not formed. If a formulation composition affects the formation of bubbles, the inspection procedure must be modified accordingly. An important factor in the velocity of a particle is the initial position on the container bottom. Those particles that are closer to the axis of rotation require greater rotational energy to achieve adequate movement during inspection. Care should be taken to avoid impact or excessive transport of the container to avoid cavitation bubbles from forming. It can be very difficult to differentiate between bubbles and solid particulates in solution. Also, it should be noted that the viscosity and container volume will affect the velocity of the particles in solution.

Inspector Training/Calibration
As stated previously, proper inspector training is critical to a properly designed visual inspection program. The handling of the sample as well as the "calibration" of the inspector is important to obtaining accurate and repeatable results. The concept of creating a calibration curve during training has been explored (11). The idea is similar to using a calibration curve for a particle counter. The rejection probability calibration curve is generated using a test set of containers, each with a single, durable, and accurately measured visible particle in a suspending fluid. Knapp has described the use of glass and stainless steel particles for the test set. Ideally, the test set should include samples that are representative of the entire particle contamination spectrum from clean to must-reject contamination. This test set can be applied to multiple inspectors at multiple sites, defining the test environment. Results should be obtained until there is sufficient data to support an analysis at the 0.05 significance level. This result set can be used to obtain a standard "visibility" reference curve. This reference curve can then be used to assure the competency of the inspectors and, ultimately, the quality of the production batch. Once the visibility reference curve is established, this information can be used to qualify new inspectors or machine inspection. A good practice would be to demonstrate that the inspection security is achieved to at least the same level as the qualified manual inspection (12).

Another aspect of training that can be easily overlooked is the requirement of an appropriate vision test. A study was performed to determine the efficiency of a group of inspectors (12). It was found that the group fell into three categories, low, middle, and high false reject rates. The high false reject rate occurrence in the high group was traced to the fact that only the standard distance eye test was specified for the inspectors. Following this observation, a close focus eye test was required to resolve the problem. It is critical to add appropriate visual acuity testing to the qualification of the inspectors.

Machine Inspection
There are numerous machines that have been developed to aid in the visual inspection process. The throughput is much higher when utilizing these systems versus manual human inspection. Machines for semiautomated inspection can perform most of the mechanical manipulations normally done by the human inspector. These manipulations include swirling

the vials, inverting samples, stopping the container, and the ability to remove units flagged as defective. These semiautomated systems can provide additional lighting, such as Tyndall or polarized light filters and adjustable container holders to change the angle of inspection or container rotation rate. The visualization process can be performed via an imaging system, which reduces eye strain to the operator. There are also completely automated systems available. Eisai, for example, has automated inspection instruments capable of inspecting vials, ampoules, and syringes (13). The AIM (Automatic Inspection Machine) is fully automatic and is capable of detecting particulate matter as well as cosmetic defects at up to 24,000 vials or ampoules per hour. Various areas of the container are inspected including the body, heel, neck, and crimp/cap area. There is also a system for the inspection of syringes. The EIS inspection system is fully automated and used primarily for prefilled syringe inspection at up to 36,000 units per hour.

These Eisai automated systems are based on transmitted light (static division) technology. Static division refers to the ability of the machine to differentiate between moving and static objects, for instance a moving particle versus a scratch on the container surface. Particles in solution will block a portion of the transmitted light passing through the container. The particle will block a portion of the light causing a shadow that is detected by an array of small diodes. Since the instrument is looking at the blockage of light, the color and reflectivity of the particle is not a factor in detection. In addition, the change in light intensity is monitored so only the signals from moving objects are recorded. During operation, the container is spun at high speeds (1000–5000 rpm) just before reaching the inspection station. At that point a brake is applied and the liquid inside the container will continue to rotate due to inertia. Any insoluble particle matter will be suspended and float past the detection system. A prespin step can be used to dislodge and remove any bubbles in the container, reducing the incidence of false rejects. Since the system utilizes transmitted light, it can be set up in any work area regardless of the external lighting. These machines are designed to perform two inspections per container. This improves both sensitivity and reproducibility. Human inspection capabilities are the benchmark used for determining the sensitivity of machine inspection. These machines claim to have a sensitivity better or equal to the human eye. Figure 6 shows an Eisai system, with vials being fed into the machine for inspection. A closer view, with a vial being illuminated, is shown in Figure 7.

Another inspection machine is made by Seidenader. The machine is used for inspection of clear liquids, suspensions, lyophilized products, all in containers that range from ampoules, cartridges, vials, or bottles up to 100 mL. Common particulates that can be detected include foreign material, floating particles, fibers, and glass shards. For lyophilized materials, inspection criteria include melt back, shrunken cake, discoloration, particulates on the cake, unlyophilized product, and fill level. The machine can also be used, with optional features, for inspecting vial and ampoule defects such as position and color of cap, missing stoppers, seal

Figure 6 A photo showing the Eisai system showing vials being fed to the instrument.

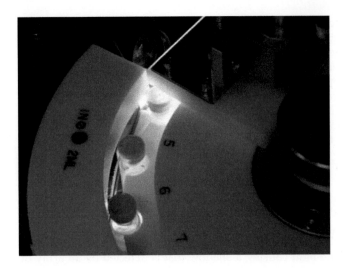

Figure 7 Vials moving into the inspection window, with a vial being illuminated.

crimp defects, scratches, dirt, cracks in sidewall or bottom, deformed ampoule tip, burn marks, and color ring code. The machines are made to analyze a wide range of products from 1 to 100 mL. The inspection mechanism is different for this machine. Each container is inspected in up to three stations. An image subtraction method is used for the detection of particulate matter. The inspection process has the following steps:

1. Container is rotated at high speed.
2. Rotation is stopped and the liquid continues to move, and particles within the liquid move.
3. A central inspection mirror moves with the transport of the container.
4. A camera acquires a sequence of images.
5. The images are sent to a processor and compared with each other via an overlay image (pixel by pixel).
6. Objects that change position between consecutive images are identified as free moving particles. Defects like scratches or glare will not change position between images.

Seidenader claims that this process of detection requires less agitation than other systems and is therefore more applicable to delicate samples such as biopharmaceuticals and viscous products. Production speeds to 36,000 vials/hr are possible. All images are stored and can be printed for further examination. The machine has some desirable options such as integration of NIR technology for product identification and residual moisture testing. An optional head space analyzer is available as well. Figures 8 and 9 show two views of a Seidenader inspection machine set up.

Important Considerations in Visual Inspection
The visual inspection process is performed not only to detect particulates within the product but also to monitor for any container and/or product defects. A useful approach to visual inspection is to create a list of criteria, categorized as critical, major, and minor defects. A typical list of inspection criteria used by the pharmaceutical and biopharmaceutical industry is shown in Table 1. Once a list of criteria has been determined, an AQL can be applied to each category of defect. The AQLs are based on historical information collected during development and manufacturing and represent the highest percentage of defective units unacceptable for releasing the batch. Having acceptance limits is a good manufacturing practice, these limits are used to trigger when an investigation should ensue. Examples of various container defects are shown in Figures 10 to 15. Most of these defects would be considered critical and easily observed by a trained inspector.

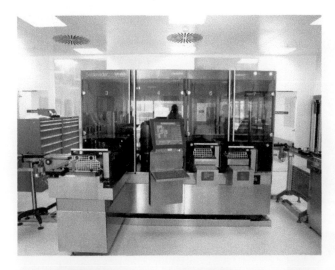

Figure 8 The Seidenader inspection machine.

Figure 9 The Seidenader inspection machine, a closer view showing the feedback module.

Table 1 Inspection Criteria, Categorized by Defect Impact

Critical defects (AQL 0.1–0.5%)	Major defects (AQL 1.0%)	Minor defects (AQL 2.5–5.0%)
Incorrect product	Extraneous color of cake	Bubbles in the glass
Melt back (lyophilized cake)	Presence of foreign material	Rough seam wave wrinkle in glass
Overfill of lyophilized product (results in superpotent dose on reconstitution)	Chipped vial	Poor appearance: dirt or specks imbedded in glass or minor scratches
Dried product on vial neck	Scratch in vial (double deep)	Crimp with poor appearance
Cracks in glass	Color variation in stopper	Presence of product on outside of container
Broken vial	Stones in glass (outside)	Broken lyophilized cake[a]
Stones in glass (inside)	Empty container	Uneven cake-cake surface on incline (lyophilized cake)
Incorrect stopper	Plastic cap from flip-off seal missing	Gross excess of product on inside shoulder
Missing/misaligned stopper	Substances on stopper	
Inadequate crimp (not tight)	Flanging incorrect	
Incorrect color of cap or seal	Amount of fill incorrect	

Example AQL limits are shown in parenthesis.
[a]For some products like antibiotics broken cake is normal and is not classified as a defect.

Figure 10   During a manual inspection, a crack along the bottom and side of a vial containing lyophilized product was detected.

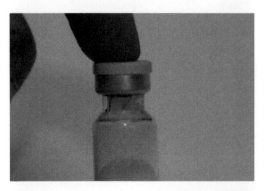

Figure 11   During a manual inspection a broken vial neck was observed.

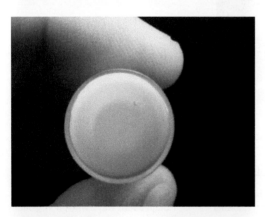

Figure 12   During manual inspection, a dark particle on a lyophilized cake is observed through the bottom of the vial.

Figure 13   Liquid product is seen below the stopper in a prefilled syringe product. This was observed during a manual inspection.

Figure 14 Photo showing vials that failed due to crimp defects, on the vial farthest right, no crimp/overseal was applied.

Figure 15 The vial on the right failed inspection due to less than acceptable fill volume (properly filled vial shown on the left).

The International Organization for Standardization (ISO) provides extensive guidelines on establishing sampling plans and determining AQLs. The international standard ISO 2859 *Sampling procedures for inspection by attributes* describes attribute sampling schemes and plans. The guidance document is divided into six parts, with part 1 being of most relevance for visual lot inspections. ISO 2859 Part 1—*Sampling schemes indexed by acceptance quality limits (AQL) for lot-by-lot inspection* provides sampling schemes indexed by AQLs. The quality measure used is percent nonconforming (or parts per 100). It was developed primarily for the inspection of a continuing series of lots that originate from the same production process. The remaining parts of the ISO guidance describe a general introduction to the series as well as sampling plans and procedures for specialized cases.

Facility Inspections/Requirements
In many cases, processing of parenteral products may be done at an external facility or contract organization. The opportunity to audit the facility is also an aspect of good process control. Some criteria that should be included in the quality audit of a contract manufacturing organization are as follows:

- Are product contact surfaces maintained under aseptic conditions?
- Are periodic evaluations of the facility performed and documentation available?

- How are out of specification results dealt with?
- How is the customer notified?
- What is the overall appearance of the facility?
- Do SOPs exist for each operation?
- Are training records current?
- Are operators trained in detecting particulate matter (visual acuity etc.)?
- Where are particulate matter investigations performed?
- Are they utilizing current/appropriate technology?
- Are they utilizing current compendial methods?

PARTICULATE MATTER CHARACTERIZATION

Particulate matter in a sample can originate from a variety of sources. Particulates can be intrinsic, coming from the product itself (as in the case of protein aggregates in a biological product, crystallization of API/excipients, precipitation of insoluble impurities or degradation products etc.) or extrinsic (sourced from a process or the environment the sample is in contact with). Any successful attempts to reduce the amount of particulate matter should include an attempt to identify particles. This will help in understanding the source of the particulate matter. Once samples have been identified as having visible particles, selecting a representative sample set and isolation of the particulate matter are the first steps in the identification process. As much as possible, the first examination of the particulate matter should be done without disrupting the integrity of the sample container, for instance, viewing the particulates in the container using a stereomicroscope. This is an opportunity to examine the population within the container—a look at the material in situ. The optical properties and morphological features of the particles should be recorded at this time.

Next, the particles should be isolated and examined by light microscopy. The use of polarized light can be very helpful in the identification of many types of foreign matter including glass, cellulosic fibers, and inorganic metallic particles. The morphological and optical properties of the particle should be recorded. If the identification is not complete, additional analysis can be performed on the isolated particles. These include spectroscopic techniques such as FTIR-microscopy and elemental techniques such as scanning electron microscopy–energy dispersive X ray (SEM-EDX).

Particle Isolation

There are several approaches that can be used to isolate the particulate matter. Filtration is a common approach, where a portion of the sample containing particulate matter is isolated on a filter membrane. The filter material should be appropriate for the type of particles being collected. For most foreign material, a Nucleopore® membrane filter, where particles are captured on the flat surface, works well. An advantage to the filtration approach is that the contents of the entire container can be captured for analysis. The process is more challenging for soft or gelatinous materials such as protein particles. These particles tend to form a film that is very difficult to recover from the filter membrane. For more challenging materials, other approaches may prove successful. Another common isolation approach involves the drawing up of particles via a microcapillary pipette. The particles are deposited onto a glass microscope slide. They can be washed, stained, or otherwise treated for further analysis. A third approach is to use centrifugation to sediment the particulate matter. This technique is useful for isolating very small particles that may be difficult to collect by other means. For larger particles, filtration or pipetting works very well. It is important to reduce the chances of introducing contamination (additional particulate matter) during the isolation process. The use of clean areas such as laminar flow hoods is encouraged. In addition, analyzing blank samples (such as neat filter membranes) can help identify any artifacts emanating from the procedure itself.

Microscopy

Light microscopy should be the first step in the identification process. A sensible approach is to first view the container in room light, then in an inspection light box. Next, the particles should be examined by low magnification using a stereomicroscope or similar instrument, recording the characteristics of the material in situ. An example of a white particle in a vial observed first by visual examination under room lighting, then inspected using a stereomicroscope is shown in Figure 16. The particle is easily detected and some additional characteristics such as the morphology can be recorded before opening the container. There are several references in the literature describing important characteristics of particulates, a comprehensive list of which is shown in Table 2 (14). Microscopy provides excellent sensitivity and can provide useful information from subnanogram amounts of material. The minimum particle length that can be resolved by microscopy varies with instruments and optical properties of the material, but is generally accepted to be 1 µm (15). The analysis can be done quickly and with a high degree of accuracy when performed by an experienced microscopist. In addition to the visual observations, microscopy can be used to aid in microchemical testing of the particles. Solubility can be an important indicator in the composition of the material, providing information on functional groups or elemental composition. Proteinaceous materials can be identified by the use of specific stains applied to the particles. Also, the use of fluorescence staining and microscopy can be advantageous for certain types of particulate matter.

Figure 16 White particle observed in a sealed vial (*circled*).

Table 2 Particle Characteristics Obtainable by Optical Microscopy with Polarized Light

Particle characteristic	Properties
Morphology	Particle shape
Size	Linear dimensions and thickness
Surface texture	Is it smooth, rough, scaly? Is there evidence of tool marks?
Hardness	Does it deform, is it brittle?
Reflectivity	Is it dull, semi-dull, or very reflective?
Transparency	Is it transparent, translucent, or opaque?
Color	Color using transmitted and/or reflected light
Magnetism	Is it magnetic?
Refractive indices	Can determine using refractive index oils? Can be very useful in identification?
Melting point	Does the particle melt when heated, and at what temperature? This requires the use of a hot stage
Chemical composition	What elements or functional groups might be present? This requires the use of microchemical or spectroscopic tests

The associated properties for each characteristic are listed in the right column.
Source: Adapted from Ref. 14.

It should be noted that there can be heterogeneity within an individual particle as well as within the particle population. It is critical that good sampling plans are utilized, and analysis of an appropriate number of particles is employed. Evaluating the sample first by optical microscopy provides the opportunity to identify heterogeneity and avoid misinterpretation of the results.

Another type of microscopy widely used in particle identification is SEM, especially when coupled with EDX analysis for elemental analysis. In SEM, a beam of high-energy electrons is rastered over a sample and an image is produced by means of low-energy secondary electrons and backscattered electrons (16). The SEM will give a topographical picture of the material, including size, shape, and detailed information on the texture. The addition of EDX provides elemental information by measuring the energy of X rays emitted from the sample when it is exposed to an electron beam. Sample preparation is minimal, usually just requiring a thin coating of gold, carbon, or other conductive material to prevent the charging by the electron beam. Instruments are available that can analyze samples under various environments and require no coating. The spatial resolution for SEM is around 0.1 μm or slightly less, making it a good technique for looking at particle homogeneity. The combination of high-resolution images and elemental composition can be a powerful characterization tool.

A fairly new microscopic technique gaining popularity is the use of flow microscopy. This technique provides particle counts as well as a digital image of each particle analyzed, using small sample amounts (as small as 500 μL) and detecting particles as small as 1 μm. Two examples of this technology are the FlowCAM and the Brightwell flow microscopes. The technique is powerful, allowing the user to view the morphology of the particles. In some cases, the morphology alone can provide critical information in identifying the type of particulate matter. Spherical particulate matter such as oil droplets and bubbles can be easily differentiated from more irregular shaped particles by evaluating shape factors associated with the particles analyzed. There has been some recent work on showing the differentiation of silicone oil droplets in the presence of protein particulates using the Brightwell flow microscope (17). A sample containing silicone oil droplets and protein particles was analyzed. Particles ≥ 5 μm were shown to be easily resolved by the flow microscope software. Because spherical particles such as oil droplets have very high aspect ratios, a filter was used in the data analysis to separate the particle population with a size ≥ 5 μm and an aspect ratio ≥ 0.85. This population was verified to be primarily comprised of silicone oil droplets. The technique is gaining popularity in the area of subvisible particle analysis since the technology claims to be more sensitive than the traditional light obscuration methodology. It is also being investigated for analysis of protein particulates since the technology also claims to be more sensitive to detecting near-transparent particles (18,19).

Within the category of dynamic imaging, another instrument, the Eyetech (Ankersmid Ltd., Antwerpen, Belgium) has been used for visualization of particulate matter. The Eyetech uses a rotating laser beam to scan individual particles within a sample zone. The instrument offers a variety of accessories for imaging wet, dry, or airborne particles. Focusing on the wet sample analysis, the instrument can provide similar information to the flow microscope. Here the detection system is based on the rotating laser beam obscuration time and signal interaction as detected by a photodiode. The duration of the laser obscuration is used in determining the particle size. Images of individual particles are recorded and can be recalled for later reprocessing. One area of concern for all of the dynamic imaging systems is the ability to only analyze particles that are in focus. The Eyetech addresses this issue by using a sophisticated algorithm that measures the angle of the laser path and particle boundary. If that angle is significantly $<90°$, it will be due to off center or out-of-focus particles. This will result in signals that have a wider pulse transition with small amplitude. These data points are then discarded, ensuring the user is not analyzing out-of-focus particles. Another feature of the Eyetech is that there are no assumptions of particle sphericity. The size measurements are solely based on the length of the cord crossed by the laser, providing an accurate length measurement for each particle. These imaging systems all provide information on particle size and shape, with some differences in how the data is presented. It is critical that the user understand the mechanism of detection and what information is being produced. This data can then be used to better understand the physical properties of the particles being analyzed.

Spectroscopic Techniques

The most useful molecular techniques include FTIR and Raman spectroscopy for particle identification. An FTIR microscope is a useful tool in particle identification. The sample can be viewed directly and in many instruments, a photomicrograph of the sample being analyzed can be obtained for reference. Variable masking options allow the analyst to measure the spectra of small areas within the field of view. There is minimal sample preparation, and many particles can be directly mounted onto potassium bromide (KBr) plates for analysis in transmission mode. Particle thickness and shape can have an effect on the spectra. Flattening the sample onto the KBr plate or using a diamond compression cell can help reduce these effects.

Raman spectroscopy can also be used for particle identification. For example, the RapID Single Particle Explorer (SPE) is an automated Raman system that allows for analysis of particles isolated on a gold filter membrane. The particles are filtered onto the membrane and it is placed into the RapID system. Run parameters are specified by the operator, including the area of the filter to be scanned. The instrument is capable of analyzing count, size, and shape of particles from 0.5 to 5000 μm. The filter is scanned and images of the scanned areas are recorded, along with the associated Raman spectra for the particles detected. The instrument compares the spectra obtained to defined libraries and reports particle composition and ID based on the libraries used. A ranking of how well spectra match library spectra is reported. An advantage of this type of system is that it can provide analysis of a great deal of particles in one run. The RapID system is also 21 CFR Part 11 compliant, making it a possible technique in a GMP setting. As with any analytical technique, care should be taken in the interpretation of the results and the analyst should use this information in conjunction with other information gathered during the identification process (e.g., optical microscopy).

In general, spectroscopy is sensitive and requires minimal sample preparation. The spectra can provide valuable information on functional groups related to the analyte. Most systems are equipped with comprehensive libraries, but care should be used when interpreting the results from automated identification software.

SOURCES AND PREVENTION

The ultimate goal of any particulate investigation is to determine the source of the particulate matter and reduce or prevent its recurrence. The procedures described thus far allow for the detection and identification of the particulate matter. This information can lead to an understanding of the source and can provide insight into methods for elimination. By understanding sources of contamination, controls can be implemented for prevention in future processes.

For biological compounds, air-water interface stress should be considered when agitating product for inspection. Such stress can create or increase particulates in these products, sometimes even with moderate agitation. Agitation can also cause the loss of visible particulates in some products (redissolution of precipitates). Visible particles that were loosely bound aggregates have been observed to break into smaller subvisible particles.

Particulate matter can come from a wide variety of sources including packaging, facilities/environment, or the formulation itself. There are numerous examples where packaging was the source of particulate matter. Glass vials have been known to delaminate, causing a haze-like contamination within liquid formulations. In addition, there are many reports of lubricants (such as silicone oil) leaching into liquid formulations, causing haze or opalescence within the sample. A list of some contaminants reported in IV solutions along with their sizes is shown in Table 3 (6). As shown, there are a large number of particles that are below visible detection (<50 μm). These particles, for the most part, will be detected during sub-visible particle analysis. However, the particle size listed is a primary particle size, and one must remember that many of these particles can be found in aggregated form putting their size in the visible region.

Environmental contamination can be the result of poor air quality and improper operator techniques during processing. It is recommended that processing and filling procedures are performed in laminar air flow devices and class-100 environments. Even in a controlled environment, particles can originate from air handling and filtration systems, room construction

Table 3   List of Contaminants Observed in IV Solutions,
Along with Associated Size Range

Contaminant	Size range (μm)
Insect parts	20–1000
Glass fragments	1–1000
Glass delamination	1–100 (extremely thin particles)
Rubber fragments	1–500
Metal particles	1–100
Cellulose fibers	1–100
Machine oil droplets	1–100
Plastic fragments	1–100
Starch	5–50
Calcium carbonate	1–10
Plasticizer droplets	1–10
Silicone oil droplets	0.01–10
Carbon black, clay, diatomaceous earth	1–5
Bacterial fragments	0.1–5
Viruses	0.05–0.1

Source: Adapted from Ref. 6.

Table 4   Theoretical Distribution of Particulate Matter Sources Under Controlled
Environmental Conditions

Area	Factor	%Contribution	
		≥10 μm	≥50 μm
Clean lab	Work area/equipment/materials	30	55
	Process flow/adjacent areas	1	5
	Personnel (worker and activity)	70	>30
Aseptic filling	Work area/equipment/materials	15	33
	Process flow/adjacent areas	<5	10
	Personnel (worker and activity)	85	40
Device production	Work area/equipment/materials	55	55
	Process flow/adjacent areas	5	5
	Personnel (worker and activity)	25	10

Source: Adapted from Ref. 6.

materials, personnel, equipment (primarily due to wear or malfunction), and instrumentation in the area. Barber, on the basis of his research, developed a theoretical distribution of particulate matter contamination (at 10 μm and 50 μm sizes) and their sources as shown in Table 4. As shown, the contribution from personnel is significant. Overall, the particle burden can be reduced by following good aseptic procedures including good personnel training, use of appropriate low shedding garb, room access control, and environmental particle monitoring.

The formulation itself may be the source of particulate matter. Excipients can be the source of particles, even if filtered before use. Degradation or interactions with materials used in processing/packaging can lead to particle formation. Storage conditions or sample-handling procedures can also induce particulate formation.

One source of particulate matter that is sometimes overlooked is filtration units. There are two types of filters commonly used, depth filters and screen filters. The choice of filter will depend on the intended application, flow rate, viscosity of the solution, and compatibility with the solution to be filtered. It has been observed that depth filters can contribute cellulose fibers, mineral fibers, animal fibers, glass fibers, or sintered steel particles into the filtered solution (5).

Figure 17 Photomicrograph showing the outside of a filled vial, with crystalline material at the vial crimp.

Screen filters generally contribute fewer particles to the solution. Screen filters are made of cellulose acetate or mixed cellulose esters with or without regenerated cellulose.

Case Study

There is considerable interest in monitoring both subvisible and visible particle load for the purpose of process and formulation evaluation. An increase in particle load during a stability study, for instance, can be an indication that the process or formulation is not well controlled. Further determination of the particle type can provide rich insight into the processing, providing information to obtain a rugged formulation or process leading to a high-quality product. For instance, during a development program, a formulation group noticed particulate matter on the outside of filled liquid product in vials as shown in Figure 17. The material was inspected by light microscopy and isolated for further analysis. The composition was found to be primarily crystalline mannitol, a main component in the drug product. The container integrity was tested on several vials and found to be acceptable. With further investigation it was determined that drug product solution was being deposited on the exterior of the vial during filling. The filling needle was not properly aligned and would drip between vial fills. This was easily corrected and this rapid investigation led to a more robust procedure for filling.

SUMMARY

An important part of particulate matter control is prevention early in the development process. The earlier controls are put in place, the less crisis situations and regulatory concerns that will have to be dealt with later. Personnel working in the processing areas should be effectively trained with regard to particle sources and particulate matter control. Since it is more difficult to control larger areas, it is advisable to create minienvironments and use isolators for various applications. Process design is critical to the control of particulate matter contamination. If particulate matter is found, techniques used in the inspection and identification of visible particulate matter have advanced to the point where investigations can be performed in a timely fashion. With a properly equipped laboratory, the detection and composition of particles can be determined, allowing for identification of the source and, ultimately, control over processes and formulations to greatly reduce or eliminate their recurrence. Many inspectors have used the visual inspection data during preapproval inspections and annual GMP inspections to identify weak links in the manufacturing process or controls. High reject rates during inspection can point to specific problems, for example,

high/low fill volumes—filling not controlled, lot checks and
lyophilized cake appearance/meltback—lyo controls, nonrobust lyo cycle.

It is important that the (bio) pharmaceutical industry utilizes this information to proactively address these potential issues. This will lead to robust and well-controlled processes for the development and manufacture of injectable drug products.

ACKNOWLEDGMENTS
The authors would like to acknowledge John Shabushnig (Pfizer Inc.) and Marie Reynolds (Pfizer Inc.) for their insight and comments on the inspection process.

REFERENCES
1. Shabushnig JG, Melchore JA, Geiger M, et al. A proposed working standard for validation of particulate inspection in sterile solutions. Paper presented at: PDA Annual Meeting. Philadelphia, 1994.
2. Douglas FG. Foreign particle embolism in drug addicts. Ann Intern Med 1971; 75(6):865–872.
3. Garvin JM, Gunner BW. The harmful effects of particles in intravenous fluids. Med J Aust 1964; 2:1–6.
4. Garvin JM, Gunner BW. Intravenous Fluids: a solution containing such particles must not be used. Med J Aust 1963; 2:140–145.
5. Thomas WH. Particles in intravenous solutions: a review. N Z Med J 1974; 80:170–178.
6. Barber TA. Control of Particulate Matter Contamination in Healthcare Manufacturing. Denver, CO: Interpharm Press, 2000.
7. Leversee RL, Shabushnig JG. A Survey of Industry Practice for the Visual Inspection of Injectable Products. PDA Visual Inspection Forum; Berlin Germany, 2008.
8. Akers MJ, Larrimore DS, Guazzo DM. Parenteral Quality Control. Sterility, Pyrogen, Particulate, and Package Integrity Testing. 3rd ed. NY: Marcell Dekker Inc, 2003.
9. Madsen RE, Cherris RT, Shabushnig JG, et al. Visible Particulates in Injections—a History and a Proposal to Revise USP General Chapter Injections <1>. Pharmacopeial Forum 2009; 35(5):1383–1387.
10. Knapp JZ. The effect of validation on non-destructive particle inspection. PDA J Pharm Sci Technol 1999; 53(3):108–110.
11. Knapp JZ. Overview of the forthcoming PDA task force report on the inspection for visible particles in parenteral products: practical answers for present problems. J Pharm Sci Technol 2003; 57(2):131–139.
12. Knapp JZ. The scientific basis for visible particle inspection. Pharm Sci Technol 1999; 53(6):291–302.
13. Eisai Machinery USA Inc. Pharmaceutical Inspection Equipment. Available at: http://www.eisaiusa.com/.
14. Delly JG. Essentials of Polarized Light Microscopy. 5th ed. Illinois: College of Microscopy, 2008.
15. McCrone WC, Delly JG. The Particle Atlas. Vol. I, 2nd ed. Principles and Techniques. Ann Arbor: Ann Arbor Science Publishers, 1973.
16. Borchert SJ, Abe A, Aldrich SD, et al. Particulate matter in parenteral products: a review. J Parenter Sci Technol 1986; 40(5):212–241.
17. Sharma DK, Oma P, Krishnan S. Silicone microdroplets in protein formulations. Pharm Technol 2009; 33(4):74–79.
18. Moore P. Automatic flow Microscopy for Pharmaceutical and Biological Analysis. Laboratory Focus 2008:12–14.
19. Sharma DK, King D, Moore P, et al. Flow microscopy for particulate analysis in parenteral and pharmaceutical fluids. Eur J Parenter Pharm Sci 2007; 12(4):97–101.

5 | Advances in parenteral injection devices and aids

Donna L. French and James J. Collins

BENEFITS OF INJECTION DELIVERY SYSTEMS

The growth of the injectable market in the last decade has led to the development of numerous injection device technologies for product preparation and administration. Injection device technologies facilitate injection preparation, ease administration, improve dose accuracy, and ensure safety, all of which contribute to improved user acceptance and compliance. Device technologies covered in this chapter include tools for injection preparation, needlestick prevention devices, and delivery devices used to administer injectable drugs.

FACTORS INFLUENCING SELECTION OF A DEVICE TECHNOLOGY

The selection of an injection device technology depends on the attributes of the patient population and drug product. Users of device technologies can be patients or health care professionals. For patient administration at home, the patient's experience with injectables, mental acuity, physical dexterity, and product storage conditions need to be considered. For the health care professional, prevention of accidental needlesticks and clinic operating procedures are the primary considerations. As for the drug product attributes, formulation factors such as preservatives, liquid versus lyophilized, and viscosity are critical. In addition, dosing factors such as the route of administration, frequency of administration, deliverable volume, and fixed versus variable dosing can have an impact on device selection. Although the selection of a device has historically been constrained by such factors, many current devices have new capabilities that overcome previous limitations and accommodate a wider range of product attributes.

Customer-Based Selection Factors

Route of Administration

The route of drug delivery—intravenous (IV), subcutaneous (SC), or intramuscular (IM)—affects the selection and design of the injection device. Devices used for IV administration must be universally compatible with clinical procedures and equipment, and most of these devices are tools that provide minor improvements to existing IV administration procedures. Injection devices are primarily designed for SC or IM use by a patient or health care provider. Key considerations in the design of devices for SC and IM use include injection depth and volume. The needles for these routes of administration vary in length and needle gauge. An injection device must be designed to ensure that the drug is injected in the appropriate SC or IM space. Incorrect administration could result in discomfort and/or alter the pharmacokinetics, efficacy, and safety of the drug.

Frequency of Administration

The frequency of drug administration impacts the selection of the type of device (e.g., cartridge pen or prefilled syringe with an autoinjector) as well as the decision to use a reusable or disposable design. Drugs may be administered frequently, such as daily or more often, or infrequently, such as once weekly or less often. For frequently administered products, devices that are portable and contain multiple doses in a compact design, such as multidose pens, are least disruptive to a patient's lifestyle and most cost effective. Infrequently administered products are likely to be supplied as a disposable single-dose device. A simple, intuitive, ready-to-use system is particularly appropriate in this setting, so that patients do not have to familiarize themselves with the instructions each time they use the device or perform a complex procedure with a product that is used only occasionally. The best example of this scenario is the use of emergency antidote devices, which may only be used once in a person's lifetime for a medical emergency. An easy, quick, and intuitive injection procedure is essential to prevent the occurrence of a serious medical condition.

The choice of a reusable or disposable device is a balance between convenience and cost. Although reusable devices require more manipulation by the end user, they are more cost effective. If a drug is frequently administered, the unit cost of disposable devices may be cost prohibitive because of the waste associated with frequent disposal.

Acute Vs. Chronic Therapies
In acute therapies, a drug is used only for a discrete period of time, whereas with chronic therapies, the drug is used for extended periods of time, and in many instances, for the duration of the patient's life. For acute therapies, ease of use is essential so that the patient does not have to be trained on a complex injection procedure when the product will be used only for a limited time. For chronic therapies, easy-to-use devices are preferred but the extended use also allows additional features to be considered. Chronic therapies may require long-term monitoring of the disease, which may require that the patient use electronic features on the injection device to record injection times, dosing, and other information useful for monitoring the compliance or other disease status indicators. Although such features are inherently more complex, they can provide significant value and are not an issue after the patient becomes accustomed to using them.

Considerations for Self-Injecting Populations
The selection and design of an injection device requires careful consideration of the needs of the end-user population—children, elderly, and physically challenged. In self-injecting populations, physical impairments, cognitive challenges, the user's degree of experience with injections, and patient age (e.g., pediatric, elderly) are important considerations in device design. For example, rheumatoid arthritis patients will require a device that is easy to grip and activate, and therefore careful study of the physical challenges is critical to ensuring the device can be used by the patient population. Elderly patients may have difficulty reading instructions or dosing information on the device. Aside from special considerations for specific end-user populations, human factors must be a key consideration in all delivery device designs. Not only are human factors a good design practice, human factors studies are required as part of the regulatory expectations for medical devices. Never assume you understand how a user will use the designed device. Experienced users will provide different feedback than inexperienced users, and the exact same device may have different challenges for each therapeutic for which it is applied. In all cases, device designs that are intuitive are less likely to have issues with end-user training and compliance.

Considerations for Health Care Professional Users
A key consideration for health care professionals is safety—prevention of cross-contamination between patients or between the patient and the health care provider. Prevention of patient to patient contamination is easily achieved with single-use disposable devices. For multiuse devices, explicit instructions not to share or reuse devices with other patients should be provided. Currently the only type of device used for multiple patients are reusable needle-free devices that have been used for mass immunization worldwide. Such devices are uncommon in other clinical settings, where single-use disposable syringes and other devices are typically used. These devices are equipped with needlestick prevention mechanisms to protect the health care provider.

Needlestick prevention devices that safeguard health care providers from contracting serious or fatal diseases from accidental sharps injuries have become a key focus area for the pharmaceutical and medical device industries within the last decade. Injuries from needles or other sharps contaminated with bloodborne pathogens such as human immunodeficiency virus, hepatitis B virus, and hepatitis C virus have been a serious problem. The Center for Disease Control and Prevention estimated that 600,000 to 800,000 needlestick and other percutaneous injuries occurred among health care workers annually, and that 62% to 88% of sharps injuries could be prevented by the use of safer medical devices (1). In 2000, the United States enacted the Needlestick Safety and Prevention Act requiring that hospitals and clinics take appropriate measures to prevent needlestick injuries to reduce the risk of transferring bloodborne pathogens to the health care provider (1). A revised Occupational Safety and Health Administration (OSHA) Bloodborne Pathogens Standard became effective in 2001. New engineering controls in this standard included the use of sharps with engineered sharps injury protections (ESIP) and needleless systems. Products that are administered by health care providers in hospital clinical settings must have a needlestick prevention mechanism, and only some device types are amenable to this design feature. Devices with engineered sharps protection are discussed in more detail in sect. "Needlestick Prevention Devices."

Product Property–Based Selection Factors

Liquid Vs. Lyophilized Products

Injectable products are supplied as liquids or lyophilized powders. A lyophilized formulation is used when the drug product is not sufficiently stable during its shelf life storage as a liquid. A disadvantage of lyophilized products is that they require reconstitution prior to injection, which is an additional step that must be performed compared with a liquid product. Some device technologies are designed solely to ease the inconvenience of the reconstitution step, and then the drug is injected with a conventional needle and syringe. When a lyophilized product is used with an injection device, the design must have a mechanism for reconstitution prior to injection. While reconstitution can be designed to be easier, a liquid formulation will always be an advantage. The more inexperienced the user, the more physically or mentally challenged the user, or the more frequent the need for reconstitution, the greater the value of a ready-to-use liquid formulation.

Viscosity

Even slightly viscous products (~5 cp) impact the delivery of a product with a device, and higher viscosity products (15–30 cp) may require significant adaptation of the drug delivery system for administration. The viscosity of the product affects the needle gauge that can be used with the injection device technology and the speed of injection. Designing a device for a viscous product will require trade-offs between the needle size, injection time, and the force applied to administer the product. To mimic a typical injection speed (less than ~10 seconds), the size of the needle must be increased or the force that the patient or device must apply to expel the product through the needle must increase. Increasing the size of the needle is potentially more painful for the patient, and the force that can manually be applied to a syringe or device to expel the product is limited. High forces can be applied using automated types of injection devices, but a specialized design may be required.

Preserved Vs. Nonpreserved Formulations

Whether the product can be stabilized with preservatives will also be a key factor in the delivery technology selection. Multiuse products such as pen and cartridges systems require the pharmaceutical product to be preserved. If the drug product is not stable with preservative, then a single-use system will be required.

Fixed Vs. Variable Dosing

Some products are dosed according to the weight or specific therapeutic needs of the patient (variable dosing). Other products are the same dose despite differences between patients (fixed or flat dosing). This is an important development parameter as some device designs are more amenable to variable dosing.

Volume of Administration

The injection volume to be administered is a key factor in device design and selection as it affects dose accuracy, ease of administration, and injection site tolerability. The administered volume is dependent primarily on the potency, physicochemical properties, and stability of the drug as a function of concentration. The administered volume of most SC products is between 0.1 and 1.0 mL, and most injection devices are designed to deliver volumes in this range.

Because of the large doses required for efficacy of some drugs and limitations in drug concentration for some products because of their stability and physicochemical properties (e.g., viscosity), an increasing number of injectable products may require administration of volumes more than 1 mL. A general clinical rule of thumb is that SC injectable volumes should be less than 1 mL to avoid injection site discomfort. However, the tolerability of higher injection volumes is not well understood, and the impact of the formulation (excipients, pH, drug properties) and the needle type (e.g., size, needle point geometry) add additional complexity to understanding the injection experience. From a technology perspective, current bolus injection devices are typically designed to deliver 1 mL or less, although technologies can be adapted to deliver larger volumes that are clinically tolerated. Infusion pumps can deliver larger volumes subcutaneously over a longer period of time (minutes to hours) than a typical bolus injection ($\leq$10 seconds). From a clinical perspective, other options to administer the product can be considered, such as IM or IV administration. Many of the options to deliver large doses are also less convenient for the patient.

Injection of small volumes poses different challenges. Because of limitations in glass-forming and filling technology, prefilled syringes and autoinjectors are not recommended for volumes below 100 μL. If a prefilled syringe is going to be utilized, the formulation should target a volume above 100 μL. If a pen cartridge device is going to be utilized, the International Standards Organization regulatory expectation (ISO 11608) for dose accuracy is an absolute value of $\pm$10 μL up to a volume of administration of 200 μL, and then $\pm$5% for volumes greater than 200 μL. Given this requirement, the system dose accuracy will be impacted by the dose volume chosen. At a dose volume of 10 μL, the dose accuracy would be $\pm$100% and the dose accuracy as a percentage would improve as the dose volume increases.

THE PRIMARY CONTAINER

The drug product is contained in a container closure system, or primary container, to prevent microbial contamination, solvent loss, or exposure to gases or water vapor. Type I borosilicate glass is the most common construction material due its excellent barrier properties and inertness. Current devices are designed to be used with three types of primary containers: vials, cartridges, and prefilled syringes.

Vials

Glass vials are the most prevalent primary container for injectable drugs. Plastic resins may be more prevalent materials for this container closure system in the future. Vials are cylindrical containers with a stopper and seal, which is crimped onto the top of the vial to maintain container closure (Fig. 1). Vials can contain either liquids or lyophilized powders. Most injection devices used with vials are tools to facilitate reconstitution (if the product is lyophilized), transfer product between vials and other containers (e.g., IV bags, syringes, or other vials), or ease removal of the product from the container.

Cartridges

Glass cartridges are used as a primary container for injection pens (described in sect. "Injection Pens" in more detail). For preserved multiuse formulations, glass cartridges are the commonly used primary container. As with other vials, plastic resins may be a more prevalent material for this primary container system in the future. A cartridge has a tubular barrel that is sealed on each end with a rubber or elastomeric closure (Fig. 1). The drug product is prefilled into the container and retained by a stopper to which a needle assembly can be attached on one end and a plunger on the other end. Dual-chamber cartridges are used for lyophilized drug products in which the dried product is contained in one chamber and the diluent in the other. A channel between the chambers allows mixing and reconstitution of the product at the time of use.

Prefilled Syringes

Prefilled syringes are syringes supplied to the patient or health care provider that already contain the drug. These systems offer a more convenient alternative to standard drug vials—the user does not need to perform the steps to prepare and administer the product from vials, such as air pressure adjustments, aspirating the drug from the vial, changing needles, and

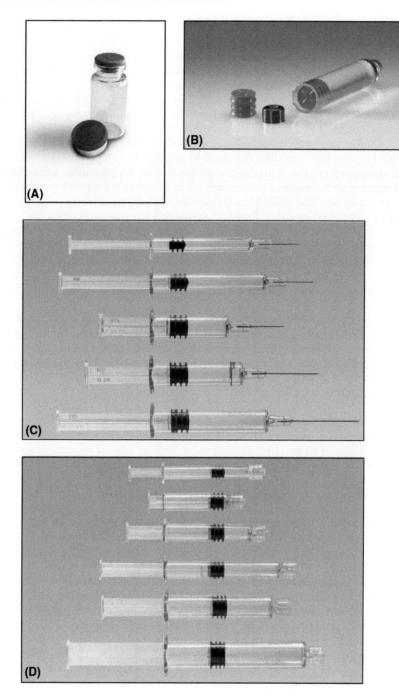

Figure 1 Primary containers: (**A**) vial, (**B**) cartridge, (**C**) staked needle prefilled syringe, and (**D**) luer-lock prefilled syringe. *Source:* Parts A and B courtesy of West Pharmaceutical Services and parts C and D courtesy of Becton, Dickinson and Company.

reconstitution and handling of diluent if the product is lyophilized. Prefilled syringes reduce or eliminate these preparation steps, improve dosing accuracy, and reduce the risk of contamination because of fewer transfers of the product prior to injection. Prefilled syringes can be used as the injection device or can be used with an autoinjector (described in sect. "Autoinjectors") to automate the injection.

Prefilled syringes consist of a cylindrical glass or plastic barrel; by far, the most common prefilled syringes are made of type I borosilicate glass. Plastic prefilled syringes are used commercially to a limited extent but may become a more prevalent technology in the future. Other components include the plunger (or stopper), plunger rod, and syringe tip, which may be a luer tip or staked needle (Fig. 1). Luer-tipped syringes have a tip cap for product containment and maintaining sterility. Prior to injection, the user removes the tip cap and attaches a needle to the syringe. Staked needle systems have the needle permanently affixed to the tip of the syringe barrel, and the needle tip is embedded into an elastomeric or rubber needle shield to maintain product containment and sterility. The preattached needle offers greater convenience because the user does not need to perform the steps to attach the needle. However, the user does not have the ability to choose the needle, which can be a disadvantage. Because of manufacturing and container closure design requirements for these systems, staked needles have different injection and glide force properties, and the injection experience may be different than that of nonstaked needles. Prefilled syringes with staked needles are intended for SC or IM administration, rather than IV use through ports. Depending on the application or preference of the user, a luer-tip system with the needle of choice may be preferable. The decision to use a staked needle or luer-tip syringe depends on various factors for which needle selection is important, such as the route of administration, product viscosity, compatibility with injection devices or aids, and market preferences for convenience versus flexibility in needle choice.

Prefilled syringes are available for lyophilized drugs. These syringes consist of a glass dual chamber container in which one chamber contains the diluent and the other contains the drug product. At the time of injection, the drug is reconstituted by moving the diluent from its chamber into the drug product chamber by pushing the plunger rod. After the reconstitution step, the product is directly injected into the skin as with any syringe. These systems are available commercially with a luer-tip system.

Drug Product and Container Compatibility

Assessment of the compatibility between the drug product, primary container, and device as a system is an essential element of product development and an area of increased scrutiny by regulatory agencies. The stability of biopharmaceutical products is impacted by the physical, chemical, and mechanical properties of the primary container and device system. Factors that must be considered in assessing compatibility include materials of construction, surface preparation or treatment, and any added excipients. Glass, plastic, elastomeric, and rubber components have inherent chemical and physical properties that can impact product stability. The integrity of the product upon actuation of the device must be also evaluated to ensure that the device is not detrimental to the product. The shear forces applied to the product by some devices have the potential to cause instability for biopharmaceuticals. The impact of manufacturing and component preparation processes including washing, sterilizing, and storage conditions must be evaluated, and understanding how these components are tested for product quality is essential. Lot-to-lot and vendor-to-vendor variability can pose unforeseen issues. In addition to drug stability issues, the functionality of the device can also be impacted by these same factors and may also vary with the drug product formulation. The compatibility and functionality of the drug-container-device system needs to be assessed under both normal and stressed conditions, and it must be understood not only initially but also as the components age. The stability of the drug and functionality of the device must be retained over the shelf life of the product.

Leachables and extractables in primary containers commonly pose chemical and physical stability challenges for injectable drug products. Primary container components can contain trace metals, plasticizers, antioxidants, accelerators, silicone, vulcanizing agents, and other chemicals that can impact product stability. For example, silicone is in most primary containers to provide lubricity for proper functionality. Silicone enables the plunger to glide smoothly through the syringe or cartridge for drug administration. However, many pharmaceutical proteins are incompatible with silicone and the product can become unstable and form particulates upon storage or shipping. Another example is tungsten, a heavy metal residual from the prefilled syringe manufacturing process, which can lead to serious stability issues for some biopharmaceuticals. Other protein compatibility issues include adherence of the drug to the glass walls of the container and delamination of glass surfaces.

Some technologies have been developed to avoid compatibility problems with primary container components. These include fluoropolymer coatings on rubber components as well as thermoplastic elastomers and fluoropolymers as materials of construction for plungers and stoppers. Improvements in silicone chemistry and application have been developed that increase adherence of the silicone to the containers and reduce the amount of free silicone available. Plastic prefilled syringes, which are composed of cyclic olefin polymer or copolymer, offer potential advantages over glass with respect to the leachable/extractable profile. In addition, plastic syringes offer the advantages of being lightweight and less prone to breakage. One type of plastic syringe is silicone-free, which is an advantage for silicone-sensitive biopharmaceuticals. These syringes are more permeable to gases than glass, which is a factor that must be considered in development of these prefilled syringe products, particularly with respect to stability of the biopharmaceuticals.

INJECTION DEVICES AND USER AIDS
Autoinjectors
Autoinjectors are spring-based systems that automatically inject drug from a prefilled syringe into the skin (Figs. 2 and 3). Insertion of the needle into the skin and delivery of the drug occurs automatically upon activation. These devices can be used for SC or IM injection. All commercial autoinjectors are fixed single-dose systems, although injectors with variable dose capabilities could be developed. Reusable, semidisposable, and disposable systems are used commercially.

Reusable systems are cost-effective options for frequently administered products in self-administering populations. However, these delivery systems require a significant amount of end-user manipulation to perform an injection, and the complexity of use has potentially limited their popularity. Numerous steps are required to operate reusable systems because the user must assemble the device for use and disassemble it after use. Typically, a user will have to separate the device into two parts, set the activation mechanism, insert the prefilled syringe, assemble the device, unlock the actuator, and activate the device for injection. After injection,

Figure 2 Single-use disposable autoinjector. *Source:* Courtesy of Scandinavian Health Limited Medical.

Figure 3 Reusable autoinjector designed for Rheumatoid Arthritis patients. *Source:* Courtesy of Owen Mumford Ltd.

the device is disassembled and the syringe is removed and discarded. Most systems are designed for staked needle prefilled syringes but some are designed to be used with a cartridge, luer-tip prefilled syringe, or a syringe that has been loaded with drug from a vial by the end user. Currently, reusable devices do not have needlestick prevention features; such a feature would require a permanent locking mechanism over the needle tip that would prevent reuse.

Single-use disposable autoinjectors were historically used for emergency medicine but are now used with pharmaceuticals. These systems are supplied preassembled with a prefilled syringe containing the drug. These systems have integrated needlestick protection mechanisms, which makes them suitable for clinical or home administration. These systems are easier to use than their reusable counterparts as they are supplied preassembled and ready to activate. The user simply removes the needle cover, unlocks the actuation mechanism, activates the device to inject, and then discards the entire device after injection. These systems are good choices for infrequently administered products or acute application because of their ease of use; disposable systems are more cost effective with short-term or infrequent use.

With semidisposable systems, the component containing the drug product is discarded after each use but the activation mechanism component is reused. All commercial disposable and semidisposable autoinjectors are used only with liquid products, although systems for use with lyophilized systems are under development.

Injection Pens

Pen injectors (Fig. 4) were first used for frequent self-administration of preserved multidose drug formulations requiring weight-based dosing. However, fixed-dose and single-use pens are now in use. Pen injectors are the most widely used injection devices and have been used in the diabetes management and human growth hormone (hGH) market for the last 20 years. Pen injectors are portable and provide greater ease of use and convenience compared with traditional vials and syringes. Cartridges are the primary drug containers used with pens. Small gauge needles designed for SC injections are used with these devices. The user is required to place a needle onto the tip of the device, set the dose, manually insert the needle into the skin, and push a button to inject the drug. Prior to the next injection, the system may need to be reset, and the needle must be replaced. The original pen injectors were reusable devices, which require the end user to insert a prefilled cartridge into the pen and to periodically replace the cartridge when it is empty. More recently, disposable pen injectors have been introduced in which a prefilled cartridge is preassembled into the pen. The entire device is discarded when the cartridge is empty. An advantage of pen injectors is that they can easily be used with liquid or lyophilized formulations. For lyophilized formulations, either the cartridge is a dual chamber system (as described in sect. "Cartridges") or an adapter is provided to reconstitute the powder in the cartridge. After the initial reconstitution, injections are identical to that of a liquid formulation. Hence, the inconvenience of reconstitution is minimized with these systems because it is performed only when a new cartridge is used, and the reconstitution process is easier with the pen than with conventional vials and syringes. More recent developments in pen devices include the use of needle safety devices, automated needle insertion and injection, smaller dosing capabilities, and electronics. Electronic capable systems allow the patient to record and review their dosing information.

Figure 4   Injection pen. *Source:* Courtesy of Y posomed.

Figure 5 Needle-free injector. *Source: Courtesy of Antares Pharma Inc.*

Needle-Free Injectors

Needle-free injection systems enable the injection of drug products without the use of a needle (Fig. 5). A high-pressure gas or spring drives the drug product through a small orifice in the device with sufficient force to create a hole in the skin and inject the drug. Needle-free injections are not pain-free but have obvious attractions for needle-phobic patients. These technologies also best meet the requirements of the U.S. Needlestick Safety and Prevention Act. Reusable needle-free devices are commercially marketed for hGH and insulin, and have historically been used for mass immunizations. These systems typically require that the drug be transferred from a vial into a cartridge that is used in the device. Either a liquid drug product or a reconstituted lyophilized drug product can be transferred into the device for injection, and then the end-user sets the dose and injects. Reusable systems offer variable dosing and formulation flexibility, but as is typical of all reusable devices, they are more difficult to use than disposable versions. Disposable needle-free devices are in development by a number of companies. As with other prefilled disposable devices, fewer steps are required for injection. Disposable devices are designed to be filled from a vial or prefilled to give a single fixed dose. Historically, disposable needle-free technologies have been limited to liquid formulations and low administrable volumes, but recent developments include the ability to dose a lyophilized formulation using a dual chamber cartridge and larger volumes.

Reconstitution Aids

Reconstitution of a lyophilized product prior to injection is an inconvenience that can be offset by the use of dual chamber syringe/cartridge systems or adaptors that enable needle-less reconstitution in vials. Dual chamber systems contain the diluent and powdered drug in the same primary container and enable mixing via a channel that connects the drug and diluent. These are available as prefilled syringes (as described in sect. "Prefilled Syringes") or as prefilled cartridges for use with an injection device.

In addition to dual chamber syringes, a variety of reconstitution aids are available to facilitate the preparation of lyophilized drug products (Fig. 6). Vial adapters provide an interface between the product and diluent that enables the reconstitution of the product by a luer or other connection rather than a needle. Some adapters connect product and diluent-filled vials with a syringe. Others connect the vial with a syringe, in which case a prefilled diluent syringe provides additional convenience. Some adapters are preattached to the product vial and diluent syringe or have preattached needles. Vial adapters for multidose formulations are also available. These systems have a mechanism in the adapter to maintain container closure between injections.

Tools to assist the patient in handling components such as removing caps on vials and syringe caps and/or handling of needles have also been marketed. Some injection device companies have developed devices that enable reconstitution and automate injection, but none have been commercialized to date.

Pumps

Pumps (Fig. 7) have been utilized for the clinical delivery of many IV products and for the SC delivery of insulin by diabetics for over 20 years. The growth of biotechnology products is increasing the opportunities for pumps, and pump technology choices for clinically and patient-administered products are likely to grow. There are three broad categories of pumps

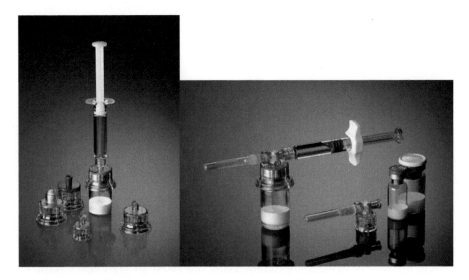

Figure 6   Reconstitution aids. *Source:* Courtesy of West Pharmaceutical Services.

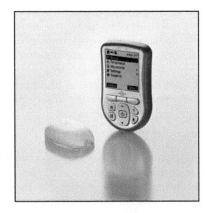

Figure 7   External injection pump. *Source:* Courtesy of Insulet Corporation.

that can be considered: IV pumps, patient-administered external pumps, and implantable pumps.

While not an option for all products, clinical IV administration with a pump is a viable option for infrequently delivered products or for very serious diseases in which IV administration provides a clinical advantage. An IV formulation provides very good systemic distribution and bioavailability, and the vial container closure system utilized to store the pharmaceutical product prior to pump administration is well established. The challenge with formulations administered by an IV pump is the end-user inconvenience.

With regard to patient-administered external pumps, there are three different types of external pumps being developed that balance ease of use with additional features and cost. These three external pump types can be categorized as disposable patch pumps, semi-disposable electromechanical pumps, and reusable electromechanical pumps. Disposable patch pumps are simple needle-based systems with an adhesive patch attached to the skin. These disposable patches are primarily designed to deliver a fixed basal rate and are utilized for one to three days. Semidisposable electromechanical pumps utilize a disposable drive system and incorporate a reusable electronic module that can provide a more advanced feature set such as a variable basal rate or bolus dosing. In some cases, the pump worn by the patient can be smaller as the electronic module that controls the pump can be carried separately. External reusable electromechanical pumps are full-featured pumps that have variable basal rates, bolus dosing, data connectivity, and dose memory options. These reusable pumps can provide very accurate dosing and can incorporate a number of features required by the therapeutic. Currently, patient-administered external pumps utilize a cartridge container that is filled by the patient from a vial. Patient-administered pumps are complex and costly, and they need to provide a clinical benefit over other options to gain customer acceptance.

The last pump category is internal implantable pumps. These pumps provide a unique benefit of being able to deliver drug product directly to specific areas such as the central nervous system (CNS) to provide a unique clinical benefit. Implantable pumps are primarily being applied in diabetes, pain, and CNS applications.

Needlestick Prevention Devices

The U.S. Federal Needlestick Prevention Act of 2001 (see sect. "Considerations for Heath Care Professional Users") has driven the commercialization ESIP systems for their injection devices, prefilled syringes, and needles. These devices are supplied with needle products designed for withdrawing body fluids, accessing veins or arteries, and administering medications in a clinical setting. Needles intended to be attached to a syringe are available with a needle cover mechanism that the user slides over the tip of the needle after use (Fig. 8). For drug products in prefilled syringes with staked needles, needlestick prevention devices can be provided preassembled with the syringe (Fig. 8). These systems typically consist of a main body attached over the syringe body with a component that slides over the tip of the syringe and needle after use. Manual, active, and passive needlestick prevention devices are commercially available. However, manual systems are becoming obsolete with the introduction of newer technologies. With manual systems, the health care provider manually slides the protective guard over the

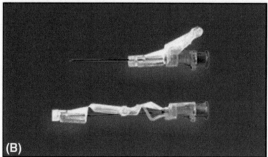

Figure 8   (**A**) Automatic needlestick prevention device with staked needle prefilled syringe and (**B**) needle-based needlestick prevention device. *Source:* Part A courtesy of Safety Syringes, Inc. and part B courtesy of Becton, Dickinson, and Company.

needle at the end of the injection. The manual design is suboptimal as users must physically pass their hand over the needle to cover it; mistakes or incorrect use of the device have the potential to cause a needlestick. With active systems, an actuation step is required to automatically activate/slide the needle protection guard, but it does not require the user to pass their hand over the needle. Passive devices are automated and safest; these systems activate the needle protection guard at the end of the injection without any additional action by the user.

DELIVERY DEVICE DEVELOPMENT QUALITY SYSTEM REQUIREMENTS

Incorporating a device into the dosage form requires the development scientist or engineer to understand the medical device regulations as well as the pharmaceutical regulations. The expectation is that a medical device incorporated into a pharmaceutical delivery system will meet the key elements of the device regulations. The two primary global governing regulations for these combination products are the ISO 13485 and the Food and Drug Administration's (FDA) Title 21 Code of Federal Regulations (CFR) Part 820. These regulations share many common requirements. In addition, the emerging Quality-by-Design initiative at the FDA contains a number of elements that are consistent with the device requirements such as working to establish patient requirements early and understanding the key patient risk in the product design.

Device Quality Management for Parenteral Delivery Device Forms

International Standards Organization—ISO 13485

This international standard outlines the requirements of a quality system for an organization to provide a delivery device that will consistently meet the customer requirements and regulatory requirements applicable to medical devices. The primary objective of this international standard is to assist in harmonizing global medical device requirements and quality systems. For this reason, ISO 13485 has many elements that are common to the FDA's Quality System Requirement (QSR) and Japan's Ministry of Health Labor and Welfare (MHLW) requirements for medical devices. In comparing the ISO 13485 to the FDA's QSR, the ISO guidance is not as prescriptive with regard to device history records, complaint handling, and device master records, but is more detailed with regard to understanding customer requirements and planning for demand realization. The ISO 13485 standard specifically focuses on medical devices, and therefore some of the requirements for ISO 9001 (Quality Management System—Requirements) have been excluded. An organization that meets the requirements of ISO 13485 will need to add the requirements for ISO 9001 if they desire to also claim conformance with that standard.

Quality System Requirements (21 CFR Part 820)

The QSR represents the FDA's requirements for medical device design, development, manufacturing, and postmarket surveillance as defined in Part 820 of the Food, Drug, and Cosmetic Act. The FDA does not expect companies that are developing combination products (containing a medical device constituent part) to operate under two separate quality systems, but they do expect that the device constituents will meet the primary device regulations and the drug components will meet the primary drug regulations. The FDA provides companies two options with regard to compliance for a combination product. A company can demonstrate that each constituent part meets the applicable regulations or demonstrate compliance with either the drug current Good Manufacturing Practices (cGMPs) or device Quality System regulations while also meeting certain conditions of the other quality system. If operating under a pharmaceutical quality system defined by Part 210/211, the FDA requires that the organization also demonstrate compliance with six specific provisions of the device regulations: Management responsibility 820.20, Design controls 820.30, Purchasing controls 820.50, Corrective and preventative actions (CAPA) 820.100, Installation 820.170, and Servicing 820.200. The use of CAPA systems is becoming more prevalent within pharmaceutical companies, but it is important to understand that a CAPA system is a critical element of the QSR and will nearly always be inspected as part of a device regulatory visit. If a company is operating under a quality system defined primarily by the device QSR contained in Part 820 and developing a product that includes a parenteral dosage form, the following elements of the Part 210/211 regulations should be included: Testing and approval or rejection of components, drug product containers, and closures 211.84; Calculation of yield 211.103; Tamper-evident packaging for over-the-counter (OTC) human drug products 211.132; Expiration dating 211.137; Testing and release for distribution 211.165; Stability testing 211.166; Special testing requirements 211.167; and Reserve samples 211.168 (2).

DESIGN AND DEVELOPMENT PLANNING

Design and development planning is a critical requirement for medical delivery devices. It includes requirements to understand the key elements of the design. The FDA requires that each company establish and maintain plans that describe the design and development activities and define responsibility for implementation. The plan needs to identify and describe the interfaces with different groups that provide input to the design and development process. The plans need to be reviewed, updated, and approved as design and development evolves (3).

Design Inputs, Design Outputs, Design Review, Design Verification, Design Transfer, and Risk Management

As devices are incorporated into a pharmaceutical dosage form, specific requirements with regard to medical device design inputs, design outputs, design reviews, design verification, design transfer, and device risk management should be included.

Design inputs are the physical and performance requirements of a device that are used as a basis for device design. The manufacturer must establish and maintain procedures to ensure that the design requirement relating to a device are appropriate and address the intended use of the device, including the needs of the user and patient. This includes a mechanism for addressing incomplete, ambiguous, or conflicting requirements. It is important to understand that market research concepts are not design inputs. The expectation is that some development will be required to transform the initial market research concepts into a more comprehensive set of documents that define the design inputs as per the QSR (3).

Design outputs are the results of a design effort at each design phase and at the end of the total design effort. The manufacturer must have procedures for defining and documenting design output in terms that allow an adequate evaluation of conformance to design input requirements. Design output procedures shall contain or make reference to acceptance criteria and shall ensure that those design outputs that are essential for the proper functioning of the device are identified. The finished delivery device design output must be the basis for the device master record. The total finished delivery device design output consists of the device, its packaging and labeling, and the device master record. Design output must include production specifications (assembly drawings, component and material specifications, production and process specifications, software machine code, work instructions, quality assurance specifications and procedures, installation and servicing procedures, packaging and labeling specifications, including methods and processes used) as well as descriptive materials that define and characterize the design (3).

Design reviews are a formal, documented, comprehensive, systematic examination of a design to evaluate the adequacy of the design requirements, to evaluate the capability of the design to meet these requirements, and to identify problems. Establishment of a formal process of reviewing the delivery system design at each stage of development and documentation of the development of the design in the design history file is required. The formal design review must have at least one qualified independent reviewer as part of the process as well as representatives of all functions concerned with the design stage being reviewed (3).

Design verification requires that an organization establish and maintain procedures for verifying the device design with objective evidence that the specified requirements of the drug delivery device have been met. Design verification shall ensure that the design outputs meet the required design inputs. These activities precede design validation, which measures whether the completed drug delivery system meets the user requirements (3).

Design technical transfer requires that each manufacturer establish procedures to ensure that the delivery device design is correctly translated into production specifications. This is not a unique requirement to delivery device technology as all pharmaceutical products must have the core elements of drug's development translated into a production specification. However, in some cases, the delivery device may be developed by a third party incorporating a unique technology platform, and in these cases a specific focus should be placed on the technical transfer of knowledge for these systems.

Design changes are common during the development of the device. Each manufacturer should have procedures for the identification, documentation, validation, or where appropriate, verification, review, and approval of design changes before their implementation (3).

Risk management is a key expectation as part of the design control process. The expectation is that an organization designing a delivery device will identify, analyze, control, and monitor the risks associated with bringing the delivery system to market. Risk to the user may be inherent in the design of the product, part of the production process, or created by the patient's use of the product. Performance of risk management identification and analysis early and throughout the design process is critical and should be part of the definition of design inputs. Risks identified late in the design process are often more challenging to mitigate and will often cause a delay in the launch of a new delivery system.

Design Validation, User Studies, and Clinical Testing

Design validation requires that an organization ensure the delivery device specifications meet the user's needs and intended use. Objective evidence that the delivery device meets the intended use of the product will typically include the design verification activities as well as evaluation of the device with the end user. The user evaluations can be completed by user studies simulating actual use or clinical testing. Design validation must be completed with delivery devices that are representative of the final product and manufactured using the same methods that will be used in final production. Labeling, packaging, and user instructions are considered part of the product, and these elements must be part of the design validation activities (3).

DESIGN HISTORY FILE

A design history file is a record of the development history of the delivery device. Unlike pharmaceutical products that have development history reports, medical devices primarily utilize a design history file to document the design and development history. The design history file is specified by the FDA QSR. A design history file is not specifically designated by ISO 13485:2003, but there is a requirement for the creation of documentation and records for design control. The primary elements of a design history file that meets the FDA requirements also meet the basic intent of the ISO 13485, so creation of a design history file or technical file is strongly recommended (3).

DELIVERY DEVICE MANUFACTURING REQUIREMENTS

The basic GMP requirements apply to delivery device manufacturing. Both the FDA and ISO requirements call for production and process controls as well as monitoring of customer feedback. An assembled mechanical medical device can have unique characteristics depending on the combination of parts used. It is very important to have well-defined component specifications and a robust test plan that will enable the manufacturer to fully characterize the production lot and ensure that any major production anomalies are detected. When a dosage form incorporates a delivery device, the need to monitor patient feedback is critical given the potential questions that may arise with regard to how the product should be used, the potential that misuse might lead to complaints, or even the possibility that an issue might cause an adverse event. Robust surveillance systems must be in place prior to launch.

PARENTERAL DEVICE REGULATORY SUBMISSIONS

510k or New Drug Application (NDA) Submission as a Combination Product

Primary oversight of device regulatory submissions is provided by the Center for Device and Radiological Health (CDRH). Drug delivery devices integrate device technology with a pharmaceutical product, and this integration of device technology is supported by the FDA's

Office of Combination Products. The Office of Combination Products helps to determine which of the FDA's centers (CDRH, CDER, CBER) will have primary jurisdiction for review of the submission. Primary oversight of a combination device regulatory submission is determined by whether the device or the pharmaceutical product is the primary mode of action. If the device is integral to the dosage form, such as a prefilled pen with a drug cartridge or a disposable autoinjector with a prefilled syringe, then the submission will likely require an NDA submission because of the primary mode of action rule. Given that the drug will most likely be the primary mode of action, the submission review will likely be led by either CDER or CBER, depending on the type of drug, with CDRH providing consultation.

European Union Regulations and CE Mark Requirements

The primary guidance with regard to medical devices in Europe is the European Union (EU) Medical Devices Directive. A conformity assessment by a notified body that will lead to a CE mark is a requirement for medical devices that are developed in compliance with the EU Medical Device Directive. Therefore, reusable delivery devices such as pen injectors, needle-free devices, or pumps would require a CE mark in Europe. Drug delivery devices that integrate device technology with a pharmaceutical product in a single prefilled unit are regulated by the EU Medicinal Products Directive. The EU Medicinal Products Directive does not require CE marking, but it is expected that the device component of the delivery system will meet the essential requirements of the Medical Device Directive.

Japan Requirements for the MHLW

The Japanese regulatory process with requirements for delivery device submissions is similar for both reusable and prefilled delivery technologies. For either a reusable or prefilled device, a separate submission is provided for review. The submission typically would include a medical device description, materials of construction, device specifications typically as defined by applicable ISO guidance, list of countries launched, address of the manufacturer, risk management process, any malfunctions reported, and a photo of the delivery device. After the medical device technology for a prefilled system is approved, it can be cross-referenced for use with other pharmaceutical products.

CONCLUDING REMARKS

Injection devices and tools are a rapidly growing and important segment of injectable products markets in the pharmaceutical and biopharmaceutical industry. The integration of a device, primary container, and drug product poses complexity for development and commercialization that are unlike the simpler presentations of the past. The rapidly evolving and unique regulatory requirements for combination products also pose additional new challenges. In the end, the most essential element for success is ensuring the combination product meets the needs of the patient in a way that provides convenience, eases discomfort or apprehension associated with injection, and enables better quality of life.

REFERENCES

1. The Needlestick Safety and Prevention Act (HR 5178), 2001.
2. Food and Drug Administration, 21 CFR Part 4, Current Good Manufacturing Requirements for Combination Products, Proposed Rule.
3. Design Control Guidance for Medical Device Manufacturers, FDA Center for Devices and Radiological Health, March 11, 1997.

6 | siRNA targeting using injectable nano-based delivery systems

Lan Feng and Russell J. Mumper

Abstract: The 2006 Nobel Prize in physiology or medicine was awarded to Andrew Fire and Craig Mello who demonstrated a fundamental control of gene expression called RNA interference (RNAi). Since the first time small interfering RNA (siRNA) was shown to knock down the expression of a target protein in mammal cells in 2001, a significant surge of interest has been focused on this promising area. This chapter will provide an overview of RNAi, siRNA, and siRNA-based therapeutics, as well as review the current state of the art of injectable siRNA nanodelivery systems and targeting strategies. The review will also discuss the chemical, physical, and biological barriers, as well as ideal criteria for effective siRNA nano-based therapeutics.

OVERVIEW
RNAi Mechanisms and siRNA

Antisense is a ubiquitous and conserved phenomenon in cells. Antisense nucleotides suppress the gene expression through several distinct mechanisms, such as RNaseH-induced degradation of complimentary mRNA through antisense oligonucleotides hybridizing to their target mRNA; sterical inhibition of mRNA translation or pretranslational splicing; cleavage of target mRNA by some ribozymes or deoxyribozymes because of their intrinsic catalytic activity; RNA-induced silencing complex (RISC)-mediated degradation of target mRNA by double-stranded RNA (dsRNA) (1–3).

RNA interference (RNAi) is the antisense effect caused by RNA (Fig. 1). dsRNAs are important regulators of gene expression in eukaryotic cells. Interfering dsRNAs cleave mRNA through several steps. First, the "DICER" enzyme and its cofactors cleave dsRNA to 21 to 23 base-pair (bp) segments, which are called small interfering RNAs (siRNAs) and assist their loading onto the RISC. RISC removes the sense strand, uses the antisense strand as a guide to seek the complimentary region in the mRNA, and pairs the antisense strand to its target. RISC contains an important protein Argonaute 2 (Ago 2) that has an RNaseH-like domain carrying the activity of RNA cleavage. After cleavage, the resulting 5' and 3' fragments are subsequently subjected to full degradation by other nucleases (2–4). Interfering dsRNA can be either endogenously produced or exogenously provided. However, exogenous dsRNAs longer than 30 bp cause severe toxic responses in mammals, which limit their applications (5). In 2001, Elbashir and colleagues published a paper in *Nature*, reporting the use of synthetic 19 bp duplexes siRNAs with 2-nucleotide (nt) 3' overhangs to mediate RNAi in mammalian cell culture systems (6). Later, researchers extended this to recombinant DNA expressing similar short interfering RNA to have longer effect in cells. siRNA has quickly become one of the most powerful and indispensable tools in molecular biology.

Therapeutic Target and Applications

Since siRNA is a highly specific tool for target gene knockdown, it has been used in the field of molecular biology to understand gene function, as well as to identify and validate genes (7–11). On the basis of knowledge of gene function, siRNA designed to target gene encoding disease-associated protein is currently under intensive investigation as a potent and specific therapeutic agent.

RNAi was found as an anti-viral defense in plants (12). Thus, siRNA as a treatment of human virus diseases may hold the greatest promise in the clinic. Recently, several groups have explored the therapeutic effects of RNAi on hepatitis B virus (HBV) (13), hepatitis C virus (14), human immunodeficiency virus type-1 (15–17), herpes simplex virus 2 (18), respiratory syncytial virus (19,20), human papillomavirus (21,22), as well as others through inhibiting viral

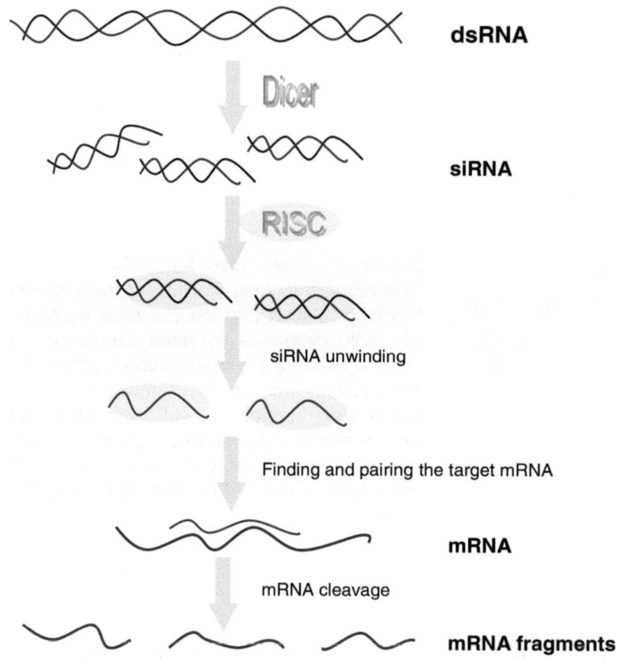

Figure 1 (*See color insert*) Mechanism of RNA interference.

replication and production mechanisms. All the studies have yielded encouraging results. Another strategy is to inhibit the host proteins for pathogen invasion or signaling pathways that initiate the inflammatory response such as cell death receptor Fas (23–25) and caspase-8 (26,27).

A second therapeutic application for RNAi is the treatment of dominant genetic diseases. Autosomal dominant diseases caused by mutant gene encoding essential proteins can be treated by siRNA targeting the mutated alleles. Studies have demonstrated that many familial neurodegenerative diseases, such as Huntington's disease, spinobulbar muscular atrophy, and slow channel congenital myasthenic syndrome (SCCMS) caused by the overexpression of mutated genes or CAG-repeat expansions that encode polyglutamine in the disease protein might be treated by siRNAs (28,29). Another example is the Cu,Zn superoxide dismutase (SOD1) gene in amyotrophic lateral sclerosis (ALS). Schwarz et al. reported that siRNA was specific enough to discriminate single-nucleotide polymorphism. Many SOD1 mutations are single-nucleotide mutations that make siRNA a promising potential therapeutic strategy for the treatment of ALS (30).

Along with the intensive research in molecular biology on cancer, the involvement of more and more signaling pathways and oncogenic genes has been demonstrated, which in turn makes RNAi anticancer therapy possible. Oncogenic genes are often important for cell survival and growth when normally expressed and strictly regulated. In addition, inhibitors of oncogenic proteins are not specific and often cause severe side effects. The high specificity of siRNA allows the selective knockdown of mutated oncogenes without influencing normal cells. Mutations of Ras are present in many cancers such as pancreatic cancers, colon cancers, leukemia, as well as others. In oncogenic K-RasV12, a point mutation results in a valine instead of a glycine in wild-type K-Ras. A viral siRNA transfection targeting this region strongly inhibited the expression of K-RasV12 and tumor formation in nude mice (31,32). Besides targeting oncogenes like Bcr-Abl (33), Bcl-2 (34), Survivin (35), some alternative strategies have also been investigated and have obtained success to some extent. Suppression of tumor angiogenesis by effectively silencing epidermal growth factor receptor (EGFR) gene and vascular endothelial growth factor (VEGF) receptors inhibited the in vivo growth of non–small lung cancer (36) and PC-3 prostate cancer cells (37), respectively. An RNAi approach also enhanced the effects of chemotherapy in resistant breast cancer cells because of the suppression of MDR1 (38,39).

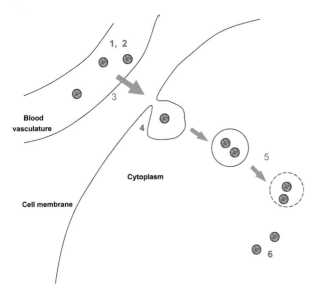

Figure 2 Challenges for siRNA delivery. The barriers include (1) susceptibility in the blood circulation and tissues after injection, (2) rapid clearance by renal excretion and RES uptake, (3) extravasation across the endothelium and to the target tissue, (4) penetration through the cell membrane, (5) endosomal escape, and (6) transient persistence in cells. *Abbreviations*: siRNA, small interfering RNA; RES, reticuloendothelial system.

Delivery Barriers and Challenges

As a potential therapeutics to treat human disease, siRNA needs to be efficiently delivered in vivo. Before designing an effective delivery system for siRNA, it is crucial to understand the six main challenges and barriers of siRNA delivery (Fig. 2).

First, siRNAs are vulnerable to nucleases in serum and tissues. Second, siRNA would be rapidly cleared from the circulation by renal excretion and the reticuloendothelial system (RES) uptake, especially delivered in a nanoparticulate formulation that was prone to RES uptake and elimination. Third, extravasation of siRNA across the endothelium and access to the target tissue is difficult because of its size and negative charge. Fourth, as hydrophilic, negatively charged macromolecules, siRNAs may have poor plasma membrane penetrating properties. Furthermore, if the siRNAs enter cells through an endocytosis mechanism, another important barrier is endosomal escape. Eventually siRNA would end up in late endosome or lysosome and be digested if they could not be released to cytoplasm where its effect takes place. Finally, the persistence of siRNA effect is not permanent because of its inability to reproduce itself.

Available Delivery Approaches

Both noncarrier and carrier strategies are available for in vivo siRNA delivery. Aimed at overcoming individual delivery barriers, various noncarrier systems or methods have been developed. Chemical modifications have been applied to improve the nuclease stability of siRNA, for example, sulfur substitution for a nonbridging oxygen in the phosphodiester linkages (40). Simple conjugation of siRNAs with ligands represents a large portion in this category. Cholesterol siRNA conjugation reduces renal excretion and increases circulation half-life by binding to plasma albumin. Long-chain fatty acid conjugation of siRNA may facilitate the cellular uptake of siRNA by receptor-mediated endocytosis (41). A considerable effort has been devoted to cell-penetrating peptide (CPP) conjugate investigation. These small polycationic peptides rich in arginine and lysine promote the cell penetration of the coupled cargo, which could be siRNA or siRNA-containing complexes. However, the mechanism of uptake and the delivery efficiency is still controversial. An intravenous injection of naked siRNA in massive volume through mouse tail vein has been performed to increase the transport of siRNA through capillary endothelial cells. This method is termed hydrodynamic injection and induces hepatic gene silencing (42). Other noncarrier methods include topical application and the gene gun, among others. Generally, these methods are less efficient and/or practical than carrier strategies.

As for carrier strategies, these can be further divided into viral and nonviral carriers. To this point, viruses are still the most efficient vehicles for gene delivery. Because of their intrinsic nature and function, they can easily penetrate capillary membranes, cell membrane, and even nuclear membranes to reach their destination. When the siRNA containing nucleic acid is inserted to the genetic DNA, it enables long-term expression, and therefore has the ability to chronically suppress gene expression. However, the disadvantages are obvious and inevitable. For example, viral carriers have the difficulties of preparation and storage, immunogenicity, and potential carcinogenicity if they either suppress tumor suppressor genes or activate oncogenes. Hence, extensive attention has been attracted to the design and study of nonviral nanoscale siRNA delivery systems. Although this is a relatively novel area, a growing number of achievements have been made in the recent years as will be discussed in detail.

Differences Between siRNA and pDNA Delivery

As double-stranded nucleic acids, siRNA and double-stranded DNA (dsDNA) share many common properties. They have similar backbone structure with the same negative charge to nucleotide ratio. They both can interact electrostatically with positively charged agents so that many delivery systems are designed on the basis of this principle. Plasmid DNA (pDNA) has been investigated and delivered for at least two decades. Considering the similarity between siRNA and pDNA, applying the knowledge from pDNA delivery systems can facilitate rationale approaches to the delivery of siRNA. However, understanding the key differences between pDNA and siRNA is critical for designing the most efficient and safe siRNA delivery systems.

First, RNA is more sensitive to enzymatic degradation than DNA. The $5'$-carbon sugar in RNA nucleotides is ribose instead of deoxyribose in DNA. This structure makes the RNA backbone more susceptible to spontaneous breakdown and hydrolysis by nucleases. Moreover, DNase and RNase are present in various environments both in vitro and in vivo. To avoid unexpected degradation during the handling and preparation process, creating a DNase/RNase-free environment is of great importance. However, DNase inhibition can be easily achieved while RNase inhibition is much more difficult. In particular, RNase A is extremely stable in an aqueous environment (43). Chemical modifications have been performed to increase the stability of dsRNA, for example, $2'$-O-methyl modification, incorporation of locked nucleic acids (LNAs), phosphorothioate, etc. (40). The greater susceptibility of RNA highlights the critical need for a protective carrier to effectively deliver siRNA.

Second, the delivery destination or intracellular location needed for pDNA and siRNA action is quite different. pDNA requires delivery into the nucleus of the host cell where it can use the transcriptional machinery of the host cell to carry out its therapeutic effect. Unlike pDNA exerting its effect in the nucleus, the target of siRNA is its complementary mRNAs that have already been released from the nucleus after transcription. Therefore, siRNA only needs to be delivered to the cytoplasm. For this reason, pDNA delivery often requires a nuclear localization mechanism such as involving the inclusion of a nuclear localization sequence or carriers that can transport their cargo to the nucleus.

Third, depending on their different action mechanisms, the duration of siRNA and pDNA effects differs as well. Naked siRNAs, unlike pDNA-expressed siRNAs, are not regenerated in cells. Thus, in rapidly dividing cells, the typical gene-silencing duration is three to seven days because of the dilution of siRNAs below a certain level. In contrast, in slowly or nondividing cells, the gene knockdown effect can last as long as three weeks depending on the stability and half-life of the suppressed protein (44). The therapeutic effects with pDNAs not only depend on their own stability but also on the strength of their promoters if they are nonintegrative. In comparison, it is well known that the therapeutic effects of integrated DNA vectors could be long term or even permanent. Hence, the contrast between the pDNA and siRNA requirements above highlight the fact that successful siRNA therapy will necessitate repeated treatment, which makes selection of the carrier with low cytotoxicity and immunogenicity even more important for siRNA.

Another obvious difference between pDNA and siRNA is the molecular weight and size of the molecules. The pDNAs used in gene therapy are usually several kilobase pairs while siRNAs are only 21 to 23 bp. In pDNA delivery, it is often complexed and condensed to nanometric-sized particles directly with cationic agents. However, it is well known that many types of cationic condensing agents (polymers, lipids, etc.) often lead to aggregation of the condensed particles. Because of its smaller size, siRNA is perhaps easier to complex with cationic condensing agents. However, these complexes with siRNA are often unstable and decomplex since the smaller siRNA is not condensed and the ionic interaction is much easier to compete off with counterions. RNA is somewhat stiffer than DNA. The persistence length, which is a basic mechanical property quantifying the stiffness of a long-chain molecule, of dsDNA is 450 to 500 Å and that of dsRNA is approximately 700 Å (45). At 2.7 Å/bp, the persistence length for RNA is 260 bp. Therefore, 21- to 23-bp siRNA behaves as a rod and is not likely to be further condensed. Thus, electrostatic interaction between siRNA and cationic agents could lead to a relatively uncontrolled interaction and form complexes of large sizes and poor stability, with the consequence of incomplete encapsulation (46).

Considering the differences discussed above, the strategy for the delivery of pDNA and siRNA should be interrogated carefully. One should not assume that a delivery system that works for pDNA could be simply transferred to an siRNA delivery system before a more thorough investigation is performed.

IDEAL INJECTABLE NANO-BASED SYSTEMS FOR siRNA DELIVERY

For an ideal injectable nano-based delivery system to efficiently deliver siRNA, regardless if it is for topical or systemic administration, certain criteria must be met. For systemic injection of siRNA, additional criteria must be considered.

Generally speaking, at the cellular level, a successful delivery vehicle must be formulated to have the following characteristics: (*i*) provide protection to siRNA against degradation in extracellular fluids, (*ii*) facilitate efficient cellular uptake, (*iii*) facilitate endosomal escape before the early endosome becomes late endosome or lysosomes in which the siRNA will be destroyed, (*iv*) be able to readily release siRNA upon arrival at the cytosol where the RNAi effect takes place, (*v*) be nontoxic to the cells, (*vi*) be stable during storage and in the vehicle for administration solution, that is, chemically and physically stable.

For a systemically administered siRNA nanocarrier, there are some additional concerns: (*i*) provide protection to siRNA against degradation not only in the extracellular fluids but also in the systemic circulation, (*ii*) be stable in the systemic circulation with limited breakdown and/or aggregation before it arrives at the target site, (*iii*) be able to extravasate blood vessels and penetrate tissues to gain access to the target site, (*iv*) maintain proper particle size and surface properties to avoid clearance and/or elimination via the kidneys and RES.

To further increase the efficiency of in vivo siRNA delivery, target strategies are widely applied. As for the targeted systemic nanocarrier, choosing a suitable targeting ligand is critical as well. First of all, the targeting should be specific enough, that is, the expression of the receptor on the target cells should be highly specific, highly expressed, and not shed, among others. Second, the targeting ligand should have high affinity with the target receptor to ensure sufficient retention time as well as trigger cellular uptake via receptor-mediated endocytosis instead of remaining bound to the receptor. Last but not least, the targeting ligand should be amenable to the required chemistries needed to attach the ligand to the nanocarrier, as well as have low or no immunogenicity.

NANO-BASED DELIVERY SYSTEMS
Complexes

RNA is a molecule consisting of a chain of nucleotide units. Each nucleotide is composed of a nitrogenous base, a ribose sugar, and a phosphate. RNA is a negatively charged molecule because of the negative charge on phosphate groups at physiological pH. siRNA molecules are dsRNA with 19 to 21 bp. Calculating the charge density gives about three negative charges per kilodalton (kDa) molecular weight of siRNA.

To date, complexes of siRNA with various positively charged materials by electrostatic interaction represent the largest portion of active research. In this category, there are two major subgroups and some others.

Lipoplex
The most often referenced formulation in this group is cationic liposomes. When cationic liposomes are mixed with negatively charged siRNA, the organized bilayer structure of the

liposome is altered by electrostatic interaction so that they are no longer referred to as liposomes but have a new name of "lipoplexes."

The DOTAP Liposomal Transfection Reagent is a commercially available liposome formulation of the monocationic lipid 1,2-dioleoyl-3-trimethylammonium-propane (DOTAP) that can be used for the transfection of nucleic acids. Mixing the DOTAP reagent with the negatively charged siRNA results in a spontaneously formed stable complex that can be directly added to the tissue culture medium with or without serum. Commercially available DOTAP is not only used as an instrumental tool for in vitro siRNA delivery to investigate gene functions in molecular biology, it has been used to deliver siRNA in mice to prove the concept and feasibility of certain therapeutic ideas (47–50). Dioleoyl-phosphatidylethanolamine (DOPE) is a neutral helper lipid usually used with DOTAP to formulate transfection reagent. It is generally believed that DOPE enhances transfection because of its tendency to form hexagonal phase structures at temperatures above 10°C, which facilitates siRNA endosomal escape (51).

On the basis of this classical liposome formulation, targeting ligands have been included to deliver siRNA to specific tissues. For example, Chang and coworkers developed a tumor targeting immunoliposome that takes advantage of elevated transferring receptor (TfR) levels on tumor cells to deliver pDNA, antisense oligonucleotides, imaging agent, or siRNA (52–54). The anti-transferrin receptor single-chain antibody fragment was incorporated into the liposomes and formed immunoliposomes. This intravenously administered immunoliposome delivered its cargo (which could be pDNA, antisense oligonucleotides, imaging agent or siRNA) specifically and efficiently to primary/metastatic tumors. In addition, a pH-sensitive histidine-lysine peptide (HoKC) was included in the complex to further increase the endosomal escape. In a recent report, the results showed increased potency of the liposome-HoKC complex and their ability to carry anti-HER2 siRNA to target and sensitize tumor cells, silencing the target gene, and inhibiting tumor growth in vivo (55). Cardoso et al. associated transferrin instead of the TfR antibody to DOTAP/cholesterol liposome, another conventional cationic liposome, to target TfR expressing cells (56). In vitro experiments by the group showed enhanced gene knockdown activity of transferrin-associated liposome compared with the conventional liposomes by anti-GFP (anti–green fluorescent protein) siRNA. Besides tumor targeting, siRNA liposomes are targeted to other tissues and organs such as the liver. Kim and his colleagues formulated anti-HBV siRNA into a complex of DOTAP/cholesterol liposome and apolipoprotein A-I (apo A-I) (57). Apolipoprotein is recognized by class B, type-1 scavenger receptor (SR-BI) that is predominantly expressed in the liver. When the liver-targeted formulation was injected intravenously into a HBV carrying mouse model, the viral protein expression was reduced to about 30% and its effect lasted up to eight days upon a single treatment.

In addition to the commercially available lipids, some cationic lipids have also been designed and synthesized to improve the transfection efficiency and reduce the cytotoxicity. It has been reported that an ether linkage containing cationic lipid, such as 1,2-dioleyloxypropyl-3-trimethylammonium chloride (DOTMA), has higher in vivo transfection efficiency than the corresponding ester analogue DOTAP (58). On the basis of the structure-activity information, Chien et al. synthesized ether-linked cationic cardiolipin analogue (CCLA) where the phosphate groups of cardiolipin were replaced with quaternary ammonium groups as shown in Figure 3A (59).

Their report showed that the transfection efficiency of the luciferase reporter gene in mice was sevenfold higher than the commercially available DOTAP-based liposome, and the CCLA-based liposome had lower toxicity than DOTAP transfection reagent. When the CCLA-based liposome was used to deliver the c-raf siRNA in mice bearing human breast cancer (MDA-MB-231) xenografts, the tumor growth was inhibited 73% as compared with free siRNA treatment. For the same reason, many groups synthesized other cationic lipids to meet the needs of in vitro and in vivo delivery such as cationic cholesterol–based polyamine lipid N'-cholesteryloxycarbonyl-3,7-diazanonane-1,9-diamine (CDAN) (46), 2-(3-[bis-(3-amino-propyl)-amino]-propylamino)-N-ditetradecylcarbamoylmethyl-acetamide (RPR209120) (60), and multivalent lipid 5 or pentavalent lipid (MVL5) (61). Their structures are shown in Figure 3B–D. Positive charges could also be incorporated by adding aminoglycoside to the lipid. Desigaux et al. synthesized a series of cationic lipids (DOST, DOSK, DOSP, DOSN)

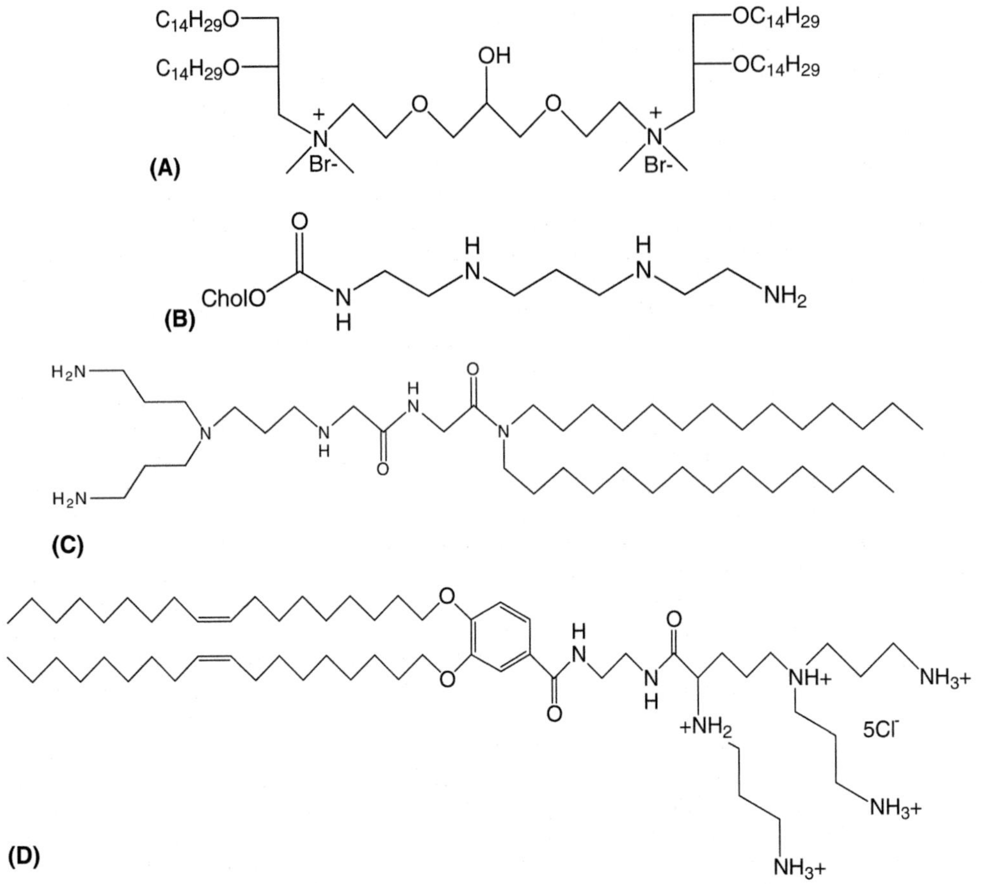

Figure 3 Structures of **(A)** CCLA, **(B)** CDAN, **(C)** RPR209120, and **(D)** MVL5. *Abbreviations*: CCLA, cationic cardiolipin analogue; CDAN, cationic cholesterol–based polyamine lipid *N'*-cholesteryloxycarbonyl-3,7-diazanonane-1, 9-diamine; RPR209120, 2-(3-[bis-(3-amino-propyl)-amino]-propylamino)-*N*-ditetradecylcarbamoylmethyl-acetamide; MVL5, multivalent lipid 5 or pentavalent lipid.

bearing various aminoglycosides (tobramycin, kanamycinA, paromomycin, and ethylthioneomycin B, respectively) linked to two dioleyl chains by a succinyl spacer for specific interaction with siRNA (62).

Besides lipid-aided cellular delivery, some positively charged CPPs have been incorporated into conventional liposomes. In a study by Mudhakir et al., liposomes composed of egg phosphatidylcholine (EPC) and cholesterol were modified by direct conjugation of a novel peptide IRQRRRR (IRQ) to the surface of liposomes (63). IRQ is a peptide ligand that targets skeletal muscle found by in vivo phage display. Since the novel peptide IRQ is rich in arginine, it not only serves as a tissue-target moiety but also triggers the cellular uptake via caveolar endocytosis.

An interesting concept called site-specific release has been applied to liposomal siRNA targeting delivery as well. It is well known that under pathological conditions the expression of many proteins are altered including intracellular receptors and enzymes, as well as others. Most of the recent studies have focused on targeting modified receptors using either an antibody or a small molecular receptor substrate. However, altered expression of enzymes in the pathological tissue could also serve as a novel target by triggering site-specific release of a therapeutic agent. For example, sPLA$_2$ is an enzyme upregulated in cancer and inflammatory tissues, but it is present at low levels in the blood circulation. Foged et al. formulated a liposome including lipid dipalmitoylphosphatidylglycerol (DPPG), which is favored by human group IIA sPLA$_2$. They hypothesized that the liposome could site-specifically release siRNA in inflammatory tissue but not in the systemic circulation or other tissues (64).

Moreover, the hydrolysis products were thought to disturb the cellular membrane and facilitate the uptake of siRNA. Although their data showed that the $sPLA_2$ degradable liposomes did not silence enhanced green fluorescent protein (EGFP) expression in HeLa cells, they did show that the siRNA from the liposomal formulation was taken up by HeLa cells and that uptake was augmented by the addition of $sPLA_2$. The concept of site-specific release with no active targeting moieties opens an alternative avenue and deserves more attention.

Polyplex

Polymers, either natural or synthetic, represent another major group of complexing agents for siRNA delivery. The formulation of nucleic acids complexed with polymers is generally called "polyplex" in this chapter even though in various literatures they are sometimes referred to as nanoparticles or micelles.

Cationic polymers, for example, polyethylenimine (PEI), polypropylenimine (PPI), poly-L-lysine (PLL), polyallylamine (PAA), cationic dextran, and chitosan are the most commonly used materials for siRNA complexation. Among them, PEI has been the most widely used polymer for complexing with siRNA.

The native branched PEI (25 kDa) is a prototype polymeric transfection agent that has gained widespread use. Branched PEI contains primary, secondary, and tertiary amines in the molar ratio of 1:2:1. The primary amines are mainly responsible for nucleic acid condensation while the secondary and tertiary amines provide buffering capacity and therefore facilitate endosomal escape via the so-called "proton-sponge" effect. The transfection efficiency of PEI, along with its cytotoxicity, strongly depends on its molecular weight. Usually, high molecular weight PEI has higher transfection efficiency but with higher toxicity as well, while low molecular weight PEI has lower cytotoxicity with reduced transfection efficiency. To enhance the gene delivery efficiency and minimize cytotoxicity of PEI, there has been a great deal of effort focused on structurally modifying PEI. For example, Dong et al. cross-linked low molecular weight PEI 800 Da with short diacrylate linkages to form higher molecular weight PEI structures (65). The modification combines the favorable low toxicity of low molecular weight PEI with the higher transfection efficiency of high molecular weight PEI. The biodegradable ester bonds are hydrolyzed under physiological conditions within the cell after delivery and convert the cross-linked high molecular weight PEI into low toxic low molecular weight PEI. In a study of pDNA transfection, an optimal cross-linked PEI, EGDMA-PEI 800-4h (the product of conjugation of amino groups of PEI 800 Da to EGDMA for four hours), resulted in a 9-fold increase in gene delivery efficiency in B16F10 cells and a 16-fold increase in 293T cells compared to with commercially available PEI 25 kDa control. Later the modified PEI was used to deliver plasmid-encoded focal adhesion kinase-1 (FAK1) siRNA in vivo and prolonged the survival of the tumor-bearing mice (66). To address the associated cytotoxicity with the use of PEI for siRNA delivery, Swami et al. cross-linked PEI with 1,4-butanediol diglycidyl ether (bisepoxide) (67). The modification converted primary amines, which are believed to be the main source of cytotoxicity, to secondary and secondary to tertiary amines. The system was found to deliver siRNA more efficiently into HEK cells as compared with native PEI 25 kDa with significantly reduced cytotoxicity.

Jere et al. conjugated low molecular weight PEI and polyethylene glycol (PEG) with biodegradable poly(β-amino ester) (PAE) (68). The high repetitive PEI units are thought to result in high delivery efficiency while PEG units and the ester linkage facilitate more rapid intracellular siRNA release and lead to enhanced polymer degradation resulting in lower cytotoxicity. As a result, PAE as a carrier was found to be less toxic and 1.5-fold more effective than standard PEI 25 kDa. Several other PEI modifications have been investigated by Zintchenko et al. (69). The group performed a number of modifications including ethyl acrylate, acetylation of primary amines, or introduction of negatively charged propionic acid or succinic acid groups to the PEI structure. All the conjugates led to reduced toxicity in comparison to the unmodified PEI. In particular, succinylation of PEI resulted in up to 10-fold lower toxicity in Neuro2A cells.

In order to facilitate release of siRNA in the cell, branches of PEI have been derivatized with ketal linkages (70). Ketal linkages are acid-degradable under mild acidic pHs (e.g., pH

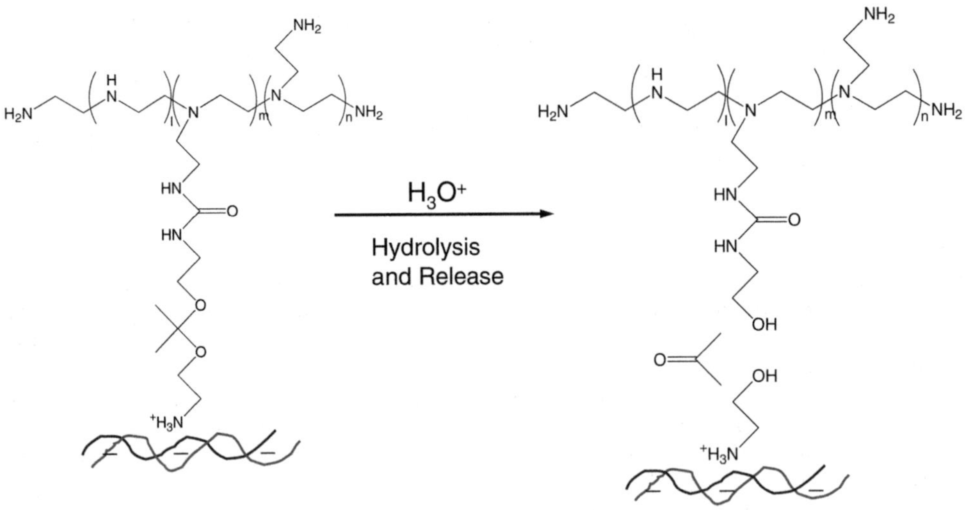

Figure 4 Dissociation of nucleic acids from ketalized PEI upon hydrolysis. *Abbreviation*: PEI, polyethylenimine. *Source*: Adapted from Ref. 70.

5.0) and facilitate the release of siRNA in the endosomal environment as shown in Figure 4. The ketalized PEI complexed with siRNA into siRNA/PEI polyplexes with a particle size range of 80 to 200 nm showed enhanced delivery efficiency with reduced cytotoxicity.

One of the primary disadvantages of the use of positively charged complexing agents is that they are prone to aggregation or disassociation in the blood when complexed to siRNA. Moreover, positively charged complexing agents tend to interact with the negatively charged proteins in the systemic circulation and are taken up by the RES. To address this potential problem, PEG has been utilized to shield the surface of the complex which serves to provide enhanced stability. PEG has either been conjugated directly to siRNA or to the cationic polymer. For example, Kim et al. conjugated PEG to siRNA via a disulfide linkage that could then be cleaved in the reductive environment in endosomes and cytoplasm. The PEG-siRNA conjugate was then complexed with PEI to form a nanoparticle (71). The resulting nanoparticle has an inner core composed of siRNA/PEI surrounded by a hydrophilic PEG shell. This kind of structure is similar to amphiphilic lipidic micelle and could be spontaneously formed, so it is called a self-assembled micelle even though the particle size is often not in the traditional micellar range. In vivo imaging results from Kim et al. showed enhanced accumulation of micelles in the tumor region following intravenous injection. Pegylated siRNA has also been complexed with other cationic polymers such as PLL (72).

PEG has also been conjugated to the cationic polymer. For example, PEG derivatized diblock or triblock copolymers have been designed and synthesized by many groups. A recent publication reported the synthesis of a triblock polymer consisting of monomethoxy PEG, poly (3-caprolactone) (PCL), and poly(2-aminoethyl ethylene phosphate) (PPEEA) (73). The polymers in an aqueous solution spontaneously formed positively charged micelles surrounded by PEG corona. siRNA was postloaded into the formed micelles resulting in complexes with an average particle size from 98 to 125 nm depending on the nitrogen to phosphate (N/P) ratio (Fig. 5).

Besides the linear copolymers, cationic graft comb-like copolymers were synthesized and used to deliver siRNA. Sato et al. prepared and evaluated a series of cationic comb-type copolymers (CCCs) consisting of a PLL backbone and PEG or dextran side chains (74). The water soluble dextran side chains of the copolymer are in abundance (>70 wt.%) and the highly dense PEG brush reinforced the electronic interaction between copolymers and siRNA instead of hindering it. The most remarkable property of the CCC with higher side-chain content (10 wt.% PLL and 90% wt.% PEG) is that it increased circulation time of siRNA in mouse bloodstream by 100-fold (74). Interestingly, even when the CCC was injected into

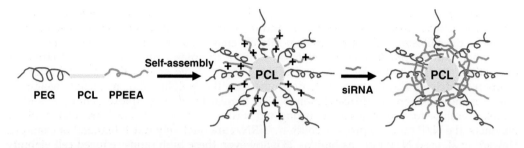

Figure 5 (*See color insert*) Self-assembling of cationic micellar nanoparticles and loading of siRNA. *Abbreviation*: siRNA, small interfering RNA. *Source*: Adapted from Ref. 73.

mouse intravenously 20 minutes prior to the injection of siRNA, the CCC still increased the half-life of the postinjected siRNA by more than 60-fold suggesting that the CCC prefers interaction with siRNA to other anions existing in blood.

While some investigators increase the stability and systemic half-life of siRNA polyplexes by incorporating PEG, others provide protection to the polycation/siRNA complex with another layer of lipid coating. Kim et al. synthesized a water soluble lipopolymer (WSLP) by conjugating cholesteryl chloroformate to PEI 1.8 kDa through a hydrophobic lipid anchor (75). The lipopolymer combined the advantages of both liposomes and cationic PEI. While the positively charged headgroup PEI complexed with siRNA and enhanced endosomal escape, the lipid coating on the complex further protected the complex from aggregation and RES clearance and increased the cell membrane permeability. The in vivo data showed that WSLP/VEGF siRNA complexes reduced tumor volume by 55% at 21 days and by 65% at 28 days relative to control tumors.

While most of the approaches discussed so far increase the transfection efficiency of PEI by reducing its cytotoxicity or provide protection against systemic clearance to some extent, a novel approach is to directly attach PEI with a membrane-active peptide. Melittin (Mel) is the major bioactive component of bee venom. Mel has been conjugated to pegylated PEI or PLL (76). To avoid its extracellular lytic activity, the amines of Mel were modified with dimethylmaleic anhydride that was cleaved under acidic pH in the endosome and enhanced the endosomolytic activity of Mel. PEG-PEI-Mel and PEG-PLL-Mel showed 70% and more than 90% in vitro luciferase gene knockdown, respectively.

To achieve targeted delivery, targeting moieties have been attached to PEI. PEI is usually pegylated with ligands conjugated to the distal end of the PEG, while direct attachment of ligands to PEI is performed as well. Schiffelers et al. targeted tumor neovasculature expressing integrins by conjugating an Arg-Gly-Asp (RGD) peptide to PEI 25 kDa (77). siRNA specific to vascular endothelial growth factor receptor-2 (VEGFR2) was complexed with the modified PEI at a N/P ratio of 2 to 6, resulting in the formation of polyplexes with average particle size of about 100 nm. The intravenous administration of these polyplexes to nude mice showed tissue-specific accumulation of PEI-PEG-RDG/siRNA. Kim et al. utilized a similar approach to complex siRNA with PEI-PEG-folate (78). Interestingly, their results showed that the delivery of siRNA led to the most pronounced gene-silencing effect compared with the delivery of antisense oligodeoxynucleotide (AS-ODN) or siRNA-expressing plasmid DNA. Another recently published paper reported on the use of hyaluronic acid (HA) as a ligand to target lymphatic vessel endothelial hyaluronan receptor-1 (LYVE-1) (79). In vitro data showed increased siRNA uptake in HA receptor expressing cells but not in nonexpressing cells, and that the gene-silencing effect was inhibited by free HA in a concentration-dependent manner.

Compared with the relative extensive investigation of PEIs and PLLs for siRNA delivery to date, studies on the use of other polymers is limited. Chitosan is one polymer being investigated for siRNA delivery. Although chitosan has been studied for more than a decade as a delivery system for pDNA, there are only few studies using it as a carrier of siRNA. Chitosan is a copolymer of N-acetyl-D-glucosamine (GlcNAc) and D-glucosamine (GlcN) produced by the alkaline deacetylation of chitin. As a natural polymer, chitosan is considered to be biocompatible and nontoxic, although this depends on various physical-chemical properties such as purity, % deacetylation, and molecular weight, among others. The primary amines in the chitosan backbone become positively charged at the pH levels below the pK_a of the primary amine (pK_a 6.5) so that chitosan forms a complex with siRNA with electrostatic interaction. Several studies of chitosan/siRNA complex have shown that the ability of the chitosan to deliver siRNA to cells is dependent on the weight ratio, molecular weight of chitosan, and the degree of deacetylation (80–83). Similar to other complexes, chitosan/siRNA complexes can be formed by a simple mixing and stirring process. Different from other synthetic polymers, the N/P ratios to prepare chitosan/siRNA are relatively much higher. For example, Howard et al. used N/P ratio as high as 285; however, these high ratios reduced cell viability (80). The in vitro data showed that chitosan/siRNA complexes formed using high molecular weight (114 and 170 kDa) and deacetylation degree (84%) at N/P 150 were most stable with particle size about 200 nm (81). The group showed that 80% EGFP gene–silencing efficiency was obtained after 24 hours in H1299 green cells in vitro. Effective in vivo gene silencing was achieved in mice bronchiole epithelial cells (37% and 43% reduction of EGFP positive cells compared with scramble siRNA and untreated control, respectively) after nasal administration. However, the ability of the complexes to deliver siRNA systemically requires further investigation.

Thiamine pyrophosphate (TPP) has been used to form salts with chitosan to improve chitosan water solubility (83). Chitosan is a weak base with a pK_a value of 6.2 to 7.0, and thus has poor solubility at neutral to alkaline pH. TPP is a zwitterionic compound, which can increase the water solubility of chitosan due to the phosphate groups. However, the amine groups of TPP together with chitosan bind to negatively charged siRNAs to form complexes. The maximal EGFP gene–silencing effect mediated by chitosan-TPP/siRNA was 70% to 73%. Another study by Katas and Alpar used sodium tripolyphosphate to ionically cross-link chitosan to form nanoparticles (82). siRNA was either mixed with sodium tripolyphosphate and then dripped into a chitosan salt solution, or adsorbed to preformed chitosan/tripolyphosphate particles. The particle size of chitosan/tripolyphosphate was 510 ± 22.9 nm and 276 ± 17.9 nm formed using chitosan glutamate 470 kDa and 160 kDa, respectively. The particle size of chitosan/tripolyphosphate was 709 ± 50.3 nm and 415 ± 44.6 nm formed using chitosan hydrochloride 270 kDa and 110 kDa, respectively.

Leng et al. synthesized several branched peptide polymers composed of histidine and lysine (HK polymer) (84). Figure 6 shows the structure of a branched HK polymer with eight terminal branches and histidine-rich domains (H³K8b). An integrin-binding ligand RGD was further added to increase the delivery efficiency of siRNA. Although the sizes of HK polymer/siRNA polyplexes were over 400 nm, the in vitro delivery efficiency was significant. The complex of H³K8b and anti-β-galactosidase (β-gal) siRNA inhibited β-gal expression by more than 80% after 48 hours in SVR-bag4 cells that stably expressed β-gal. The H³K8b/anti-luciferase siRNA complex inhibited more than 90% luciferase activity in MDA-MB-435 cells, which were cotransfected with a luciferase expression plasmid.

Others
In addition to the two larger families of cationic complexing reagents, lipids and polymers, there are several other molecules that have been proposed to make nano-based siRNA delivery systems.

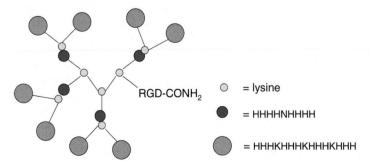

RGD-CONH₂   ○ = lysine

              ● = HHHHNHHHH

              ◐ = HHHKHHHHKHHHKHHH

Figure 6 (*See color insert*) Schematic structure of H³K8b polymer.

Positively charged natural proteins are a pool of convenient reagents in terms of their potential to complex and deliver siRNA. In a broad sense, proteins are also a group of polymers. To date, atelocollagen is the only protein used alone to deliver siRNA both in vitro and in vivo (85). Atelocollagen is a highly purified decomposition product of type I collagen derived from the dermis of cattle with a molecular weight of 300 kDa. The amino acid sequence at both N- and C-terminal of a collagen called telopeptide is the main source of the immunogenicity. Therefore, since atelocollagen obtained by pepsin treatment of collagen lacks immunogenic telopeptides, atelocollagen itself has low relative immunogenicity. It is a rod-like molecule with a length of 300 nm and a diameter of 1.5 nm. Atelocollagen, which is positively charged, interacts with the negatively charged siRNA to form an atelocollagen/siRNA complex with a diameter of 100 to 300 nm. An interesting property of atelocollagen is that it is soluble at a lower temperature but solidifies at a temperature over 30°C. Therefore, the atelocollagen/siRNA complexes were prepared and stored at 4°C. Once introduced into animals, the complex becomes solidified and releases siRNA in a controlled manner for a period of time due to the biodegradable nature of atelocollagen. Direct intratumoral injection of human HST-1/FGF-4 (fibroblast growth factor) siRNA complexed with atelocollagen resulting in about 12-fold and 8-fold tumor growth inhibition compared with atelocollagen alone and control siRNA, respectively, in an orthotopic xenograft of a human non-seminomatous germ cell tumor at 21 days after treatment.

On the basis of the barriers that must be overcome to deliver siRNA, some innovative carriers have been synthesized to fulfill multiple functions in one system. 1,4,7-Triazanonylimino-bis [*N*-(oleicyl-cysteinyl-histinyl)-1-aminoethyl) propionamide] (THCO) (Fig. 7) and (1-aminoethyl)imino-bis [*N*-(oleicyl-cysteinyl-histinyl-1-aminoethyl)propionamide] (EHCO) are two molecules containing a protonatable amine head group of different pK_as, two cysteine residues and two 8-heptadecenyl tails (86,87). They form stable complexes with siRNA through charge and hydrophobic interaction. The protonatable amino head group consists of primary, secondary, and tertiary amines having different pK_as (the pK_a values of primary, secondary, and tertiary amines are approximately 6.5, 7.0, and 6.0, respectively), which is similar to branched PEI. Thus, these molecules not only complex siRNA but also facilitate endosomal escape. The dithiol groups in the molecules can be polymerized by forming disulfide bonds to further provide stability to the formed siRNA complex. The disulfide bonds may be reduced in the endosome and cytoplasm resulting in the dissociation and release of siRNA. The multifunctional compounds mediated 40% to 88% silencing of luciferase expression with 100 nM siRNA in U87-luc cells.

Additionally, there are some interesting carriers that are quite unique in terms of geometry and other physical-chemical properties. For example, a cone-shaped macrocyclic

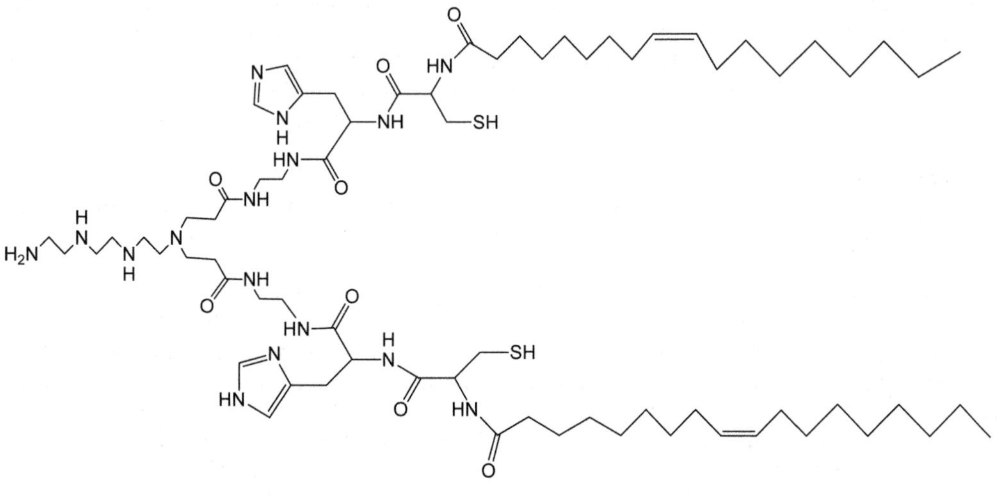

Figure 7 Structure of THCO. *Abbreviation*: THCO, 1,4,7-triazanonylimino-bis [*N*-(oleicyl-cysteinyl-histinyl)-1-aminoethyl) propionamide].

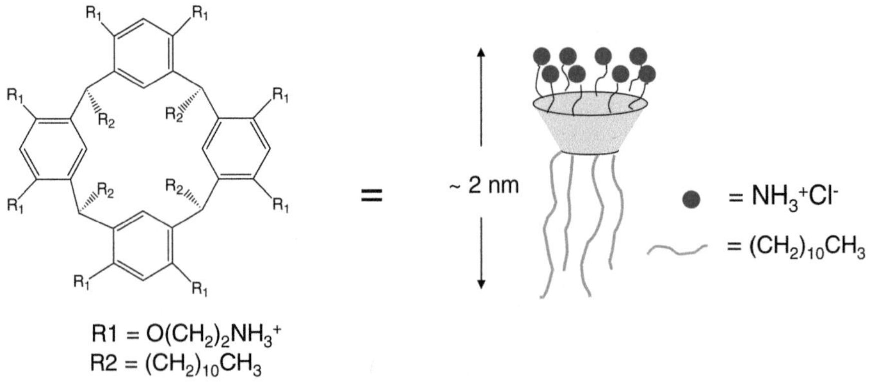

R1 = O(CH$_2$)$_2$NH$_3$$^+$
R2 = (CH$_2$)$_{10}$CH$_3$

Figure 8 (*See color insert*) Cone-shaped structure of macrocyclic octaamine. *Source*: Adapted from Ref. 88.

octaamine as shown in Figure 8 has been proposed by Matsui et al. (88). The novel carrier has four long alkyl chains and eight amino groups on the opposite side of the calix[4]resorcarene macrocycle. What makes the macrocyclic octaamine different from other cationic lipids or polymers is that being a small and single molecule (molecular weight 1740), the compound unimolecularly presents a positive charge cluster motif with a well-defined geometry. Like amphiphilic micelle-forming polymers, the macrocyclic octaamine itself may form small micelle-like particles, with hydrophilic amino groups outside and lipophilic chain inside as illustrated in Figure 8. As a result, the cone-shaped macrocyclic octaamine formed complexes with siRNA in a compact size of approximately 10 nm. Although the in vitro delivery of macrocyclic octaamine/siRNA complex occurred with 90% to 95% knockdown of luciferase expression in HeLa, HepG2, and HEK293 cells at 48 hours, its in vivo performance remains to be investigated.

The KALA peptide (WEAK LAKA LAKA LAKH LAKA LAKA LKAC EA) is a well-known cationic, amphiphilic, and fusogenic peptide, which has been popularly studied as an endosomal escaping peptide complexing with various nucleic acids. However, it was reported that KALA/siRNA complexes did not show sufficient gene-silencing effect in the presence of serum proteins. In a recent study, two cysteine residuals were added to both terminals of KALA (89). The cysteine-KALA-cysteine peptide (CWEAK LAKA LAKA LAKH LAKA LAKA LKAC) self-cross-linked through reducible disulfide linkage. The cross-linked KALA (cl-KALA) formed more stable and compact complexes with siRNA. To further improve the colloidal stability, siRNA was modified with PEG. According to a previous report of the same group, direct PEG conjugation to siRNA could form more stable complexes than those by PEG-modified cationic polymers (71). Although cl-KALA/siRNA and cl-KALA/siRNA-PEG only showed 23.6% and 47% knockdown of GFP expression in MDA-MB-435-GFP cells at the N/P ratio of 64, the data showed their potential as a nano-based delivery system for siRNA.

Nanoparticles

In a broad sense, all particles in the nanoscale range are called nanoparticles. However, in this chapter, nanoparticles are differentiated from nanocomplexes by their more organized structures, that is, well-defined shell and core structures.

Huang et al. has developed a targeted nanoparticle formulation for siRNA systemic delivery to metastatic tumors overexpressing the sigma receptor (90,91). The core of the nanoparticle is a complex of siRNA, calf thymus DNA, and protamine, a highly positively charged peptide. The shell of the nanoparticle is a reorganized liposome structure consisting of DOTAP and cholesterol (1:1 molar ratio). Thus, the nanoparticle is referred to as "LPD," or liposome-polycation-DNA (Fig. 9). The nanoparticles are formed spontaneously by mixing the core complexes with preformed cationic liposomes. To create a sterically stabilized particle and for subsequent targeting, DSPE-PEG$_{2000}$ (1,2-distearoyl-sn-glycerol-3-phosphoethanolamine-N-[methoxy(polyethylene glycol)$_{2000}$]) or DSPE-PEG-anisamide was postinserted into the preformed LPD. The in vitro results showed that the delivery efficiency of the targeted

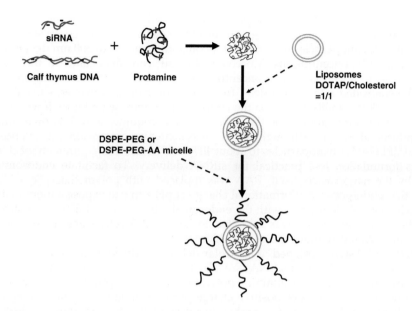

Figure 9 (*See color insert*) Preparation of PEGylated LPD. *Abbreviation*: LPD, liposome-polycation-DNA. *Source*: Adapted from Ref. 90.

nanoparticles was four- to seven-fold higher than the nontargeted nanoparticles. The in vivo tissue distribution results suggested that LPD surface-modified by PEG delivered a therapeutic dose to the tumor and avoided substantial accumulation in the liver with either targeted or untargeted LPD. These results suggest that the tumor accumulation of LPD with particle size around 100 nm is primarily due to the EPR (enhanced permeability and retention) effect as compared with targeting. After a single IV injection of 150 μg/kg anti-luciferase siRNA, 70% to 80% luciferase activity was silenced in a metastatic mouse tumor model. To avoid potential immunogenicity and inflammatory responses with calf thymus DNA, the calf thymus DNA has also been replaced with HA to produce LPH (liposome-polycation-hyaluronic acid) nanoparticles (92). The results showed that while the gene-silencing effect of LPH nanoparticles is comparable to LPD nanoparticles, the immunotoxicity of LPH is much lower.

A similar structure to LPH has also been reported by Peer et al. to develop leukocyte-directed nanoparticles to deliver anti-cyclin D1 siRNA (93). The preformed liposome is composed of phosphatidylcholine (PC), dipalmitoylphosphatidylethanolamine (DPPE), and cholesterol (Chol). High molecular weight hyaluronan (850 kDa), was attached to the outer surface of the liposomes by covalent linkage to DPPE to provide steric stabilization. The resulting nanoparticles were equipped with targeting function by covalently conjugating to the hyaluronan, a monoclonal antibody FIB504 against β7 integrins, which are highly expressed in gut mononuclear leukocytes. Anti-cyclin D1 siRNA loaded nanoparticles were formed by rehydrating lyophilized liposomes with water containing protamine-condensed siRNA. In an experimentally induced colitis mouse model, the β7 integrin–targeted nanoparticles knocked down the cyclin D1 expression to the normal level and ameliorated the colitis score.

An organic-inorganic hybrid nanoparticle was developed by Kakizawa et al. (94,95). The organic-inorganic shell-core structure is a core composed of nanocrystals of siRNA/CaP (calcium phosphate) complexes surrounded by a hydrophilic shell of a PEG-PAA block copolymer (polyethylene glycol-aspartic acid). Because of its potential biocompatibility, CaP is widely applied in various biomedical applications. Its binding affinity to a variety of molecules including proteins, nucleic acid, and small-molecule drugs makes it a potential controlled release material. However, one of the difficulties in using CaP to form nanoparticles is the relatively rapid crystallization rate of CaP. In the absence of other materials, the growth of siRNA/CaP complex crystals is rapid and precipitates are formed within minutes after mixing siRNA and CaP solutions. However, in the presence of a PEG-polycarboxylate block copolymer, such as PEG-PAA, the rapid crystal growth is controlled or even prevented through the absorption of the PAA segment of PEG-PAA on the formed crystal surface. The resulting complex nanoparticles have diameters ranging from 100 to 300 nm depending on the PEG-PAA and CaP concentrations. Moreover, the CaP core dissociates in the intracellular environment with lower calcium concentration compared with the extracellular fluids, allowing the controlled release of siRNA from the core matrix. However, since the complex nanoparticles lack the ability to escape the endosomes, in vitro gene knockdown experiments are performed by pretreatment of the cells with chloroquine, a well-known adjuvant to provide endosomal escape. Although the in vitro luciferase expression was silenced by the siRNA/CaP/PEG-PAA nanoparticles to about 40% in 293 cells, the requirement of chloroquine makes this formulation less practical for siRNA delivery. To facilitate endosomal escape provided by the nanoparticle itself, PAA was replaced with polymethacrylic acid, another polyanion that undergoes a conformational change at pH 4 to 6 to expose a more hydrophobic structure which is able to interact with the endosomal membrane and disturb its structure. As a result, the luciferase activity was inhibited to 20% in 293 cells using as low of a siRNA concentration of 25 nM without the use of chloroquin.

Bartlett and Davis designed a modular-delivery vehicle that utilized an inclusion complex for targeted delivery of siRNA (96). The inclusion complex was comprised of siRNA and a synthesized cyclodextrin-containing polycation (CDP) that provided two functions. First, the polycation contains 2 mol of positive charge per CDP monomer, which complexes with negatively charged siRNA and self-assembles to nanoparticles. Second, the cyclodextrin motifs on the surface of the nanoparticle serve as a "loading dock" to incorporate PEGs and target ligands. PEG molecules containing adamantane (AD) on the proximal end and either methoxy (AD-PEG) or a targeting ligand such as transferrin (AD-PEG-Tf) on the distal end was mixed with CDP at a 1:1 AD-PEG/β-CD (mol/mol) ratio in water. AD-PEG or AD-PEG-Tf was attached to the polymer via inclusion complex formation between AD and the β-CD motifs on the polycation backbone. A calculation of the stoichiometry of each particle estimated that a 70-nm particle contained about 2000 siRNA molecules and around 100 AD-PEG-Tf molecules. Thus, each CDP nanoparticle could theoretically deliver a large payload of siRNA with a large ratio of siRNA to targeting ligand (20:1). The functional efficiency of CDP nanoparticles was demonstrated through knockdown of luciferase reporter protein expression. HeLa cells treated with CDP nanoparticles containing both pGL-3 plasmid DNA expressing firefly luciferase and siRNA against luciferase showed 50% lower expression of luciferase than cells that received either the plasmid alone or the plasmid plus control siRNA.

While most of the nanoparticle designs tend to entrap or hide siRNA in the nanoparticle core, thus providing siRNA protection against degradation, a few groups have attempted to adsorb siRNA on the surface of solid nanoparticles. For example, Kim et al. developed cationic solid lipid nanoparticles consisting of natural components of protein-free low-density lipoprotein (LDL) to deliver siRNA (97). LDLs are natural nanocarriers abundant in the bloodstream, transporting lipids, cholesterol, proteins, and hydrophobic drugs throughout systemic circulation. Solid lipid nanoparticles, mimicking natural LDL, have been shown to be very stable and behave similarly to native LDL when injected into the bloodstream. The solid lipid nanoparticles were composed of 45% (wt/wt) cholesteryl ester, 3% (wt/wt) triglyceride, 10% (wt/wt) cholesterol, 14% (wt/wt) DOPE, and 28% (wt/wt) 3β-[N-(N', N'-dimethylamino-ethane)-carbamoyl]-cholesterol (DC-chol). The function of the cationic DC-chol was to make the surface of the nanoparticles positively charged with a zeta potential of about +40 mV. siRNA was conjugated to PEG via a disulfide linkage and anchored onto the surface of cationic solid lipid nanoparticles through charge interaction. Under an optimal weight ratio of DC-chol and siRNA-PEG conjugate, the LDL-like nanoparticles silenced the expression of GFP and VEGF to 40% and showed much less cytotoxicity than PEI 25 kDa in MDA-MB-435 cells. Although work with the LDP-like particles has only progressed to in vitro studies, it is expected that the LDL-like nanoparticle may be useful for in vivo tumor-targeting delivery of siRNA since elevated levels of low-density lipoprotein receptor (LDLR) are reported in various cancer cells such as myeloid leukemic cells, colon, kidney, and brain tumor cells.

Finally, like DOTAP liposomes, nanoparticles for nucleic acid delivery including siRNA are also patented and commercially available for the purpose of scientific research. Bioalliance

(Paris, France) patented a chitosan-coated polyisohexylcyanoacrylate (PIHCA) nanoparticle in 2004. The nanoparticle was directly utilized by Pille et al. to deliver anti-RhoA (Ras homologous A) siRNA in mice and to prove the therapeutic potential of the strategy to treat aggressive breast cancers (98).

Nanocapsules

Nanocapsules are functionally similar to nanoparticles except for having a liquid-filled core instead of a solid core. To date, there are just a few publications on the use of nanocapsules as siRNA delivery carriers. The following will discuss two such nanocapsules that have novel properties as potential siRNA delivery systems.

Ideally, to entrap siRNA in the internal core of a nanocapsule, the core should be aqueous to accommodate the hydrophilic siRNA. The preparation of nanocapsules usually involves the preparation of an emulsion. An oil-in-water emulsion is unable to encapsulate the hydrophilic siRNA alone. In addition, a water-in-oil emulsion leads to nanocapsules suspended in an oil phase, which may not be desirable for intravenous administration or would have to be removed prior to injection. To facilitate the formulation of siRNA in a nanocapsule, Toub et al. developed a nanocapsule with an aqueous core that also could suspended in an aqueous vehicle (99). A water-in-oil nanoemulsion was first prepared by adding an aqueous phase containing siRNA to an oil phase composed of Miglyol and Span 80. Then, isobutylcyanoacrylate (IBCA) monomer was added to the nanoemulsion under mechanical stirring. When IBCA polymerized, it formed a shell structure surrounding the aqueous core containing entrapped siRNA. Later, the oil phase and surfactant were removed by ultracentrifugation. The resulting pellet was resuspended in water to produce a nanosuspension with a particle size of 350 ± 100 nm. In vitro studies in NIH/3T3 cells stably transfected with human EWS-Fli1 gene showed that siRNA against EWS-Fli1 oncogene delivered in the nanocapsules inhibited the EWS-Fli1 mRNA level to 40%. When tested in vivo in xenograft mice bearing EWS-Fli1-expressing tumors, the nanocapsules were found to inhibit 80% of the tumor growth after intratumoral injection when compared with the saline treated control mice. This was the first study reporting on the use of aqueous core nanocapsules for the delivery of siRNA with resulting efficacy in vivo.

To facilitate endosomal escape and release siRNA to the cytosol where RNAi events take place, various endosomal escaping agents have been utilized, such as fusogenic lipids and peptides, polymers exerting proton-sponge effect, etc. A novel endosomal breaking formulation called thermosensitive hydrogel nanocapsules were developed by Lee et al. (100). The thermosensitive Pluronic F-127/PEI 2 kDa nanocapsules were synthesized by interfacial cross-linking reaction between preactivated Pluronic F-127 and low molecular weight PEI 2 kDa at the oil-in-water interface. The resulting Pluronic/PEI 2 kDa nanocapsules had an interior structure filled with aqueous fluid surrounded by a cross-linked Pluronic/PEI 2 kDa shell. Most pluronic copolymers have the critical micelle temperature (CMT) ranging from 25°C to 40°C. Above the CMT, the pluronic copolymers self-assemble to form a spherical micellar structure by dehydration of the poly-(propylene oxide) (PPO) moieties within the structure. The average particle size of Pluronic/PEI 2 kDa nanocapsules was 118.9 ± 15.3 nm at 37°C and 412.3 ± 83.2 at 15°C, respectively. According to the temperature-dependent property of pluronic, the collapse of the nanocapsules with increasing temperature is primarily caused by enhanced hydrophobic interactions between the PPO blocks in the Pluronic F-127 copolymers. PEG-conjugated siRNA was anchored to the surface of Pluronic/PEI 2 kDa nanocapsules through charge interaction. During in vitro cell transfection experiments, three hours after the cells treated with the nanoparticles at 37°C, 15 minutes of 15°C cold shock was given to the cells. The increased particle size under 15°C caused a 41.7-fold volume change, which disrupts the endosomal membrane by physical strength. With cold shock treatment, the expression of GFP in HeLa cells and VEGF in PC-3 cells was reduced to 37.3% and 3.2%, respectively.

Dendrimers

Polycationic dendrimers such as poly-(amidoamine) (PAMAM) dendrimers have long been used to deliver DNA. Recent studies have shown that PAMAM may also serve as siRNA delivery carriers (101). PAMAM dendrimers contain primary amine groups on the surface and tertiary amine groups in the internal architecture. The primary amines bind siRNA, whereas the tertiary amines act as a proton-sponge and facilitate the endosomal release of siRNA into the cytoplasm. The siRNA-PAMAM complexes are very stable, which could only have been dissociated under very strong ionic strength conditions. PAMAM dendrimers are termed as Gn with n denoting dendrimer generation number. As the generation number increases, the number of terminal amines increases. Thus, similar to DNA-PAMAM affinities, an increase in PAMAM generation leads to stronger interactions between the dendrimer and the siRNA. Zhou et al. showed that GL3Luc siRNA-G$_7$ complex reduced the expression of luciferase to 20% in A549Luc cells in vitro (102). To lower the cytotoxicity of G$_7$ PAMAM dendrimers while maintaining the siRNA binding affinity, surface PAMAM-NH$_2$ was acetylated with acetic anhydride and internal PAMAM-OH was quaternized with methyl iodide (103). Both modifications generate neutral outer surface with internal positive charges. It was found that the modifications did not interfere with the binding ability but significantly decrease the cytotoxicity of G$_7$ PAMAM dendrimers. An effort was also made to further increase the cellular uptake of siRNA-PAMAM complex by conjugating the cell-penetrating peptide, Tat; however, the conjugation of Tat did not improve the efficiency of the dendrimer (104).

The terminal groups of G$_3$ PAMAM dendrimer have been partially conjugated with α-cyclodextrin (α-CDE) to deliver siRNA (105). CDE, at high concentration, disturbs the cellular membrane components such as phospholipids and cholesterol, leading to increased membrane permeability. Moreover, the α-CDE has low cytotoxicity even at high charge ratio of α-CDE/nucleic acid. Thus, the G$_3$ PAMAM dendrimer/α-CDE conjugate was developed to reduce the cytotoxicity and increase the delivery efficiency for nucleic acids. A pilot study showed that siRNA against pGL3 luciferase delivered by G$_3$ PAMAM dendrimer/α-CDE conjugate suppressed the luciferase gene expression level in vitro by about 50% in NIH3T3-luc cells.

Dendritic poly(L-lysine) generation 6 (KG$_6$) was used to deliver several siRNAs by Inoue et al. (106). KG$_6$ was used in combination with the amphiphilic weak-base peptide Endo-Porter (EP), which is a commercially available cellular delivery reagent available from Gene Tools. Neither KG$_6$ nor EP could efficiently deliver glyceraldehyde 3-phosphate dehydrogenase (GAPDH) siRNA when KG$_6$ or EP was used alone. However, when KG$_6$ and EP were used together, GADPH was efficiently knocked down both protein levels and mRNA levels in H4IIEC3 cells.

Other Novel Carriers

In addition to the traditional or conventional siRNA delivery carriers discussed above, there are several highly innovative new strategies that are being developed and tested as potential delivery systems for siRNA.

Quantum dots (QDs) are nanoscaled semiconductor inorganic materials that have provided greatly enhanced capabilities for medical imaging and diagnostics. Yezhelyev et al. developed a class of dual-functional nanoparticle for both siRNA delivery and imaging based on the use of QDs (107). Highly luminescent QDs were first synthesized and encapsulated in the poly-(maleic anhydride-alt-1-tetradecene) bearing surface carboxylic acid groups. The carboxylic acid groups were then partially converted to tertiary amines (Fig. 10). It was found that by balancing the ratio of the carboxylic acid and tertiary amine moieties, the proton-sponge effect could be precisely controlled. The resulting polymer-coated QDs were suitable

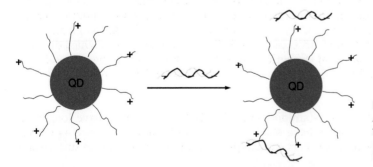

Figure 10 (*See color insert*) Adsorption of siRNA onto surface-modified QDs. *Abbreviations*: siRNA, small interfering RNA; QDs, quantum dots.

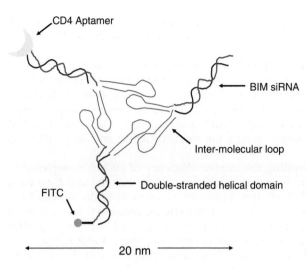

CD4 Aptamer

BIM siRNA

Inter-molecular loop

FITC

Double-stranded helical domain

20 nm

Figure 11 (*See color insert*) Schematic structure of engineering pRNA nanoparticle containing siRNA, aptamer, and fluorescent label. *Abbreviations*: pRNA, packing RNA; siRNA, small interfering RNA.

for siRNA binding, penetrating the cell, and for providing a mechanism for endosomal escape. In comparison to cationic lipids and polymer-based siRNA delivery systems, the QD-based nanoparticles have much smaller size and more uniform size distribution. A QD core size of 6 nm yielded polymer-coated dots with sizes of 13 and 17 nm before and after siRNA binding, respectively. The QD nanoparticles efficiently delivered siRNA against cyclophilin B in a human breast cancer cell line and led to nearly complete suppression of cyclophilin B expression, which was superior to three most commonly used transfection reagents (Lipofectamine^{TM}, TransIT-TKO^{TM}, and JetPEI^{TM}). Another advantage of the QD-siRNA particles is that they afford simultaneous delivery with imaging allowing for real-time tracking and intracellular localization of QDs during delivery.

Khaled et al. have also engineered protein- and lipid-free multifunctional RNA nanoparticles to deliver siRNA and combine targeted therapy and imaging in a natural modality, pRNA (packing RNA), by utilizing RNA nanotechnology (108). pRNA is a vital component of molecular motor, which uses ATP as energy to package DNA into the procapsid during the replication of linear dsDNA viruses. The 117-nucleotide pRNA monomer contains two functional domains: the intermolecular-interacting domain and the double-stranded helical DNA packaging domain. The intermolecular-interacting domain contains left and right loops like two arms that interlock with other pRNA monomers via base-pairing to form dimer, trimer, or hexamers of size 10 to 30 nm. Figure 11 shows the structure of a pRNA trimer. According to their study, the replacement of pRNA helical region with siRNA, or connection of the RNA aptamer, or connection of other chemical components did not interfere either with the folding and trimering of the pRNA, as long as the two strands were paired, or with the function of siRNA and other connected moieties. Therefore, they tried to replace the helical region with small RNA fragments and connect RNA aptamer or other chemical components to this region to engineer a variety of chimeric pRNAs. The pRNA trimers with the size of about 20 nm are extremely compact and versatile nanoparticles with lots of advantages. For example, as shown in Figure 11, a trimeric complex composed of pRNA/aptamer (CD4), pRNA/siRNA (BIM), and pRNA/fluorescein isothiocyanate (FITC) could target CD4 positive cells and simultaneously deliver siRNA against proapoptosis factor Bcl-2 interacting mediator of cell death (BIM) and imaging molecule FITC to these cells. In addition, more than one siRNA could be constructed to the pRNA nanoparticles to inhibit the expression of multiple oncogenes. RNA aptamers, compared with antibodies and phage-displaying peptides, have very low immunogenicity. Furthermore, the size, shape, stoichiometry, and the functions of the final product are highly controllable.

FUTURE PERSPECTIVE

The promise of siRNA applications as a powerful therapeutic agent relies on a successful delivery vehicle. In this chapter, a series of criteria of an "ideal" nano-based siRNA delivery system were addressed and can be summarized as efficient, specific, and safe. It is obvious that a great deal of effort has been devoted to pursuing the ideal nano-based systems for siRNA delivery, and the field is developing rapidly. However, all current-reported formulations have recognizable gaps.

Delivery efficiency depends on many factors. First, the structures of carrier materials are critical. Currently, although there are some general rules to design and synthesize siRNA complexing agents (e.g., the presence of positive charges), the investigation of structure-efficiency relationship is still under a trial-and-error mode. In the future, when a large amount of compounds have been studied, databases could be built, and thus computer simulation and modeling would be performed to rationally design the delivery agent and to predict the binding and assembly with siRNA.

Particle size is another factor controlling the in vivo efficiency of siRNA nanoparticles. Nanoparticles with a broad range of particle size (from 20 nm up to about 800 nm) have been reported in the literature to deliver siRNA in vitro. However, since most of the studies have stopped at in vitro experiments, the in vivo efficacy of the siRNA nanoparticles remains a question. What is not fully understood yet is how the particle size and surface properties minimize systemic clearance and optimize target tissue penetration, and cellular uptake.

In addition to particle size and size distribution, other properties such as shape, mechanical properties, and surface texture and morphology are also important factors affecting siRNA delivery efficiency of nanoparticles both in vitro and in vivo. While chemical modifications of carrier materials are the major strategy to increase the efficiency of siRNA delivery nowadays, the influence of physical properties of the nanoformulation has been underestimated. Together with particle size, these physical properties and their influence on the nanoparticle behavior in circulation, tissue distribution, cell penetration, and cellular trafficking require more attention.

The specificity of siRNA delivery primarily depends on the selection of a target and ligand, both of which would benefit from progress and advances in other fields. The advances in molecular biology would help find more specifically expressed targets such as receptors, integrins, or enzymes in pathological tissues as well as more specific and high-affinity ligand via, for example, in vivo phage display.

Years ago, the incorporation of PEG in various nanoformulations dramatically decreased their nonspecific RES clearance and increased their circulation half-life. As the nonspecific RES clearance decreases, the accumulation of nanoparticles in target organ or tissue increases. Hence, to increase the delivery specificity, active targeting using a targeting ligand is preferred; however, improved delivery by passive targeting may also have therapeutic potential and utility.

In terms of the safety for nano-based siRNA delivery systems, on one hand, efforts need to be made to further decrease the cytotoxicity of carrier materials and to look for less immunogenic targeting ligands. On the other hand, the toxicity of different formulations is mostly identified and/or estimated by in vitro experiments. However, cytotoxicity is often cell-type dependent. Thus, the field also is in need of improved, predictive, in vitro models to more accurately reflect the in vivo environment.

There is no doubt that delivering siRNA safely and efficiently is a challenging task. The field is in need of a breakthrough.

REFERENCES

1. Meister G, Tuschl T. Mechanisms of gene silencing by double-stranded RNA. Nature 2004; 431 (7006):343–349.
2. Hannon GJ. RNA interference. Nature 2002; 418(6894):244–251.
3. Valencia-Sanchez MA, Liu J, Hannon GJ, et al. Control of translation and mRNA degradation by miRNAs and siRNAs. Genes Dev 2006; 20(5):515–524.
4. Ameres SL, Martinez J, Schroeder R. Molecular basis for target RNA recognition and cleavage by human RISC. Cell 2007; 130(1):101–112.
5. Reynolds A, Anderson EM, Vermeulen A, et al. Induction of the interferon response by siRNA is cell type- and duplex length-dependent. RNA 2006; 12(6):988–993.
6. Elbashir SM, Harborth J, Lendeckel W, et al. Duplexes of 21-nucleotide RNAs mediate RNA interference in cultured mammalian cells. Nature 2001; 411(6836):494–498.

7. Gurney AM, Hunter E. The use of small interfering RNA to elucidate the activity and function of ion channel genes in an intact tissue. J Pharmacol Toxicol Methods 2005; 51(3):253–262.
8. Zaffaroni N, Pennati M, Folini M. Validation of telomerase and survivin as anticancer therapeutic targets using ribozymes and small-interfering RNAs. Methods Mol Biol 2007; 361:239–263.
9. Morgan-Lappe SE, Tucker LA, Huang X, et al. Identification of Ras-related nuclear protein, targeting protein for xenopus kinesin-like protein 2, and stearoyl-CoA desaturase 1 as promising cancer targets from an RNAi-based screen. Cancer Res 2007; 67(9):4390–4398.
10. Tyner JW, Walters DK, Willis SG, et al. RNAi screening of the tyrosine kinome identifies therapeutic targets in acute myeloid leukemia. Blood 2008; 111(4):2238–2245.
11. Gust TC, Neubrandt L, Merz C, et al. RNA interference-mediated gene silencing in murine T cells: in vitro and in vivo validation of proinflammatory target genes. Cell Commun Signal 2008; 6:3.
12. Vance V, Vaucheret H. RNA silencing in plants—defense and counterdefense. Science 2001; 292 (5525):2277–2280.
13. McCaffrey AP, Nakai H, Pandey K, et al. Inhibition of hepatitis B virus in mice by RNA interference. Nat Biotechnol 2003; 21(6):639–644.
14. McCaffrey AP, Meuse L, Pham TT, et al. RNA interference in adult mice. Nature 2002; 418(6893): 38–39.
15. Song E, Zhu P, Lee SK, et al. Antibody mediated in vivo delivery of small interfering RNAs via cell-surface receptors. Nat Biotechnol 2005; 23(6):709–717.
16. Park WS, Hayafune M, Miyano-Kurosaki N, et al. Specific HIV-1 env gene silencing by small interfering RNAs in human peripheral blood mononuclear cells. Gene Ther 2003; 10(24):2046–2050.
17. Novina CD, Murray MF, Dykxhoorn DM, et al. siRNA-directed inhibition of HIV-1 infection. Nat Med 2002; 8(7):681–686.
18. Palliser D, Chowdhury D, Wang QY, et al. An siRNA-based microbicide protects mice from lethal herpes simplex virus 2 infection. Nature 2006; 439(7072):89–94.
19. Bitko V, Musiyenko A, Shulyayeva O, et al. Inhibition of respiratory viruses by nasally administered siRNA. Nat Med 2005. 11(1):50–55.
20. Zhang W, Yang H, Kong X, et al. Inhibition of respiratory syncytial virus infection with intranasal siRNA nanoparticles targeting the viral NS1 gene. Nat Med 2005; 11(1):56–62.
21. Jiang M, Milner J. Selective silencing of viral gene expression in HPV-positive human cervical carcinoma cells treated with siRNA, a primer of RNA interference. Oncogene 2002; 21(39):6041–6048.
22. Yoshinouchi M, Yamada T, Kizaki M, et al. In vitro and in vivo growth suppression of human papillomavirus 16-positive cervical cancer cells by E6 siRNA. Mol Ther 2003; 8(5):762–768.
23. Song E, Lee SK, Wang J, et al. RNA interference targeting Fas protects mice from fulminant hepatitis. Nat Med 2003; 9(3):347–351.
24. Wang J, Li W, Min J, et al. Fas siRNA reduces apoptotic cell death of allogeneic-transplanted hepatocytes in mouse spleen. Transplant Proc 2003; 35(4):1594–1595.
25. Perl M, Chun-Shiang Chug, Joanne Lomas-Neira, et al. Silencing of Fas, but not caspase-8, in lung epithelial cells ameliorates pulmonary apoptosis, inflammation, and neutrophil influx after hemorrhagic shock and sepsis. Am J Pathol 2005; 167(6):1545–1559.
26. Zender L, Hutker S, Liedtke C, et al. Caspase 8 small interfering RNA prevents acute liver failure in mice. Proc Natl Acad Sci U S A 2003; 100(13):7797–7802.
27. Wesche-Soldato DE, Chung CS, Lomas-Neira J, et al. In vivo delivery of caspase-8 or Fas siRNA improves the survival of septic mice. Blood 2005; 106(7):2295–2301.
28. Caplen NJ, Taylor JP, Statham VS, et al. Rescue of polyglutamine-mediated cytotoxicity by double-stranded RNA-mediated RNA interference. Hum Mol Genet 2002; 11(2):175–184.
29. Miller VM, Xia H, Marrs GL, et al. Allele-specific silencing of dominant disease genes. Proc Natl Acad Sci U S A 2003; 100(12):7195–7200.
30. Schwarz DS, Ding H, Kennington L, et al. Designing siRNA that distinguish between genes that differ by a single nucleotide. PLoS Genet 2006; 2(9):e140.
31. Brummelkamp TR, Bernards R, Agami R. A system for stable expression of short interfering RNAs in mammalian cells. Science 2002; 296(5567):550–553.
32. Brummelkamp TR, Bernards R, Agami R. Stable suppression of tumorigenicity by virus-mediated RNA interference. Cancer Cell 2002; 2(3):243–247.
33. Wohlbold L, van der Kuip H, Moehring A, et al. All common p210 and p190 Bcr-abl variants can be targeted by RNA interference. Leukemia 2005; 19(2):290–292.
34. Manka D, Spicer Z, Millhorn DE. Bcl-2/adenovirus E1B 19 kDa interacting protein-3 knockdown enables growth of breast cancer metastases in the lung, liver, and bone. Cancer Res 2005; 65 (24):11689–11693.
35. Kami K, Doi R, Koizumi M, et al. Downregulation of survivin by siRNA diminishes radioresistance of pancreatic cancer cells. Surgery 2005; 138(2):299–305.

36. Zhang M, Zhang X, Bai CX, et al. Silencing the epidermal growth factor receptor gene with RNAi may be developed as a potential therapy for non small cell lung cancer. Genet Vaccines Ther 2005; 3:5.

37. Takei Y, Kadomatsu K, Yuzawa Y, et al. A small interfering RNA targeting vascular endothelial growth factor as cancer therapeutics. Cancer Res 2004; 64(10):3365–3370.

38. Wu H, Hait WN, Yang JM. Small interfering RNA-induced suppression of MDR1 (P-glycoprotein) restores sensitivity to multidrug-resistant cancer cells. Cancer Res 2003; 63(7):1515–1519.

39. Stierle V, Laigle A, Jolles B. Modulation of MDR1 gene expression in multidrug resistant MCF7 cells by low concentrations of small interfering RNAs. Biochem Pharmacol 2005; 70(10):1424–1430.

40. Manoharan M. RNA interference and chemically modified small interfering RNAs. Curr Opin Chem Biol 2004; 8(6):570–579.

41. Wolfrum C, Shi S, Jayaprakash KN, et al. Mechanisms and optimization of in vivo delivery of lipophilic siRNAs. Nat Biotechnol 2007; 25(10):1149–1157.

42. Lewis DL, Wolff JA. Systemic siRNA delivery via hydrodynamic intravascular injection. Adv Drug Deliv Rev 2007; 59(2–3):115–123.

43. Maruyama T, Sonokawa S, Matsushita H, et al. Inhibitiory effects of gold(III) ions on ribonuclease and deoxyribonuclease. J Inorg Biochem 2007; 101(1):180–186.

44. Bartlett DW, Davis ME. Insights into the kinetics of siRNA-mediated gene silencing from live-cell and live-animal bioluminescent imaging. Nucleic Acids Res 2006; 34(1):322–333.

45. Kebbekus P, Draper DE, Hagerman P. Persistence length of RNA. Biochemistry 1995; 34(13): 4354–4357.

46. Spagnou S, Miller AD, Keller M. Lipidic carriers of siRNA: differences in the formulation, cellular uptake, and delivery with plasmid DNA. Biochemistry 2004; 43(42):13348–13356.

47. Kornek M, Lukacs-Kornek V, Limmer A, et al. 1,2-dioleoyl-3-trimethylammonium-propane (DOTAP)-formulated, immune-stimulatory vascular endothelial growth factor a small interfering RNA (siRNA) increases antitumoral efficacy in murine orthotopic hepatocellular carcinoma with liver fibrosis. Mol Med 2008; 14(7–8):365–373.

48. Cardoso AL, Simões S, de Almeida LP, et al. Tf-lipoplexes for neuronal siRNA delivery: A promising system to mediate gene silencing in the CNS. J Control Release 2008; 132(2):113–123.

49. Sioud M, Sorensen DR. Cationic liposome-mediated delivery of siRNAs in adult mice. Biochem Biophys Res Commun 2003; 312(4):1220–1225.

50. Sorensen DR, Leirdal M, Sioud M. Gene silencing by systemic delivery of synthetic siRNAs in adult mice. J Mol Biol 2003; 327(4):761–766.

51. Zhang Y. Cationic Liposome-Mediated Gene Delivery. Cambridge Healthtech Institute's 4th Annual Meeting, October 12–14, 1997.

52. Xu L, Tang WH, Huang CC, et al. Systemic p53 gene therapy of cancer with immunolipoplexes targeted by anti-transferrin receptor scFv. Mol Med 2001; 7(10):723–734.

53. Xu L, Huang CC, Huang W, et al. Systemic tumor-targeted gene delivery by anti-transferrin receptor scFv-immunoliposomes. Mol Cancer Ther 2002; 1(5):337–346.

54. Pirollo KF, Dagata J, Wang P, et al. A tumor-targeted nanodelivery system to improve early MRI detection of cancer. Mol Imaging 2006; 5(1):41–52.

55. Pirollo KF, Rait A, Zhou Q, et al. Materializing the potential of small interfering RNA via a tumor-targeting nanodelivery system. Cancer Res 2007; 67(7):2938–2943.

56. Cardoso AL, Simões S, de Almeida LP, et al. siRNA delivery by a transferrin-associated lipid-based vector: a non-viral strategy to mediate gene silencing. J Gene Med 2007; 9(3):170–183.

57. Kim SI, Shin D, Choi TH, et al. Systemic and specific delivery of small interfering RNAs to the liver mediated by apolipoprotein A-I. Mol Ther 2007; 15(6):1145–1152.

58. Song YK, Liu F, Chu S, et al. Characterization of cationic liposome-mediated gene transfer in vivo by intravenous administration. Hum Gene Ther 1997; 8(13):1585–1594.

59. Chien PY, Wang J, Carbonaro D, et al. Novel cationic cardiolipin analogue-based liposome for efficient DNA and small interfering RNA delivery in vitro and in vivo. Cancer Gene Ther 2005; 12 (3):321–328.

60. Khoury M, Louis-Plence P, Escriou V, et al. Efficient new cationic liposome formulation for systemic delivery of small interfering RNA silencing tumor necrosis factor alpha in experimental arthritis. Arthritis Rheum 2006; 54(6):1867–1877.

61. Bouxsein NF, McAllister CS, Ewert KK, et al. Structure and gene silencing activities of monovalent and pentavalent cationic lipid vectors complexed with siRNA. Biochemistry 2007; 46(16):4785–4792.

62. Desigaux L, Sainlos M, Lambert O, et al. Self-assembled lamellar complexes of siRNA with lipidic aminoglycoside derivatives promote efficient siRNA delivery and interference. Proc Natl Acad Sci U S A 2007; 104(42):16534–16539.

63. Mudhakir D, Akita H, Tan E, et al. A novel IRQ ligand-modified nano-carrier targeted to a unique pathway of caveolar endocytic pathway. J Control Release 2008; 125(2):164–173.
64. Foged C, Nielsen HM, Frokjaer S. Liposomes for phospholipase A2 triggered siRNA release: preparation and in vitro test. Int J Pharm 2007; 331(2):160–166.
65. Dong W, Jin GH, Li SF, et al. Cross-linked polyethylenimine as potential DNA vector for gene delivery with high efficiency and low cytotoxicity. Acta Biochim Biophys Sin (Shanghai) 2006; 38 (11):780–787.
66. Li S, Dong W, Zong Y, et al. Polyethylenimine-complexed plasmid particles targeting focal adhesion kinase function as melanoma tumor therapeutics. Mol Ther 2007; 15(3):515–523.
67. Swami A, Kurupati RK, Pathak A, et al. A unique and highly efficient non-viral DNA/siRNA delivery system based on PEI-bisepoxide nanoparticles. Biochem Biophys Res Commun 2007; 362 (4):835–841.
68. Jere D, Xu CX, Arote R, et al. Poly(beta-amino ester) as a carrier for si/shRNA delivery in lung cancer cells. Biomaterials 2008; 29(16):2535–2547.
69. Zintchenko A, Philipp A, Dehshahri A, et al. Simple modifications of branched PEI lead to highly efficient siRNA carriers with low toxicity. Bioconjug Chem 2008; 19(7):1448–1455.
70. Shim MS, Kwon YJ. Controlled delivery of plasmid DNA and siRNA to intracellular targets using ketalized polyethylenimine. Biomacromolecules 2008; 9(2):444–455.
71. Kim SH, Jeong JH, Lee SH, et al. Local and systemic delivery of VEGF siRNA using polyelectrolyte complex micelles for effective treatment of cancer. J Control Release 2008; 129(2):107–116.
72. Oishi M, Nagasaki Y, Itaka K, et al. Lactosylated poly(ethylene glycol)-siRNA conjugate through acid-labile beta-thiopropionate linkage to construct pH-sensitive polyion complex micelles achieving enhanced gene silencing in hepatoma cells. J Am Chem Soc 2005; 127(6):1624–1625.
73. Sun TM, Du JZ, Yan LF, et al. Self-assembled biodegradable micellar nanoparticles of amphiphilic and cationic block copolymer for siRNA delivery. Biomaterials 2008; 29(32):4348–4355.
74. Sato A, Choi SW, Hirai M, et al. Polymer brush-stabilized polyplex for a siRNA carrier with long circulatory half-life. J Control Release 2007; 122(3):209–216.
75. Kim WJ, Chang CW, Lee M, et al. Efficient siRNA delivery using water soluble lipopolymer for anti-angiogenic gene therapy. J Control Release 2007; 118(3):357–363.
76. Meyer M, Philipp A, Oskuee R, et al. Breathing life into polycations: functionalization with pH-responsive endosomolytic peptides and polyethylene glycol enables siRNA delivery. J Am Chem Soc 2008; 130(11):3272–3273.
77. Schiffelers RM, Ansari A, Xu J, et al. Cancer siRNA therapy by tumor selective delivery with ligand-targeted sterically stabilized nanoparticle. Nucleic Acids Res 2004; 32(19):e149.
78. Hwa Kim S, Hoon Jeong J, Chul Cho K, et al. Target-specific gene silencing by siRNA plasmid DNA complexed with folate-modified poly(ethylenimine). J Control Release 2005; 104(1):223–232.
79. Jiang G, Park K, Kim J, et al. Hyaluronic acid-polyethyleneimine conjugate for target specific intracellular delivery of siRNA. Biopolymers 2008; 89(7):635–642.
80. Howard KA, Rahbek UL, Liu X, et al. RNA interference in vitro and in vivo using a novel chitosan/siRNA nanoparticle system. Mol Ther 2006; 14(4):476–484.
81. Liu X, Howard KA, Dong M, et al. The influence of polymeric properties on chitosan/siRNA nanoparticle formulation and gene silencing. Biomaterials 2007; 28(6):1280–1288.
82. Katas H, Alpar HO. Development and characterisation of chitosan nanoparticles for siRNA delivery. J Control Release 2006; 115(2):216–225.
83. Rojanarata T, Opanasopit P, Techaarpornkul S, et al. Chitosan-thiamine pyrophosphate as a novel carrier for siRNA delivery. Pharm Res 2008; 25(12):2807–2814.
84. Leng Q, Scaria P, Zhu J, et al. Highly branched HK peptides are effective carriers of siRNA. J Gene Med 2005; 7(7):977–986.
85. Minakuchi Y, Takeshita F, Kosaka N, et al. Atelocollagen-mediated synthetic small interfering RNA delivery for effective gene silencing in vitro and in vivo. Nucleic Acids Res 2004; 32(13):e109.
86. Wang XL, Nguyen T, Gillespie D, et al. A multifunctional and reversibly polymerizable carrier for efficient siRNA delivery. Biomaterials 2008; 29(1):15–22.
87. Wang XL, Ramusovic S, Nguyen T, et al. Novel polymerizable surfactants with pH-sensitive amphiphilicity and cell membrane disruption for efficient siRNA delivery. Bioconjug Chem 2007; 18 (6):2169–2177.
88. Matsui K, Horiuchi S, Sando S, et al. RNAi silencing of exogenous and endogenous reporter genes using a macrocyclic octaamine as a "compact" siRNA carrier. Studies on the nonsilenced residual activity. Bioconjug Chem 2006; 17(1):132–138.
89. Mok H, Park TG. Self-crosslinked and reducible fusogenic peptides for intracellular delivery of siRNA. Biopolymers 2008; 89(10):881–888.

90. Li SD, Huang L. Surface-modified LPD nanoparticles for tumor targeting. Ann N Y Acad Sci 2006; 1082:1–8.
91. Li SD, Huang L. Targeted delivery of antisense oligodeoxynucleotide and small interference RNA into lung cancer cells. Mol Pharm 2006; 3(5):579–588.
92. Chono S, Li SD, Conwell CC, et al. An efficient and low immunostimulatory nanoparticle formulation for systemic siRNA delivery to the tumor. J Control Release 2008; 131(1):64–69.
93. Peer D, Park EJ, Morishita Y, et al. Systemic leukocyte-directed siRNA delivery revealing cyclin D1 as an anti-inflammatory target. Science 2008; 319(5863):627–630.
94. Kakizawa Y, Furukawa S, Ishii A, et al. Organic-inorganic hybrid-nanocarrier of siRNA constructing through the self-assembly of calcium phosphate and PEG-based block aniomer. J Control Release 2006; 111(3):368–370.
95. Kakizawa Y, Furukawa S, Kataoka K. Block copolymer-coated calcium phosphate nanoparticles sensing intracellular environment for oligodeoxynucleotide and siRNA delivery. J Control Release 2004; 97(2):345–356.
96. Bartlett DW, Davis ME. Physicochemical and biological characterization of targeted, nucleic acid-containing nanoparticles. Bioconjug Chem 2007; 18(2):456–468.
97. Kim HR, Kim IK, Bae KH, et al. Cationic solid lipid nanoparticles reconstituted from low density lipoprotein components for delivery of siRNA. Mol Pharm 2008; 5(4):622–631.
98. Pille JY, Li H, Blot E, et al. Intravenous delivery of anti-RhoA small interfering RNA loaded in nanoparticles of chitosan in mice: safety and efficacy in xenografted aggressive breast cancer. Hum Gene Ther 2006; 17(10):1019–1026.
99. Toub N, Bertrand JR, Tamaddon A, et al. Efficacy of siRNA nanocapsules targeted against the EWS-Fli1 oncogene in Ewing sarcoma. Pharm Res 2006; 23(5):892–900.
100. Lee SH, Choi SH, Kim SH, et al. Thermally sensitive cationic polymer nanocapsules for specific cytosolic delivery and efficient gene silencing of siRNA: swelling induced physical disruption of endosome by cold shock. J Control Release 2008; 125(1):25–32.
101. Shen XC, Zhou J, Liu X, et al. Importance of size-to-charge ratio in construction of stable and uniform nanoscale RNA/dendrimer complexes. Org Biomol Chem 2007; 5(22):3674–3681.
102. Zhou J, Wu J, Hafdi N, et al. PAMAM dendrimers for efficient siRNA delivery and potent gene silencing. Chem Commun (Camb) 2006(22):2362–2364.
103. Patil ML, Zhang M, Betigeri S, et al. Surface-modified and internally cationic polyamidoamine dendrimers for efficient siRNA delivery. Bioconjug Chem 2008; 19(7):1396–1403.
104. Kang H, DeLong R, Fisher MH, et al. Tat-conjugated PAMAM dendrimers as delivery agents for antisense and siRNA oligonucleotides. Pharm Res 2005; 22(12):2099–2106.
105. Tsutsumi T, Hirayama F, Uekama K, et al. Evaluation of polyamidoamine dendrimer/alpha-cyclodextrin conjugate (generation 3, G3) as a novel carrier for small interfering RNA (siRNA). J Control Release 2007; 119(3):349–359.
106. Inoue Y, Kurihara R, Tsuchida A, et al. Efficient delivery of siRNA using dendritic poly(L-lysine) for loss-of-function analysis. J Control Release 2008; 126(1):59–66.
107. Yezhelyev MV, Qi L, O'Regan RM, et al. Proton-sponge coated quantum dots for siRNA delivery and intracellular imaging. J Am Chem Soc 2008; 130(28):9006–9012.
108. Khaled A, Guo S, Li F, et al. Controllable self-assembly of nanoparticles for specific delivery of multiple therapeutic molecules to cancer cells using RNA nanotechnology. Nano Lett 2005; 5(9): 1797–1808.

7 | Excipients for parenteral dosage forms: regulatory considerations and controls

Sandeep Nema and Ronald J. Brendel

INTRODUCTION

A survey of commercial parenteral products confirms an interesting observation—the active drug molecule typically comprises only a small percentage of the drug product formulation whereas the excipients make up the primary components. Excipients provide the enhanced vehicle for the active pharmaceutical ingredient (API) and are typically referred to as inactive or inert ingredients, where "inactive" or "inert" indicates the compound does not directly contribute to the intended therapeutic or diagnostic activity of the drug product. Pharmaceutical excipients or additives are compounds added to the finished drug product with a specific functional role [other than that defined for API or in case of biologics, drug substance (DS)]. These functions include increasing the bulk to aid in manufacturing, stabilizing of the active drug, improving delivery and targeting, and modifying the safety or pharmacokinetic profile of the active drug. Compounds considered excipients also encompass ingredients that are used for the production of dosage forms, but may or may not be present in the finished dosage forms. Examples are water for lyophilized product and inert gases in the head space of container (1). It is clear that excipients have a "functionality" in the dosage form synonymous to the pharmacological activity of API/DS, which is currently being acknowledged by various pharmacopoeias (2). It is noted, however, that many of the existing excipient monographs do not address this functionality aspect and its control.

With the current Quality-by-Design (QbD) initiative, pharmaceutical companies are achieving a better understanding of how the functionality and performance of excipients can influence the drug product (3). Concurrently, steps are being initiated to rectify the failure of existing pharmacopoeial excipient monographs to directly address excipient functionality. These activities include the proposal of an USP (United States Pharmacopoeia) general chapter on functionality of excipients and Ph. Eur. (European Pharmacopoeia) nonmandatory Functionality-Related Characteristics section in the monographs of some excipients. The QbD approach will also provide better insight into the potential impact of excipient variability on product quality. Design space studies of excipient variability and functional performance will provide a higher level of assurance that excipient standards accurately reflect excipient quality (4).

While the functional role of the inert or inactive excipient does not include the therapeutic or diagnostic activity of the active drug ingredient, it may have some level of pharmacological activity. Therefore, restrictions have been placed on the type or amount of excipient that can be included in the formulation of parenteral drug products because of safety issues. For example, Japan, the United States, and the European Union (EU) prohibit or discourage the use of amino mercuric chloride or thimerosal, yet these excipients are still widely used in several products in rest of the world.

A typical definition of parenteral is *not oral or not through the alimentary canal*. As defined in the Ph. Eur. and the British Pharmacopoeia (BP), "Parenteral preparations are sterile preparations intended for administration by injection, infusion, or implantation into the human or animal body"; however, for the purposes of this chapter, only sterile preparations for administration by injection or infusion into the human body will be surveyed (5,6). Injectable formulations are subject to a strict set of requirements. The formulated product has to be sterile, pyrogen-free, and in the case of solution, free of particulate matter. Coloring agents added solely for the purpose of coloring the parenteral preparation are not allowed. An isotonic formulation is preferred, and depending on the route of administration, some excipients may be prohibited. Certain drugs administered by injection, rather than orally, may pose a higher risk for an adverse event or the drug's effect may be especially difficult to reverse because the injected drug bypasses natural defense barriers and is quickly distributed throughout the body. The excipient must be able to withstand the rigors of the sterilization process such as the very high temperatures required for terminal steam sterilization, or filtration and lyophilization in aseptic processing. All of the above factors can limit the choice of excipients available to the formulator.

When choosing acceptable excipients for a parenteral formulation, a formulator should first look for those excipients already used in approved parenteral drug products and/or designated for parenteral use in regulatory listings such as the FDA's inactive ingredient database. Using these excipients provides increased assurance to the formulator that they will probably be safe for their new drug product. However, it should be understood that as excipients are combined with other additives and/or with a new drug molecule, unforeseen potentiation or synergistic toxic effects could result. Utilizing excipients previously approved in an injectable product will often ease the regulatory scrutiny and may require less safety data. In contrast, a new additive will certainly require additional studies adding to the cost, time, and risk to product development. It is important to note that inclusion of an excipient in the GRAS (Generally Recognized as Safe) list or pharmacopoeia such as the USP-NF (United States Pharmacopoeia–National Formulary) does not mean that the excipient has been deemed safe by the regulatory agencies for use in parenteral products.

In Japan, if the drug product contains an excipient with no precedence of use in Japan, then the quality and safety attributes of the excipient must be evaluated by the Subcommittee on Pharmaceutical Excipients of the Central Pharmaceutical Affairs Council concurrently with the evaluation of the drug product application (7). Precedence of use means that the excipient has been used in a drug product in Japan, which is administered via the same route and in a dose level equal to or greater than the excipient in question in the new application.

This chapter offers a comprehensive review of excipients that have been included in injectable products marketed in the United States, Europe, and Japan. A review of the literature indicates that only limited articles have been published which specifically deal with the selection of parenteral excipients (8–17). However, excipients included in other sterile dosage forms not administered parenterally, such as solution for irrigation, ophthalmic or otic drops, and ointments, are not covered in this chapter.

Several sources of information were used to summarize the information in this chapter (9–22). The tables are categorized on the basis of the excipient's primary function in the formulation. For example, ascorbates are categorized as antioxidants, although they can also serve as buffers. This classification system minimizes redundancy and results in a simplified format. Excipient concentration is expressed as percent weight by volume (wt/vol) or volume by volume (vol/vol). For lyophilized or powder products, the percentages were calculated on the basis of the most commonly used reconstitution volume.

TYPES OF EXCIPIENTS
Solvents and Cosolvents

Table 1 list solvents and cosolvents used in parenteral products. Water for injection (WFI) is the most common solvent that can be produced by a variety of technologies. The Ph. Eur. recognizes WFI produced by the distillation process only, even though WFI produced by reverse osmosis will meet all the specifications (23,24). WFI may be combined or substituted with a cosolvent to improve the solubility or stability of drugs (25,26). The dielectric constant and solubility parameters are among the most common polarity indices used for solvent blending (27,28). For more than 50% of parenteral cosolvent systems, ethanol and propylene glycol are used either alone or in combination with other solvents. Interestingly, propylene glycol is used more often than polyethylene glycols (PEGs) in spite of its higher myotoxicity and hemolyzing effects (29–32). A review of toxicity for commonly used parenteral cosolvents is summarized in an article by Mottu et al. (33). The hemolytic potential of cosolvents is as follows (31):

Dimethyl acetamide < PEG 400 < ethanol < propylene glycol < dimethylsulfoxide

Degradation of the drug in the cosolvent system may result due to the possible presence of residual peroxide from the bleaching of PEG or the generation of peroxides in PEG. Hence, it is important to use unbleached and/or low-peroxide PEGs in the formulation.

Table 1 Solvents and Cosolvents

Excipient	Frequency	Range	Example
Benzyl benzoate	3	20–44.7% wt/vol	Delestrogen® 40 mg/mL (Bristol Myers) 44.7% wt/vol
Castor oil	2	11.50%	Delestrogen 40 mg/mL (Bristol Myers)
Cottonseed oil	1	73.6–87.4% wt/vol	Depo® Testosterone (Pfizer) 73.6% wt/vol
N,N dimethylacetamide	2	6–33% wt/vol	Busulfex® (Orphan Medical) 33%
Ethanol/ethanol dehydrated	34	0.6–100%	Prograf® (Fujisawa) 80% vol/vol, Alprostadil (Bedford Lab) 100%
Glycerin (glycerol)	17	1.6–70% wt/vol	Multitest CMI® (Pasteur Merieux) 70% wt/vol
N-methyl-2-pyrrolidone	1	a	Eligrad 7.5 mg (Sanofi)
Peanut oil	1	a	Bal in Oil® (Becton Dickinson)
Polyethylene glycol			
PEG[b]	5	0.15–50%	Secobarbital sodium (Wyeth-Ayerst) 50%
PEG 300	4	50–65%	VePesid® (Bristol Myers) 65% wt/vol
PEG 400	4	11.2–67% vol/vol	Busulfex® (Orphan Medical) 67%
PEG 600	1	5% wt/vol	Persantine® (Dupont-Merck)
PEG 3350	4	0.3–3%	Depo-Medrol® (Upjohn) 2.95% wt/vol
PEG 4000	1	0.3–3%	Invega Sustenna® (Janssen)
Poppyseed oil	1	a	Ethiodol® (Savage)
Propylene glycol	32	0.0025–80%	Ativan® (Wyeth-Ayerst) 80%
Safflower oil	2	5–10%	Liposyn II® (Abbott) 10%
Sesame oil	7	100%	Solganal Inj.® (Schering)
Soybean oil	1	10% wt/vol	Diprivan Inj. (Zeneca)
Vegetable oil	2	a	Virilon IM Inj.® (Star Pharmaceuticals)

[a]Not applicable or no data available.
[b]PEG molecular weight not specified.

Oils such as cottonseed, castor, safflower, and soybean have additional specifications if they are used in parenterals. These specifications include saponification value, iodine number, test for unsaponifiable matter, test for free fatty acid, solid paraffin test at 10°C, and acid value. Oils are used to dissolve drugs with low aqueous solubility and provide a mechanism to slowly release drug over a long period of time. In total parenteral nutrition (TPN) products, oils serve as a fat source and as carriers for fat-soluble vitamins. Two important concerns when using fixed oils in injectable products are (*i*) oil degradation, which leads to rancidity and production of free fatty acids, and (*ii*) presence of mineral oil or paraffin that the body cannot metabolize.

Polymeric and Surface Active Compounds

Table 2 includes a broad category of excipients whose functions include the following:

1. Viscosity enhancing and suspending agents such as carboxymethylcellulose, sodium carboxymethylcellulose, acacia, Povidone, hydrolyzed gelatin, sorbitol
2. Solubilizing, wetting, or emulsifying agents such as Cremophore EL, sodium desoxycholate, Polysorbate 20 or 80, PEG 40 castor oil, PEG 60 castor oil, sodium dodecyl sulfate, lecithin, or egg yolk phospholipid
3. Gelling agent such as aluminum monostearate that is added to fixed oil to form a viscous or gel-like suspension medium.
4. Complexing agents such as cyclodextrins

Polysorbate 80 is the most common and versatile solubilizing, wetting, and emulsifying agent. Again, the level of residual peroxides present in polysorbates and protecting them from light and air to prevent further oxidation is an important concern (34). Polysorbate 80 is a polyoxyethylene sorbitan ester of oleic acid (unsaturated fatty acid) while Polysorbate 20 is a polyoxyethylene sorbitan ester of lauric acid (saturated fatty acid). Thus, stability differences in the drug product formulated with Polysorbate 80 versus Polysorbate 20 may occur in some

Table 2 Solubilizing, Wetting, Suspending, Emulsifying, or Thickening Agents

Excipient	Frequency	Range	Example
Acacia	2	7%	Tuberculin Old Test® (Lederle) 7%
Aluminum monostearate	1	2%	Solganal Inj.® (Schering) 2%
Carboxymethylcellulose	4	0.50–0.55%	Bicillin® (Wyeth-Ayerst) 0.55%
Carboxymethylcellulose, sodium (Croscarmellose sodium)	21	0.15–3.0%	Nutropin Depot® (Genentech) 3%
Cremophor EL[a]	3	50–65% wt/vol	Sandimmune® (Sandoz) 65% wt/vol
Cyclodextrin-γ	1	5.0%	Cardiotec (BMS)
Cyclodextrin-α	1	0.14%	Edex (Schwartz)
Hydroxypropyl-β-cyclodextrin	2	16–40%	Sporanox (Janssen)
Sulfobutylether cyclodextrin sodium	3	15–29.4%	Geodon (Pfizer)
Desoxycholate sodium	1	0.4% wt/vol	Fungizone® (Bristol Myers) 0.41% wt/vol
Egg yolk phospholipid	3	1.2%	Cleviprex® (The Medicines Co.)
Gelatin, hydrolzyed	1	16% wt/vol	Cortone® (Merck) 16% wt/vol
Lecithin	8	0.4–1.2% wt/vol	Diprivan® (Zeneca) 1.2% wt/vol
Polyoxyethylated fatty acid	1	7% wt/vol	AquaMephyton® (Merck) 7% wt/vol
Polysorbate 80 (Tween 80)	72	0.001–100%	Taxotere® (Aventis) 100%
Polysorbate 20 (Tween 20)	22	0.001–0.4%	Calcijex® (Abbott) 0.4% wt/vol
PEG 40 castor oil[b]	1	11.5% vol/vol	Monistat® (Janssen) 11.5% vol/vol
PEG 60 castor oil[c]	1	20% wt/vol	Prograf® (Fujisawa) 20% wt/vol
Poloxamer-188 (Pluronic F68)	5	0.005–0.3%	Norditropin (NovoNordisk) 0.3%
Povidone (Polyvinyl pyrrolidone, Crospovidone)	7	0.5–0.6% wt/vol	Bicillin® (Wyeth-Ayerst) 0.6% wt/vol
Sodium dodecyl sulfate (sodium lauryl sulfate)	1	0.018% wt/vol	Proleukin® (Cetus) 0.018% wt/vol
Sorbitol	3	25–50%	Aristrospan® (Fujisawa) 50% vol/vol
Triton X-100 (Octoxynol-9)	1	0.0085% wt/vol	Fluarix® (GSK)

[a]Cremophor EL, Etocas 35, polyethoxylated castor oil, polyoxyethylene 35 castor oil.
[b]PEG 40 castor oil, polyoxyl 40 castor oil, castor oil POE-40, Croduret 40, polyoxyethylene 40 castor oil, Protachem CA-40.
[c]PEG 60 hydrogenated castor oil, Cremophor RH 60, hydrogenated castor oil POE-60, Protachem CAH-60.

cases. This has been noted with Neupogen®, which when exposed to a high concentration of Polysorbate 20 exhibited substantially less oxidation than when exposed to similar concentration of Polysorbate 80 (35). In many other formulation studies with proteins no such advantage of polysorbate could be confirmed as the stability is molecule dependent.

Several new excipients, such as cyclodextrins, are being evaluated to increase the solubility or improve the stability of parenteral drugs. Currently, there are two FDA-approved parenteral products that utilize α- and γ-cyclodextrins. β-Cyclodextrin is unsuitable for parenteral administration because it causes necrosis of the proximal kidney tubules upon IV and SC administration (36). Hydroxypropyl-β-cyclodextrin (HPβCD) and sulfobutylether-β-cyclodextrin (SBE-7-β-CD) have shown the most promise. Captisol™ is the trade name of SBE-7-β-CD and is anionic. Currently, three Captisol-based drug formulations have been approved in the United States. One parenteral formulation is in phase II/III clinical trial that utilizes HPβCD (Cavitron®) and another (Sporanox) has been approved by the FDA. Manufacturers of HPβCD and SBE-7-β-CD have established a Drug Master File (DMF) with the FDA. A detailed review of cyclodextrins has been recently published (37,38). A caution when using cyclodextrin is it can accelerate drug product degradation (39) and can sequester preservatives rendering them ineffective (40).

Chelating Agents

Table 3 lists the relatively few chelating agents that are used in parenteral products. These agents complex heavy metals allowing for improved efficacy of antioxidants or preservatives. Citric acid, tartaric acid, and some amino acids can also act as chelating agents. There has been some misunderstanding that ethylenediaminetetraacetic acid (EDTA) (as calcium or sodium salt) has

Table 3 Chelating Agents

Excipient	Frequency	Range	Example
Calcium disodium EDTA	11	0.01–0.1%	Wydase® (Wyeth-Ayerst) 0.1% wt/vol
Disodium EDTA	48	0.01–0.11%	Calcijex® (Abbott) 0.11% wt/vol
Sodium EDTA	1	0.20%	Folvite® (Lederle) 0.20%
Calcium versetamide sodium	1	2.84% wt/vol	OptiMARK® (Mallinckrodt)
Calteridol	1	0.023% wt/vol	Prohance® (Bracco Diagnostics, Inc.)
DTPA	3	0.04–1.2%	Omniscan™ (GE Healthcare) 1.2%

Abbreviations: EDTA, ethylenediaminetetraacetic acid; DTPA, diethylenetriaminepentaacetic acid, pentetic acid.

not been used in an approved injectable product in Japan. There are some drug products that contain calcium disodium EDTA on market currently in Japan, and this excipient is also listed as an official excipient in Japan (refer to sect. "Special Additives" for details). One possible advantage calcium EDTA has over the tetrasodium salt is that it does not contribute sodium and does not chelate calcium from the blood.

Complexing agents should not be used in metalloprotein formulations, where the protein subunits are held by the metal (41). EDTA, in rare instances, can increase the oxidation rate due to binding of EDTA-metal complex to protein, resulting in site-specific generation of radicals (42).

Antioxidants
Antioxidants are used to prevent the oxidation of active substances and excipients in the finished product and may be categorized into three groups:

1. True antioxidants: They react with free radicals via a chain termination mechanism for example, butylated hydroxytoluene.
2. Reducing agents: They have a lower redox potential than the drug and get preferentially oxidized, for example, ascorbic acid. Thus, they can be consumed during the shelf life of the product.
3. Antioxidant synergists: These enhance the effect of antioxidants, for example, EDTA.

Table 4 summarizes the antioxidants, their frequency of use, concentration range, and examples of product containing them. Sulfite, bisulfite, and metabisulfite constitute the majority of antioxidants used in parenteral products despite several reports of incompatibility and toxicity (43,44). Butylated hydroxy anisole, butylated hydroxy toluene, α-tocopherol, and propyl gallate are primarily used in semi-/nonaqueous vehicles because of their low aqueous solubility (45). Ascorbic acid/sodium ascorbate may serve as an antioxidant, a buffer, and a chelating agent in the same formulation. Some amino acids such as methionine and cysteine also function as effective antioxidants.

The Committee for Proprietary Medicinal Products (CPMP) guideline calls for a full explanation and justification for including antioxidants in the formulation (46,47). Specific evidence must be provided that their use cannot be avoided and its concentration must be justified in terms of efficacy and safety. Thus, it is imperative to first try an inert gas such as nitrogen or argon in the head space to prevent oxidation. Antioxidants such as sulfites and metabisulfites are especially undesirable.

Some antioxidants such as propyl gallate and butylated hydroxy anisole possess antimicrobial properties. Compatibility of antioxidants with the drug, packaging system, and the body should be studied carefully. For example, tocopherols may be absorbed on to plastics; ascorbic acid is incompatible with alkalis and oxidizing materials such as phenylephrine; propyl gallate forms complexes with metal ions such as sodium, potassium, and iron.

Preservatives
Benzyl alcohol is the most common antimicrobial preservative present in parenteral formulations (Table 5). This observation is consistent with other surveys (12,48). Parabens are the second most common preservatives. Surprisingly, thimerosal is also common,

Table 4 Antioxidants and Reducing Agents

Excipient	Frequency	Range	Example
Acetone sodium bisulfite	4	0.2–0.4% wt/vol	Novocaine® (Sanofi-Winthrop) 0.4% wt/vol
Argon	–	100%	Used to fill headspace of lyophilized or liquid products. TechneScan MAG3® (Covidien)
Ascorbyl palmitate	1		Visudyne® (QLT)
Ascorbate (sodium/acid)	10	0.1–4.8% wt/vol	Vibramycin® (Pfizer) 4.8% wt/vol
Bisulfite sodium	31	0.02–0.66% wt/vol	Amikin® (Bristol Myers) 0.66% wt/vol
Butylated hydroxy anisole (BHA)	3	0.00028–0.03% wt/vol	Aquasol A® (Astra) 0.03% wt/vol
Butylated hydroxy toluene (BHT)	4	0.00116–0.03% wt/vol	Aquasol A (Astra) 0.03% wt/vol
Cystein/Cysteinate HCl	4	0.07–1.3% wt/vol	Acthrel® (Ferring) 1.3% wt/vol
Dithionite sodium (sodium hydrosulfite, sodium sulfoxylate)	1	0.10%	Numorphan® (Endo Lab) 0.10%
Gentisic acid	1	0.02% wt/vol	OctreoScan® (Mallinckrodt) 0.02% wt/vol
Gentisic acid ethanolamine	1	2%	M.V.I. 12® (Astra) 2%
Glutamate monosodium	1	0.1% wt/vol	Varivax® (Merck) 0.1% wt/vol
Glutathione	1	0.01% wt/vol	Advate® (Baxter) 0.01% wt/vol
Formaldehyde sulfoxylate sodium	10	0.075–0.5% wt/vol	Terramycin solution (Pfizer) 0.5% wt/vol
Metabisulfite potassium	1	0.10%	Vasoxyl® (Glaxo-Wellcome) 0.10%
Metabisulfite sodium	33	0.02–1% wt/vol	Intropin® (DuPont) 1% wt/vol
Methionine	5	0.01–0.15%	Depo-subQ provera 104 (Upjohn)
Monothioglycerol (thioglycerol)	8	0.1–1%	Terramycin solution (Pfizer) 1%
Nitrogen	–	100%	Used to fill headspace of lyophilized or liquid products
Propyl gallate	3	0.02%	Navane® (Pfizer) 0.02%
Sulfite sodium	8	0.05–0.2% wt/vol	Enlon® (Ohmeda) 0.2% wt/vol
α-Tocopherol	2	0.005–0.075%	Torisel (Wyetth) 0.075%
α-Tocopherol hydrogen succinate	1	0.02% wt/vol	Fluarix® (GSK) 0.02% wt/vol
Thioglycolate sodium	1	0.66% wt/vol	Sus-Phrine® (Forest) 0.66% wt/vol

Table 5 Antimicrobial Preservatives

Excipient	Frequency	Range	Example
Benzalkonium chloride	1	0.02% wt/vol	Celestone Soluspan® (Schering) 0.02% wt/vol
Benzethonium chloride	4	0.01%	Benadryl® (Parke-Davis) 0.01% wt/vol
Benzyl alcohol	90	0.75–5%	Dimenhydrinate Inj., USP (APP Pharmaceuticals) 5%
Chlorbutanol	19	0.25–0.5%	Codine phosphate (Wyeth-Ayerst) 0.5%
m-Cresol	13	0.1–0.315%	Humalog® (Lilly) 0.315%
Myristyl γ-picolinium chloride	2	0.0195–0.169% wt/vol	Depo-Provera® (Pharmacia-Upjohn) 0.169% wt/vol
Paraben methyl	55	0.05–0.18%	Inapsine® (Janssen) 0.18% wt/vol
Paraben propyl	45	0.005–0.1%	Xylocaine w/epinephrine (Astra) 0.1% wt/vol
Phenol	55	0.15–0.5%	Calcimar® (Rhone-Poulanc) 0.5% wt/vol
2-Phenoxyethanol	4	0.50%	Havrix® (SmithKline Beecham) 0.50% wt/vol
Phenyl mercuric nitrate	3	0.001%	Antivenin® (Wyeth-Ayerst) 0.001%
Thimerosal	50	0.003–0.012%	Atgam® (Pharmacia-Upjohn) 0.01%

especially in vaccines, even though some individuals are sensitive to mercurics. Several preservatives can volatilize easily (e.g., benzyl alcohol and phenol) and therefore should not be used in a lyophilized dosage form. Chlorocresol is purported to be a good preservative for parenterals, but our survey did not find any examples of commercial products containing chlorocresol. The British Pharmaceutical Codex and Martindale list chlorocresol to be used as a

Table 6 Maximum Permissible Amount of Preservatives and Antioxidants

Excipient	Maximum limit in USP (%)
Mercurial compounds	0.01
Cationic surfactants	0.01
Chlorobutanol	0.50
Cresol	0.50
Phenol	0.50
Sulfur dioxide or an equivalent amount of the sulfite, bisulfite, or metabisulfite of potassium or sodium	0.20

preservative in multidose aqueous injections at concentration of 0.1% but no examples of injectable products have been provided (49,50).

Antimicrobial preservatives are allowed in multidose injections to prevent growth of microorganisms that may accidentally enter the container during withdrawal of the dose. In the United States, preservatives are discouraged from being used in single-dose injections, while Ph. Eur. and BP allow aqueous preparations that are manufactured using aseptic techniques to contain suitable preservatives. However, Ph. Eur. and BP prohibit antimicrobials from single-dose injections where the dose volume is greater than 15 mL or if the drug product is to be injected via intracisternal or any route (e.g., retro-ocular) that gives access to the cerebrospinal fluid (CSF). It is imperative that preservatives should never be used as a substitute for inadequate cGMP (current Good Manufacturing Practices). The primary reason for minimizing the use of antimicrobial preservatives is toxicity. For example, many individuals are allergic to mercury preservatives, and benzyl alcohol is contraindicated in children under the age of two years. The USP has also placed some restrictions on the maximum concentration of preservatives allowed in the formulation to address toxicity and allergic reactions (Table 6). The World Health Organization has set an estimated total acceptable daily intake for sorbate (as acid, calcium, potassium, and sodium salts) as not more than 25 mg/kg body weight. Recently, concerns have been raised on the safety of parabens in pediatric formulations based primarily on reports by one Japanese laboratory between 2001 and 2004, which indicated effects on reproductive apparatus of juvenile male rats given propyl (51) or butyl paraben (52), but lack of effects for methyl and ethyl parabens (53). However, toxicological data suggests otherwise (54). Until a comprehensive assessment is done, formulators should take into account current view of the regulatory agencies (e.g., Agence Francaise de Securite Sanitaire des Produits de Sante (AFSSAPS), Scandinavian, etc.) and may opt for other preservative options for pediatric products.

Preservative efficacy is assessed during product development using Antimicrobial Preservative Effectiveness Testing, PET (55–57). Satisfactory PET results on finished aqueous-preserved parenteral product in the commercial package can be used up to a maximum of 28 days after the container has been opened (58). Unpreserved product should preferably be used immediately following opening, reconstitution, or dilution.

Similar to antioxidants, addition of preservatives in medicinal products requires justification. Wherever possible, their use should be avoided, particularly for pediatric products, but if required, minimal concentrations should be determined (47).

Buffers

Buffers are added to a formulation to adjust the pH to optimize solubility and stability. For parenteral preparations, it is desirable to target the pH of the product to physiological pH. Consideration of the buffer concentration (ionic strength) and the buffer species is important. For example, citrate buffers in the range of 5 to 15 mM are used in the formulations, but increasing the buffer concentration to >50 mM will result in excessive pain on subcutaneous injection and toxic effects because of the chelation of calcium in the blood if large volumes of product are injected.

Buffers have maximum buffer capacities near their pK_a. It is important to select buffers with a small $\Delta pK_a/°C$ for products that may be subjected to excessive temperature fluctuations during processing such as steam sterilization, thermal cycling, or lyophilization. Tris, whose

$\Delta pK_a/^\circ C$ is large ($-0.028/^\circ C$), the pH of buffer, made at 25°C will change from 7.1 to 5.0 at 100°C, which could dramatically alter the stability or solubility of the drug. Similarly, the preferred buffers for a lyophilized product may be those that show the least pH change upon cooling, do not crystallize out, and will remain in the amorphous state protecting the drug. For example, replacing succinate with glycolate buffer improves the stability of lyophilized interferon-γ (59). During the lyophilization of mannitol containing succinate buffer at pH 5, monosodium succinate crystallizes reducing the pH and resulting in the unfolding of interferon-γ. This pH shift is not seen with glycolate buffer.

Table 7 lists buffers and chemicals used to adjust the pH of parenteral formulations and maintain the product pH range. The most common buffers used are phosphate, citrate, and acetate. Mono- and di-ethanolamines are added to adjust pH and form corresponding salts. Hydrogen bromide, sulfuric acid, benzene sulfonic acid, and methane sulfonic acids are added to drugs that are salts of bromide (Scopolamine HBr, Hyoscine HBr), sulfate (Nebcin, Tobramycin Sulfate), besylate (Tracrium Injection, Atracurium besylate), or mesylate (DHE 45 Injection, dihydroergotamine mesylate). Glucono-δ-lactone is used to adjust the pH of Quinidine Gluconate. Benzoate buffer, at a concentration of 5%, is used in Valium Injection. Citrates are a common buffer that can serve a dual role as chelating agents. Amino acids, histidine, arginine, aspartic, and glycine, function as buffers and stabilize proteins and peptide formulations. These amino acids are also used as lyo-additives and may prevent cold denaturation. Lactate and tartrate are occasionally used as buffer systems. Acetates are good buffers at low pH, but they are not generally used for lyophilization because of the potential sublimation of acetates.

Bulking Agents, Protectants, and Tonicity Adjusters

Table 8 lists additives that are osmolality adjusters and bulking or lyo-/cryoprotective agents. The most common tonicity adjusters are dextrose and sodium chloride. Additives that serve as lyophilization bulking agents and also as stabilizers and/or as buffers include some amino acids such as glycine, alanine, histidine, imidazole, arginine, asparagine, and aspartic acid. Other commonly used lyo-additives are monosaccharides (dextrose, glucose, maltose, lactose), disaccharides (sucrose, trehalose), polyhydric alcohols (inositol, mannitol, sorbitol), glycols (PEG 3350), Povidone (polyvinylpyrrolidone, PVP), and proteins (albumin, gelatin). Hydroxyethyl starch (hetastarch) and pentastarch, which are currently used as plasma expanders in commercial injectable products such as Hespan and Pentaspan, are also being evaluated as protectants during freeze-drying of proteins.

PVP has been used in injectable products (except in Japan) as a solubilizing agent, a protectant, and as a bulking agent. Only pyrogen-free grade, with low molecular weight (K-value less than 18), should be used in parenteral products to allow for rapid renal elimination. PVP not only solubilizes drugs such as rifampicin, but it may also reduce the local toxicity as seen in case of Oxytetracycline Injection.

Protein stabilization in the lyophilized state can be achieved if the stabilizer and protein do not phase separate during freezing or the stabilizer does not crystallize out. For Neupogen (GCSF), mannitol was replaced with sorbitol in the formulation to prevent the loss of activity of the liquid formulation upon accidental freezing (35). If the solution freezes, mannitol crystallizes while sorbitol will remain in an amorphous state protecting GCSF. However, a recent report suggests that sorbitol can also crystallize under certain conditions (60). Similarly, it is preferred that the drug remains dispersed in the stabilizer upon freezing of the solution to maximize protection. For example, cefoxitin, a cephalosporin, is more stable when freeze-dried with sucrose than with trehalose. Although the glass transition temperature and structural relaxation time is much greater for trehalose than sucrose (61), FTIR data indicates that the trehalose-cefoxitin system phase separates into two nearly pure components resulting in no protection (stability). Similarly, sucrose was found to be a better cryoprotectant than dextran for protein because dextran and protein underwent phase segregation as the solution started to freeze. The mechanism of cryoprotection in the solution has been explained by the preferential exclusion hypothesis (62).

Trehalose is a nonreducing disaccharide composed of two D-glucose monomers. It is found in some plants and animals that can withstand dehydration (anhydrobiosis) and

Table 7 Buffers and pH Adjusting Agents

Excipient	pH range	Example
Acetate		
Sodium	2.5–7.0	Syntocinon® (Novartis)
Acetic acid	2.5–7.2	Syntocinon (Novartis)
Glacial acetic acid	3.5–7.0	Brevibloc® (Ohmeda)
Ammonium	6.8–7.8	Bumex Injection® (Roche)
Ammonium sulfate	–	Innovar® (Astra)
Ammonium hydroxide	–	Triostat® (Jones Medical)
Arginine	7.0–7.4	Retavase® (Boehringer)
Aspartic acid	5.0–5.6	Pepcid® (Merck)
Benzene sulfonic acid	3.25–3.65	Nimbex® (Glaxo Wellcome)
Benzoate sodium/acid	3.5–6.9	Valium® (Roche)
Bicarbonate, sodium	5.5–11.0	Cenolate® (Abbott)
Boric acid/sodium		Comvax® (Merck)
Carbonate, sodium	4.0–11.0	Hyperab® (Bayer)
Carbon dioxide	–	Serentil® (Boehringer) Used to fill headspace
Citrate		
Acid	2.5–9.0	DTIC-Dome® (Bayer)
Sodium	3.0–8.5	Amikin® (Bristol Myers)
Disodium	6.1	Cerezyme® (Genzyme)
Trisodium	6.1	Cerezyme® (Genzyme)
Diethanolamine	9.5–10.5	Bactim IV® (Roche)
Glucono δ-lactone	5.5–7.0	Quinidine Gluconate (Lilly)
Glycine/Glycine HCl	2.5–10.8	Hep-B Gammagee® (Merck)
Histidine/Histidine HCl	5.0–6.5	Doxil® (Sequus)
Hydrochloric acid	Broad range	Amicar® (Immunex)
Hydrobromic acid	3.5–6.5	Scopolamine (UDL)
Lactate sodium/acid	2.7–5.8	Innovar® (Janssen)
Lysine (L)	–	Eminase® (Roberts)
Maleic acid	3.0–5.0	Librium® (Roche)
Meglumine	6.5–11.0	Magnevist® (Berlex)
Methanesulfonic acid	3.2–4.0	DHE-45® (Novartis)
Monoethanolamine	8.0–9.0	Terramycin (Pfizer)
Phosphate		
Acid	6.5–8.5	Saizen® (Serono Labs)
Monobasic potassium	6.7–7.3	Zantac® (Glaxo-Wellcome)
Dibasic potassium	6.7–7.3	Aminosyn® (Hospira)
Monobasic sodium[a]	2.5–8.0	Pregnyl® (Organon)
Dibasic sodium[b]	2.5–8.3	Zantac® (Glaxo-Wellcome)
Tribasic sodium	–	Synthroid® (Knoll)
Sodium hydroxide	Broad range	Optiray® (Mallinckrodt)
Succinate sodium/disodium	5.0–6.0	AmBisome® (Fujisawa)
Sulfuric acid	3.0–7.0	Nebcin® (Lilly)
Tartrate sodium/acid	2.5–6.2	Methergine® (Novartis)
Tromethamine (Tris)	6.5–9.0	Optiray® (Mallinckrodt)

[a]Sodium biphosphate, sodium dihydrogen phosphate, or sodium dihydrogen orthophosphate.
[b]Sodium phosphate, disodium hydrogen phosphate.

therefore had been suggested to stabilize drugs that undergo denaturation during spray or freeze-drying (63). Herceptin® (trastuzumab) is a recombinant DNA-derived monoclonal antibody (MAb) used for treating metastatic breast cancer. The MAb is stabilized in the lyophilized formulation using α,α-trehalose dihydrate. Trehalose is also used as a cryoprotectant to prevent liposomal aggregation and leakage. In the dried state, carbohydrates such as trehalose and inositol exert their protective effect by acting as a water substitute (64).

Formulations may require additives for specific gravity adjustments, especially for drugs which, upon administration, may come in contact with CSF. CSF has a specific gravity of 1.0059 at 37°C. Solutions with the same specific gravity as that of CSF are termed isobaric while those with a specific gravity greater than that of CSF are called hyperbaric. Upon

Table 8   Bulking Agents, Protectants, and Tonicity Adjusters

Excipient	Example
Alanine	Thrombate III® (Bayer)
Albumin	Bioclate® (Arco)
Albumin (human)	Botox® (Allergan)
Amino acids	Havrix® (Smith Kline Beecham)
Arginine (L)	Activase® (Genentech)
Aspargine	Tice BCG® (Organon)
Aspartic acid (L)	Pepcid® (Merck)
Calcium chloride	Xyntha® (Wyeth)
Cyclodextrin-α	Edex® (Schwartz)
Cyclodextrin-γ	Cardiotec® (Squibb)
Dextran 40	Etopophos® (Bristol Myers)
Dextrose	Betaseron® (Berlex)
Gelatin (cross-linked)	Kabikinase® (Pharmacia-Upjohn)
Gelatin (hydrolyzed)	Acthar® (Rhone-Poulanc Rorer)
Lactic and glycolic acid copolymers	Lupron Depot® (TAP)
Glucose	Iveegam® (Immuno-US)
Glycerine	Tice BCG® (Organon)
Glycine/glycine hydrochloride	Atgam® (Pharmacia-Upjohn)
Histidine	Antihemophilic Factor, human (American Red Cross)
Imidazole	Helixate® (Armour)
Inositol	OctreoScan® (Mallinckrodt)
Lactose	Caverject® (Pharmacia-Upjohn)
Magnesium chloride	Terramycin Solution (Pfizer)
Magnesium sulfate	Tice BCG® (Organon)
Maltose	Gamimune N® (Bayer)
Mannitol	Elspar® (Merck)
Polyethylene glycol 3350	Bioclate® (Arco)
Polylactic acid	Lupron Depot® (TAP)
Potassium chloride	Varivax® (Merck)
Povidone	Alkeran® (Glaxo-Wellcome)
Sodium chloride	WinRho SD® (Univax)
Sodium cholesteryl sulfate	Amphotec® (Sequus)
Sodium succinate	Actimmune® (Genentech)
Sodium sulfate	Depo-Provera® (Pharmacia-Upjohn)
Sorbitol	Panhematin® (Abbott)
Sucrose	Prolastin® (Bayer)
L-Threonine	Temodar® (Schering)
Trehalose (α, α)	Herceptin® (Genentech)

administration of a hyperbaric solution in the spinal cord, the injected solution will settle and affect spinal nerves at the end of the spinal cord. For example, dibucaine hydrochloride solution (Nupercaine® 1:200) is isobaric, while Nupercaine 1:500 is hypobaric (specific gravity of 1.0036 at 37°C). Nupercaine heavy solution is made hyperbaric by addition of 5% dextrose solution. This solution will block (anesthetize) the lower spinal nerves as it settles down in the spinal cord.

Special Additives
Table 9 lists special additives that have been included in pharmaceutical formulations to serve specific functions. Some of the special additives are summarized in the following along with their intended use:

1. Calcium D-saccharate tetrahydrate 0.46% wt/vol is used in Calcium Gluconate Injection, a saturated solution of 10% wt/vol, to prevent crystallization during temperature fluctuations.

Table 9 Special Additives

Excipient	Example
Acetyl tryptophanate	Human Albumin (American Red Cross)
Aluminum hydroxide	Recombivax HB® (Merck)
Aluminum phosphate	Tetanus Toxoid Adsorbed (Wyeth)
Aluminum potassium sulfate	TD Adsorbed Adult (Pasteur Merieux)
Amino acids [leucine, isoleucine, lysine (as acetate or HCl salt), valine, phenylalanine, threonine, tryptophan, alanine, aspartic acid, glutamic acid, proline, serine, tyrosine, taurine]	Travasol 10% Injection (Baxter) Aminosyn-PF10% (Hospira)
ε-Aminocaproic acid	Eminase® (Roberts)
Calcium D-saccharate	Calcium Gluconate (American Regent)
Caprylate sodium	Human Albumin (American Red Cross)
8-Chlorotheophylline	Dimenhydrinate® (Steris)
Creatine	Dalalone DP® (Forest)
Creatinine	Decadron® (Merck)
Cholesterol	Doxil® (Sequus)
Cholesteryl sulfate sodium	Amphotec® (Sequus)
Cyclohexanedione dioxime	Cardiotec® (BMS)
Diethanolamine	Bactrim® IV Infusion (Roche)
Distearyl phosphatidylcholine	DaunoXome® (Nexstar)
Distearyl phosphatidylglycerol	MiKasome® (NeXstar)
L-α-Dimyristoylphosphatidylcholine	Abelcet® (The Liposome Co.)
L-α-Dimyristoylphosphatidylglycerol	Abelcet (The Liposome Co.)
Dioleoylphosphatidylcholine (DOPC)	DepoCyt® (Chiron)
Dipalmitoylphosphatidylglycerol (DPPG)	DepoCyt (Chiron)
(R)-hexadecanoic acid, 1-[(phosphonoxy)methyl]-1,2-ethanediyl ester, monosodium salt (DPPA)	Definity® (Lantheus Medical Imaging)
(R)-4-hydroxy-N,N,N-trimethyl-10-oxo-7-[(1oxohexadecyl)oxy]-3,4,9-trioxa-4-phosphapentacosan-1-aminium, 4-oxide, inner salt (DPPC)	Definity (Lantheus Medical Imaging)
(R)-[6-hydroxy-6-oxido-9-[(1-oxohexadecyl)oxy]-5,7,11-trioxa-2aza-6-phosphahexacos-1-yl]-ω-methoxypoly(ox-1, 2-ethanediyl), monosodium salt (MPEG5000 DPPE)	Definity (Lantheus Medical Imaging)
MPEG-distearoyl phosphoethanolamine	Doxil® (Sequus)
Ethyl lactate	Ergotrate maleate (Lilly)
Ethylenediamine	Aminophylline (Abbott)
L-Glutamate sodium	Kabikinase® (Pharmacia-Upjohn)
Hyaluronate sodium	Trivaris® (Allergan)
Hydrogenated soy phosphatidylcholine	Doxil® (Sequus)
Iron ammonium citrate	Tice BCG® (Organon)
Lactic acid	Cipro IV® (Bayer)
D,L-Lactic and glycolic acid copolymer	Zoladex® (Zeneca)
Meglumine	Magnevist® (Berlex)
Methyl boronic acid	Cardiotec® (BMS)
Niacinamide	Estradurin® (Wyeth-Ayerst)
Paraben methyl	Adriamycin RDF® (Pharmacia-Upjohn)
Phosphatidylglycerol, egg (EPG)	Visudyne® (QLT)
Potassium sodium tartrate	CEA-Scan® (Immunomedics)
Protamine (as sulfate)	Insulatard NPH® (Novo Nordisk)
Simethicone	Premarin Injection® (Wyeth-Ayerst)
Saccharin sodium	Compazine Injection® (Smith Kline Beecham)
Sodium D-gluconate	Myoview® (Amersham)
Sodium hypochlorite	Ultratag™ RBC (Covidien)
Sodium sulfate	Depo-Provera® (Pfizer)
Stannous chloride	Myoview® (Amersham)
Sulfosalicylate disodium	Myoview® (Amersham)
Tin chloride (stannous and stannic)	Ultratag™ RBC (Covidien)
Tri-n-butyl phosphate	Venoglobulin® (Alpha Therapeutic)
Tricaprylin	DepoDur® (SkyePharma)
Triolein	DepoCyt® (Chiron)
von Willebrand factor	Bioclate® (Arco)
Zinc	Lente Insulin® (Novo Nordisk)
Zinc acetate	Nutropin Depot® (Genentech)
Zinc carbonate	Nutropin Depot® (Genentech)
Zinc oxide	Humalog® (Lilly)

2. Lactic acid is used in Cipro IV® as a solubilizing agent for the antibiotic.
3. Simethicone is used in the lyophilized product Premarin Injection® to prevent the formation of foam during reconstitution.
4. Creatine or creatinine are used in the dexamethasone formulations Dexamethasone acetate and Dexamethasone sodium phosphate, which are available as a suspension or a solution.
5. Methyl paraben (0.2 mg/mL) is used in Adriamycin RDF® to increase dissolution (65).
6. 0.1% Ethyl lactate is used in Ergotrate maleate as a solubilizing agent.
7. Niacinamide (12.5 mg/mL) is used in Estradurin Injection® as a solubilizing agent. Hydeltrasol® also contains niacinamide. The concept of hydrotropic agents to increase water solubility has been tried on several compounds including proteins (66,67).
8. Aluminum, in the form of aluminum hydroxide, aluminum phosphate, or aluminum potassium sulfate, is used as adjuvant in various vaccine formulations to elicit an increased immunogenic response.
9. Lupron Depot® Injection is lyophilized microspheres of gelatin and glycolic-lactic acid for intramuscular injection. Nutropin Depot® consists of polylactate-glycolate microspheres.
10. Sodium caprylate (sodium octoate) has antifungal properties, but it is also used to improve the stability of albumin solution against the effects of heat. Albumin solution can be heat pasteurized by heating at 60°C for 10 hours in the presence of sodium caprylate. Acetyl tryptophanate sodium is also added to albumin formulations.
11. Meglumine (*N*-methylglucamine) is used to form in situ salt. For example, diatrizoic acid, an X-ray contrast agent, is more stable when autoclaved as meglumine salt than as sodium salt (68). Meglumine is also added to Magnevist®, a magnetic resonance contrast agent.
12. Tri-*n*-butyl phosphate is present as an excipient in human immune globulin solution (Venoglobulin®). Its exact function in the formulation is not known, but it may serve as a scavenging agent.
13. von Willebrand factor is used to stabilize recombinant antihemophilic factor (Bioclate®).
14. Epsilon amino caproic acid (6-amino hexanoic acid) is used as a stabilizer in anistreplase (Eminase Injection®).
15. Zinc, zinc acetate, zinc carbonate, and protamine have been added to growth hormone and insulin to form complexes and control the duration of action.
16. Lipids (natural or modified) and cholesterols are used in liposomes and lipid complexes (e.g., PC, DMPC, DMPG, DOPC, DPPG, DSPC, DSPG, DPPA, DPPC, MPEG5000DPPE).
17. Several amino acids are used as either stabilizers or as part of amino acid solution for parenteral nutrition.

The FDA has published the "Inactive Ingredient Guide," which lists excipients in alphabetical order (20). The Inactive Ingredient database is reasonably comprehensive and is updated on a quarterly basis, but it does not include several excipients used in recently approved drug products. Each listed ingredient is followed by the route of administration, the CAS #, the UNII #, and in some cases, the range of concentration used in the approved drug product. However, this list does not provide the name of commercial product(s) corresponding to each excipient. Table 10 lists excipients that are included in the FDA database but were not found in our survey.

Similarly, in Japan the "Japanese Pharmaceutical Excipients Directory (JPED)" is published by the Japanese Pharmaceutical Excipients Council, with the cooperation and

Table 10 Excipients Listed in the FDA Inactive Ingredient Guide but Not Found in Our Survey

Excipient	Route/dosage form	Conc. (%)
Acetic anhydride	IV; injection	
Acetylated monoglycerides	IV; injection	
Activated charcoal	IM; injectable	0.3
Adipic acid	IM; injection	1
Alcohol, denatured	IV; injection	
Benzyl chloride	IV; injection	1.00E-03
Bibapcitide	IV; injection	0.01
Brocrinat	IV(infusion); injection	
Calcium gluceptate	IV; injection	5
Calcium hydroxide	IV; injection	0.37
Caldiamide sodium	IV; solution, injection	1.2
Calteridol calcium	IV; injection	0.02
Cellulose, microcrystalline	IV; injection	14.9
Corn oil	IM; injection	
Deoxycholic acid	IV(infusion); powder, for injection solution	
Diatrizoic acid	IM; injection	59.7
Dimethyl sulfoxide	IV(infusion); powder, for injection suspension, lyophilized	
Dimyristoyl lecithin	IV; powder, for injection solution, lyophilized	7.05
Disofenin	IV(infusion); injection	2
Docusate sodium	IM; injection	0.01
Edetic acid	Submucosal; solution, injection	0.05
Ethanolamine hydrochloride	IV; injection	0.15
Ethyl acetate	IM; injection	
Exametazime	IV; injection	
Fampridine	IV; injection	
Ferric chloride	IV; injection	6.05
Fructose	IV(infusion); powder, for injection solution, lyophilized	5
Gadolinium oxide	IV; injection	
Gluceptate sodium	IV; powder, for injection solution	20
Gluceptate sodium dihydrate	IV; injection	7.5
Glucuronic acid	IV; injection	
Glycocholic acid	IV; powder, for injection solution, lyophilized	14
Guanidine hydrochloride	IV; injection	0.25
Hydroxyethylpiperazine ethane sulfonic acid	IV; injection	
Insulin beef	SC; injection	0.1
Insulin pork	SC; injection	0.1
Iodine	IV; injection	
Iodoxamic acid	IV; injection	31
Iofetamine hydrochloride	IV; injection	
Isopropyl alcohol	IV; injection	
Lactobionic acid	IV(infusion); powder, for injection solution	
Lecithin, egg	IV; injectable	1.2
Lidofenin	IV; injection	
Magnesium stearate	Implantation; injection	1.50E-03
Mebrofenin	IV; injection	
Medronate disodium	IV; injection	1
Medronic acid	IV; injection	2.5
Metaphosphoric acid	IV(infusion); injection	0.13
Methylcellulose	Intra-articular; injection	0.1
Methylene blue	IV; injection	1
N-(carbamoyl-methoxy PEG-40)-1,2-distearoyl-cephalin sodium	IV; injection, suspension, liposomal	0.31
Nioxime	IV; injection	0.2
Octanoic acid	IV; injection	0.01
Oxidronate sodium	IV; injection	0.2
Oxyquinoline	IV; injection	5.00E-03
PEG sorbitan isostearate	IM; injection	
PEG vegetable oil	IM, SC; injection	7

(continued)

Table 10   Excipients Listed in the FDA Inactive Ingredient Guide but Not Found in Our Survey (*Continued*)

Excipient	Route/dosage form	Conc. (%)
Pentetate calcium trisodium	Intrathecal; injection	
Pentetate pentasodium	IV; injection	0.5
Perflutren	IV; injection	
Polysiloxane	IV; injectable	
Polysorbate 40	IM, IV; injection	
Polyvinyl alcohol	IM; injection, microspheres	
Potassium bisulfite	IV; injection	
Potassium hydroxide	IV; injection	
Silicone	IM, IV; injection	
Sodium bisulfate	IM, IV; injection	0.32
Sodium chlorate	IV(infusion); injection	15.4
Sodium cysteinate hydrochloride	Intradiscal; powder, for injection solution	
Sodium iodide	IV; powder, for injection solution	5
Sodium pyrophosphate	IV; injection	1.2
Sodium thiomalate	IM, IV; injection	
Sodium thiosulfate	IV; solution	
Sodium thiosulfate anhydrous	IV; solution	0.19
Sodium trimetaphosphate	IV; powder, for injection solution	
Sorbitan monopalmitate	IM; injection	
Stannous fluoride	IV; injection	0.07
Stannous tartrate	IV; injection	8.00E-03
Starch	IM; injection	0.6
Succimer	IV; injection	
Succinic acid	IM, IV; injection	
Sulfur dioxide	IV(infusion); solution, injection	0.15
Sulfurous acid	IM; injection	
Tetrakis(1-isocyano-2-methoxy-2-methyl-propane)-copper(i) tetrafluoroborate	IV; injection	0.1
Tetrofosmin	IV(infusion); powder, for injection solution, lyophilized	0.02
Theophylline	IV(infusion); injection	
Trifluoroacetic acid	IV(infusion); powder, for injection solution, lyophilized	
Urea	IM; injection	
Zinc chloride	Intradermal; injection	0.04

guidance of the Ministry of Health, Labour and Welfare (69). This directory divides the excipients into the following two categories:

1. Official excipients—those that have been recognized in the Japanese Pharmacopoeia (JP), Japanese Pharmaceutical Codex, and Japanese Pharmaceutical Excipients (JPE) and for which testing methods and standards have been determined.
2. Nonofficial excipients—those that have been used in pharmaceutical products sold in Japan and are planned to be included in the official book or in supplemental editions.

JPED lists the excipients, the route of administration, and the maximum amount or concentration that has been approved. An excipient used within the listed concentration limits is considered "precedented" and no additional data is needed. If the excipient concentration is outside of the limits, additional safety info may be needed (experimental or published literature). Unprecedented or novel excipients may have to be placed on stability and quality standards developed. If the excipient is listed in JP, JPE, or JPED, it must meet the specifications listed in the monograph; however, if the excipient is not listed in any of the above three books then USP, Ph. Eur., or other pharmacopoeial standards should be used. For excipients not listed in pharmacopoeias, in-house specifications are used.

REGULATORY PERSPECTIVE

The International Pharmaceutical Excipients Council (IPEC) has classified excipients into four classes on the basis of the safety testing information available (70).

1. New chemical excipients: require a full safety evaluation program. EU directive 75/318/EEC states that new chemical excipients will be treated in the same way as new actives. Safety studies for a new chemical excipient have been estimated to cost about $3.5 million over four to five years. In the United States, relevant information from these safety studies will need to be filed with the FDA in a DMF, and in Europe a dossier needs to be established. The IPEC Europe has issued a guideline (Compilation of Excipient Masterfiles Guidelines) that provides guidance to excipient producers on how to construct a dossier that will support MAA (Marketing Authorization Application) while maintaining the confidentiality of the data.

2. Existing chemical excipient—first use in man: implies that animal safety data exist since it may have been used in some other application. Additional safety information may have to be gathered to justify its use in humans.

3. Existing chemical excipient: indicates that it has been used in humans but change in route of administration (say from oral to parenteral), new dosage form, higher dose, etc. may require additional safety information.

4. New modifications or combinations of existing excipients: a physical interaction NOT a chemical reaction. No safety evaluation is necessary in this case.

Formulators should understand that just because an excipient is listed as GRAS does not mean that it can be used in a parenteral dosage form. The GRAS list includes materials that have been demonstrated as safe for food (oral administration) but have not necessarily been deemed safe for use in an injectable product. Therefore, additives included in this list are of very limited value for selecting excipients for parenteral formulations.

The USP, JP, Ph. Eur., BP, and other pharmacopoeias may have monographs for identical excipients, which differ considerably with regards to specifications, test criteria, and analytical methods. This presents a significant testing burden on a pharmaceutical manufacturer intending to supply a global product because they will have to perform testing on the same excipient numerous times to meet the various compendial specifications. Under the auspices of the Pharmacopoeial Discussion Group (PDG), there is ongoing harmonization of excipient monographs. PDG has been working on several commonly used excipients to achieve a single monograph for each excipient. Presently, 26 General Chapters and 40 excipient monographs have been harmonized (stage 6 of the process). For example, benzyl alcohol undergoes degradation by a free radical mechanism to form benzaldehyde and hydrogen peroxide. The degradation products are much more toxic than the parent molecule. The USP, JP, and Ph. Eur. require three different chromatographic systems to test for organic impurity (mainly benzaldehyde). The harmonized monograph of benzyl alcohol has eliminated unnecessary repetition, which does not contribute to the overall quality of the product (71).

In addition to testing and specifications, regulatory bodies are also focusing their attention on excipient manufacturing processes. There have been major initiatives on the part of IPEC to improve the quality of additives, which has resulted in a publication titled "Good Manufacturing Practices Guide for Bulk Pharmaceutical Excipients" (72). The excipients may be manufactured for food, cosmetic, chemical, agriculture, or pharmaceutical industries, and the requirement for each industry is different. The purpose of this guide is to develop a quality system framework that may be used for excipient suppliers, which will be acceptable to the pharmaceutical industry, and to harmonize the requirements in the United States, Europe, and Japan.

The United States and Europe require all excipients to be declared, along with their quantity, on the label if the product is an injectable preparation. In Japan, only the names of excipients are required in the labeling (information that is included with the product like the package insert); however, information of the quantity of each excipient is not required on the label. EU Article 54(c) requires that all excipients need to be declared on the labeling if the medicinal product is an injectable or a topical or an eye preparation. The European guide for the label and package leaflet also lists excipients, which have special issues, and are addressed in an Annex (73). Table 11 is a summary of some of these ingredients that are commonly used as parenteral excipients and the corresponding safety information that should be included in the leaflet. The package leaflet must include a list of information on those excipients, knowledge of which is important for the safe and effective use of the medicinal product.

Table 11   Excipients for Label and Corresponding Information for Leaflet

Name	Threshold level	Information for the package leaflet
Arachis oil	Zero	Whenever arachis oil appears, peanut oil should appear besides it.
		If you are allergic to peanut or soya, do not use this medicinal product
Benzoic acid and benzoates	Zero	It may increase the risk of jaundice in newborn babies
Benzyl alcohol	Exposures less than 90 mg/kg/day	Must not be given to premature babies or neonates. May cause toxic reactions and allergic reactions in infants and children up to 3 yr old.
	90 mg/kg/day	Must not be given to premature babies or neonates.
		Due to the risk of fatal toxic reactions arising from exposure to benzyl alcohol in excess of 90 mg/kg/day, this product should not be used in infants and children up to 3 yr old.
Castor oil polyoxyl and castor oil polyoxyl hydrogenated	Zero	May cause severe allergic reactions
Chlorocresol	Zero	May cause allergic reactions
Ethanol	<100 mg/dose	This medicinal product contains small amounts of ethanol (alcohol), <100 mg/dose.
	100 mg–3 g/dose	This medicinal product contains . . . vol % ethanol (alcohol), i.e., up to . . . mg/dose, equivalent to . . . mL beer, . . . mL wine per dose.
		Harmful for those suffering from alcoholism.
		To be taken into account in pregnant or breast-feeding women, children and high-risk groups such as patients with liver disease or epilepsy.
	3 g/dose	This medicinal product contains . . . vol % ethanol (alcohol), i.e., up to . . . mg/dose, equivalent to . . . mL beer, . . . mL wine per dose.
		Harmful for those suffering from alcoholism.
		To be taken into account in pregnant or breast-feeding women, children and high-risk groups such as patients with liver disease or epilepsy.
		The amount of alcohol in this medicinal product may alter the effects of other medicines.
		The amount of alcohol in this medicinal product may impair your ability to drive or use machines.
Fructose	Zero	If you have been told by your doctor that you have intolerance to some sugars, contact your doctor before taking this medicinal product. Patients with rare hereditary problems of fructose intolerance should not take this medicine.
	5 g	Contains *x* g fructose per dose.
		This should be taken into account in patients with diabetes mellitus.
Galactose	Zero	If you have been told by your doctor that you have intolerance to some sugars, contact your doctor before taking this medicinal product.
		SPC proposal: patients with rare hereditary problems of galactose intolerance, e.g., galactosemia should not take this medicine.
	5 g	Contains *x* g galactose per dose.
		This should be taken into account in patients with diabetes mellitus

Table 11 (*Continued*)

Name	Threshold level	Information for the package leaflet
Glucose	5 g	Contains *x* g glucose per dose. This should be taken into account in patients with diabetes mellitus.
Heparin (as an excipient)	Zero	May cause allergic reactions and reduced blood cell counts that may affect the blood clotting system. Patients with a history of heparin-induced allergic reactions should avoid the use of heparin-containing medicines.
Organic mercury compounds (like thiomerosal phenylmercuric nitrate, acetate, borate)	Zero	This medicinal product contains (thiomerosal) as a preservative and it is possible that you/your child may experience an allergic reaction. Tell your doctor if you/your child have/has any known allergies. Tell your doctor if you/your child have/has experienced any health problems after previous administration of a vaccine.
Parahydroxybenzoates and their esters	Zero	May cause allergic reactions (possibly delayed), and exceptionally, bronchospasm.
Phenylalanine	Zero	This medicine contains phenylalanine. May be harmful for people with phenylketonuria.
Potassium	<1 mmol/dose	This medicine contains potassium, <1 mmol (39 mg) per dose, i.e., essentially "potassium-free."
	1 mmol/dose	This medicine contains *x* mmol (or *y* mg) potassium per dose. To be taken into consideration by patients with reduced kidney function or patients on a controlled potassium diet.
	30 mmol/L	May cause pain at the site of injection.
Propylene glycol and esters	400 mg/kg adults 200 mg/kg children	May cause alcohol-like symptoms.
Sesame oil	Zero	May rarely cause severe allergic reactions.
Sodium	<1 mmol/dose	This medicinal product contains <1 mmol sodium (23 mg) per dose, i.e., essentially "sodium-free."
	1 mmol/dose	This medicinal product contains *x* mmol (or *y* mg) sodium per dose. To be taken into consideration by patients on a controlled sodium diet.
Sorbitol	Zero	If you have been told by your doctor that you have intolerance to some sugars, contact your doctor before taking this medicinal product. SPC proposal: Patients with rare hereditary problems of fructose intolerance should not take this medicine.
Soya oil (and hydrogenated soya oil)	Zero	Medicinal product contains soya oil. If you are allergic to peanut or soya, do not use this medicinal product.

Abbreviation: SPC, Summary of Product Characteristics.
Source: From Ref. 73.

Similarly, 21 CFR 201.22 requires prescription drugs containing sulfites to be labeled with a warning statement about possible hypersensitivity. An informational chapter in USP <1091> Labeling of Inactive Ingredients provides guidelines for labeling of inactive ingredients present in dosage forms.

According to the Notes for Guidance on Pharmaceutical Development (CHMP/ICH/167068/04), the choice of excipients, their grade, compatibility, concentration, and function should be described in the P2 section of the Common Technical Document. It is necessary to justify inclusion of all the ingredients in the drug product and describe their intended function. Generally, a specification of ±10% at the end of shelf life is acceptable except for antioxidant and preservatives where performance data from PET or stability data may justify broader limits.

The bioburden and endotoxin limits of excipients used in the manufacture of sterile medical products shall be stated. However, this individual testing of excipients may be omitted if bioburden and endotoxin testing of the solution is checked prior to sterilization.

If an excipient is present in Ph. Eur. or other major pharmacopoeias, the monograph specifications are generally acceptable in the registration file. However, excipients that are not described in any pharmacopoeia, specifications should include physical characterization, identification tests, purity test, assay, and impurity tests. A certification is also included to confirm that excipients are of non-animal (specifically non-ruminant) origin. If this is not the case, a regulatory agency will require documentation to demonstrate freedom from viral and transmissible spongiform encephalopathies (TSE) risks (74).

CRITERIA FOR THE SELECTION OF EXCIPIENT AND SUPPLIER

Excipient selection during formulation development of parenteral dosage forms is focused on providing a safe, stable, efficacious, and functional product. The choice and the characteristics of excipients should be appropriate for the intended purpose.

> An explanation should be provided with regard to the function of all constituents in the formulation, with justification for their inclusion. In some cases, experimental data may be necessary to justify such inclusion e.g. preservatives. The choice of the quality of the excipient should be guided by its role in the formulation and by the proposed manufacturing process. In some cases it may be necessary to address and justify the quality of certain excipients in the formulation (75).

Normally a pharmaceutical development report is written in the United States, which should be available at the time of preapproval inspection (PAI). The development report captures the choice of excipients, their purpose and level in the drug product, their compatibility with other excipients, drug or package system, and how they may influence the stability and efficacy of the finished product. This information is similar to that included in the P2 section of the Common Technical Document submission.

The following key points should be considered in selecting an excipient and its supplier for parenteral products:

1. Influence of the excipient on the overall quality, stability, and effectiveness of drug product.
2. Physical, chemical, and biological compatibility of the excipient with drug and the packaging system (76).
3. Compatibility of the excipient with the manufacturing process; for example, preservatives may be adsorbed by rubber tubes or filters, acetate buffers will be lost during lyophilization process, etc.
4. The amount or percentage of excipients that can be added to the drug product. Table 6 summarizes the maximum amount of preservatives and antioxidants allowed by various pharmacopoeias.
5. Route of administration. The USP, Ph. Eur., and BP do not allow preservatives to be present in injections intended to come in contact with brain tissues or CSF. Thus, intracisternal, epidural, and intradural injections should be preservative-free. Also, it is preferred for a drug product to be administered via IV route to be free of particulate matter. However, if the size of the particle is well controlled, as in fat emulsion or colloidal albumin or amphotericin B dispersion, they can be administered by IV infusion.
6. Dose volume. All large-volume parenterals (LVPs) and those small-volume parenterals (SVPs) where the single-dose injection volume can be greater than 15 mL are required by the Ph. Eur./BP to be preservative-free (unless justified). The USP recommends that special care be observed in the choice and the use of added substances in preparations for injections that are administered in volumes exceeding 5 mL (77).
7. Whether the product is intended for single or multiple-dose use. According to USP, single-dose injections should be preservative-free. The FDA takes the position that even though a single-dose injection may have to be aseptically processed, the manufacturer should not use a preservative to prevent microbial growth. European agencies have taken a more lenient attitude on this subject.

8. The length or duration of time that the drug product will be used once the multidose injection is opened.
9. How safe is the excipient? Does it cause tissue irritation, hemolysis, or other toxic effects on cells, tissues, or organs?
10. Does the parenteral excipient contain very low levels of lead, aluminum, or other heavy metals?
11. Does a dossier or DMF exist for the excipient?
12. Has the excipient been used in humans? Has it been used via a parenteral route and in the amount and concentration that is being planned?
13. Has the drug product containing this excipient been approved throughout the world?
14. What is the cost of the excipient and is it readily available?
15. Is the excipient vendor following the GMP guide? Is the vendor ISO 9000 certified?
16. Will the excipient supplier certify the material to meet USP, BP, Ph. Eur., JP, and other pharmacopoeias?
17. Has the supplier been audited by the FDA or the company's audit group? How did they fare?

Presence of impurities in excipients can have a dramatic influence on the safety, efficacy, or stability of the drug product. Monomers or metal catalysts used during a polymerization process are toxic and can also destabilize the drug product if present in trace amounts. Because of safety concerns, the limit of vinyl chloride (monomer) in polyvinyl pyrrolidone is ≤ 10 ppm and for hydrazine (a side product of polymerization reaction) is ≤ 1 ppm. Monomeric ethylene oxide is highly toxic and can be present in ethoxylated excipients such as PEGs, ethoxylated fatty acids, etc.

An FDA guidance suggests that the animal-derived materials (e.g., egg yolk lecithin, egg phospholipid) used in drug product, originating from Belgium, France, and Netherland, between January to June, 1999, should be investigated for the presence of dioxin and polychlorinated biphenyls. Contaminated animal feed is the likely source of contamination in the animal-derived product.

Excipients such as dextrose, citric acid, mannitol, and trehalose are manufactured by fermentation processes and should be specially controlled for endotoxin levels. Mycotoxin (highly toxic metabolic products of certain fungi species) contamination of an excipient derived from natural material has not been specifically addressed by regulatory authorities. The German health authority has issued a draft guideline in 1997 where a limit has been specified for aflotoxins M_1, B_1, and the sum of B_1, B_2, G_1, and G_2 in the starting material for pharmaceutical products.

Heavy metal contamination of an excipient is a concern, especially for sugars, phosphate, and citrate. Several rules have been proposed or established. For example, Ph. Eur. sets a limit of ≤ 1 ppm of nickel in polyols. California Proposition 65 specifies a limit of ≤ 0.5 µg of lead per day per product (78). Similarly, the USP and FDA have issued guidelines that limit the aluminum content for all LVPs used in TPN therapy to 25 µg/L (79). Further, it requires that the maximum level of aluminum in SVPs intended to be added to LVPs and pharmacy bulk packages, at expiration date, be stated on the immediate container label.

An excipient's physical and chemical stability will determine the frequency for retesting. Because of the relatively small amount of active ingredient compared with the amount of excipients in most parenteral formulations, the degradation of even a small percentage of excipient can lead to levels of impurities sufficient to react or degrade a large percentage of active material. For example, in the presence of light and oxygen, benzyl alcohol decomposes via a free radical mechanism to form benzaldehyde ($x\%$ of benzaldehyde is approximately equivalent to $1/3$ $x\%$ of hydrogen peroxide). Hydrogen peroxide can rapidly oxidize sulfhydryl groups of amino acids such as cysteine present in peptides or proteins.

Thorough due diligence and risk analysis should be conducted in the selection of a pharmaceutical excipient supplier. Because excipients are often commodity (low value–high volume) products, suppliers focus on improving manufacturing efficiency to reduce cost, which frequently results in manufacturing process changes that potentially could impact the quality or characteristic profile of the excipient. Generally, the pharmaceutical industry is a relatively small customer (in terms of volume of material purchased) of these suppliers and has limited business leverage. For example, the pharmaceutical industry uses approximately 20% of gelatin produced. Of this 20%, most is for production of oral dosage forms. The parenteral portion is approximately 5% of this 20%. Therefore, it is imperative that the drug manufacturer negotiates a detailed contract with the excipient supplier, which strictly prohibits the supplier from making any changes in the process or quality of the material without informing the customer well in advance. Also, the pharmaceutical manufacturer should investigate, and even consider qualifying, alternate suppliers who could be used in case of an emergency. A change in the supplier should not be made without consulting the pertinent regulatory bodies, since such an event may require prior regulatory approval.

The pharmaceutical manufacturer should have an active Vendor Certification Program and assure that the vendor is ISO 9000 certified. An audit of the excipient manufacturer is essential since the pharmaceutical industry is ultimately responsible for the quality of the drug product that includes the excipient(s) as one of the components. A useful audit tool is the IPEC GMP guide that is written in the format of ISO 9000 using identical nomenclature and paragraph numbering. The audit should determine and ensure that the quality is being built into the excipient, which may be difficult to measure by incoming quality control assessment of the material. This is especially true for parenteral excipients where not only chemical, but also microbiological, attributes are critical. Bioburden and endotoxin limits may be needed for each of the excipients, and several guidelines are available to establish the specifications (80,81).

There are no legal requirements for excipient GMPs in Europe (82). The Qualified Person (QP) is responsible to assure that the quality of excipients is appropriate on the basis of pharmacopoeial specifications or a company's quality systems.

Unfortunate events in Haiti highlight the importance of assuring the quality of excipients to the same degree that one normally does for active ingredients. From November 1995 through June 1996, acute anuric renal failure was diagnosed in 86 children. This was associated with the use of diethylene glycol contaminated glycerin used to manufacture acetaminophen syrup (83). The FDA is advising pharmaceutical companies to test for melamine down to a 2.5 ppm level in certain nitrogen-rich drug ingredients (raw materials that contain more than 2.5% nitrogen and those for which purity or strength is determined on the basis of nitrogen content) (84). The list of excipients includes albumin, amino acids derived from casein protein hydrolysates, ammonium salts, protamine sulfate, povidone, lactose, gelatin, etc. This guidance is in response to the incidents of pet food and Chinese milk doped with melamine.

The FDA recognizes the importance of excipients in the product for performance and safety. An injectable generic product should have identical nonexceptional excipients (qualitatively and quantitatively) as that of reference listed drug if the generic drug product is to follow the simplest path of registration, otherwise additional data must be submitted to demonstrate that the differences do not affect the safety or performance (85). For parenteral products, nonexceptional excipients are ingredients other than preservatives, pH adjuster, antioxidant, and buffers.

SAFETY ISSUES

Clinical experience with many of the excipients has resulted in some safety watch outs. For example, sensitization reactions have been reported for the parabens, thimerosal, and propyl gallate. Sorbitol is metabolized to fructose and can be dangerous when administered to fructose intolerant patients. Table 11 lists safety concerns that need to be included on the labeling.

Progress in drug delivery systems and new proteins/peptides being developed for parenteral administration has created a need to expand the list of excipients that can be safely used. The informational chapter in the USP presents a scientifically based approach for a safety assessment of new pharmaceutical excipients (86). A new or novel excipient is defined as one that has not been previously used in a pharmaceutical preparation or has not been fully qualified by existing safety data with respect to the proposed level of exposure, duration of exposure, or route of administration (87). Besides the baseline toxicity data (either through literature or experimentation), if the drug (excipient) will be administered short term

(<14 days of consecutive days per treatment episode), intermediate term (14–90 days), or long term (>3 months), additional safety information on the excipient is needed (87).

Currently, there are concerns regarding TSE via animal-derived excipients such as gelatin (88). TSE are caused by prions that are extremely resistant to heat and normal sterilization processes. Hence, a risk assessment is done at early stage of product development to make sure that the excipients do not contribute to this risk.

Several guidelines are available that address the issue of animal-derived excipients and scientific principles to minimize the possible transmission of TSE via medicinal products (89,90). The current situation indicates that there are negligible concerns for lactose, glycerol, fatty acids, and their esters, but the situation is less clear for gelatin. Gelatin is still a necessary ingredient for some medicinal products, and the European Medicines Agency (EMEA) has updated its guidance to allow gelatin from category I and II countries if gelatin is produced by the acid process and category I, II, and III countries if produced by the alkali process (91). Additional information on risk assessment of ruminant materials originating from United States, Canada, and other countries can be found in Refs. 90, 92, and 93.

In the current regulatory environment, if feasible, it may be beneficial to select non-animal-derived excipient. There have always been concerns in using bovine serum albumin (BSA) or human serum albumin (HSA) because they may have been possibly derived from virus contaminated blood. Recombinant human erythropoietin and darbepoetin alfa formulations were changed to replace albumin with Polysorbate 80. Currently, recombinant HSA is available from several companies, which reduces probability of TSE (94).

European Commission directive EMEA/410/01/rev. 2 requires manufacturers to provide a "Certificate of Suitability" or the underlying "scientific information" to attest that their pharmaceuticals are free of TSEs.

Vegetable origin polysorbate should be used. If older products contain animal sourced polysorbate then a switch should be made. The EMEA has stated that such a change of a Polysorbate 80 source will not result in reperforming viral inactivation studies (95). It is also important to know the vegetable source (e.g., is trehalose being made from corn or tapioca) and if during the manufacturing of the excipient any processing aids (e.g., enzymes during production of lactose) are being used that are derived from an animal source.

FUTURE DIRECTION

Biodegradable polymeric materials such as polylactic acid, polyglycolic acid, and other poly-α-hydroxy acids have been used as medical devices and also as biodegradable sutures since the 1960s (96). Currently, the FDA has approved for marketing only devices made from homopolymers or copolymers of glycolide, lactide, caprolactone, p-dioxanone, and trimethylene carbonate (97). Such biopolymers are finding increased application as a matrix to deliver parenteral drugs for prolonged delivery (98). At least four drug products—Lupron Depot®, Decapeptyl®, Nutropin Depot®, and Zoladex®—have been approved. All four drug products are microspheres in polyglycolic acid(PLG), polylactic acid(PLA), or the co-polylactic-glycolic acid(PLGA) matrix. Decapeptyl is approved in France and is a microsphere for IM administration. It contains drug in a matrix of PLGA and carboxymethylcellulose with mannitol and Polysorbate 80.

Several phospholipid-based excipients are finding increased application as solubilizing agents, emulsifying agents, or as components of a liposomal formulation. The phospholipids occur naturally and are biocompatible and biodegradable, for example, egg phosphatidylcho-line, soybean phosphatidylcholine, hydrogenated soybean phosphatidylcholine (HSPC), DMPC, DSPC, DOPC, DSPE, DMPG, DPPG, and DSPG. Spartaject™ technology uses a mixture of phospholipids, to encapsulate poorly water soluble drug, to form a micro-suspension that can be injected intravenously. Busulfan drug product uses this technology and is currently undergoing phase I clinical trials. Many liposomal and liposomal-like formulations (DepoFoam®) are either approved (DepoCyt®) or are undergoing clinical trials to reduce drug toxicity, improve drug stability, prolong the duration of action, or to deliver drug to the central nervous system (99). Two amphotericin formulations have been approved in the United States, which are a liposomal or lipid complex between the antifungal drug and the positively charged lipid. Amphotec® is a 1:1 molar ratio complex of amphotericin B and cholesteryl sulfate, while Abelcet® is a 1:1 molar complex of amphotericin B with phospholipids (7 parts of L-α-dimyristoylphosphatidylcholine and L-α-dimyristoylphosphatidyl glycerol).

Poloxamer or Pluronic are block copolymers composed of polyoxyethylene and polyoxypropylene segments. They exhibit reverse thermal gelation and are being tried as solubilizing, emulsifying, and stabilizing agents. Thus, a depot drug delivery system can be created using Pluronics whereby the product is a viscous injection that gels upon intramuscular injection (100). Pluronics can prevent protein aggregation or ad/absorption and can help in the reconstitution of lyophilized products. Pluronic F68 (Polaxamer-188), F38 (Poloxamer-108), and F127 (Poloxamer-407) are the most commonly used Pluronics. For example, a liquid formulation of human growth hormone and Factor VIII can be stabilized using Pluronics. Fluosol® is a complex mixture of perfluorocarbons, with a high oxygen-carrying capacity, emulsified with Pluronic F68 and various lipids. It was recently approved by the FDA for adjuvant therapy to reduce myocardial ischemia during coronary angioplasty. A highly purified form of Poloxamer-188 (Flocor™), intended for IV administration, is undergoing phase III clinical trials for various cardiovascular diseases. Purification of Poloxamer-188 has been shown to reduce nephrotoxicity. Another nonionic surfactant, Solutol HS 15 (Macrogol-15-Hydroxystearate), has been approved by the Health Protection Branch (Canada) in vitamin K1 formulation for human application.

Polymeric materials such as Poloxamer and albumin may coat micro- or nanoparticles, alter their surface characteristics, and reduce their phagocytosis and opsonization by reticuloendothelial system following IV injection. Such surface modifications often result in prolongation in the circulation time of intravenously injected colloidal dispersions (101). Poloxamers have also been used to stabilize suspensions such as NanoCrystal™ (102).

Fluosol-DA®, manufactured by Green Cross Corporation in Japan, was the first successfully developed injectable perfluorocarbons-based commercial product. It is a dilute (20% wt/vol) emulsion based on perfluorodecalin and perfluorotripropylamine emulsified with potassium oleate, Pluronic F68, and egg yolk lecithin. These perfluorocarbons are inert and can also be used to formulate nonaqueous preparations of insoluble proteins and small molecules (103). Perfluorocarbons have also been approved by the FDA in one ultrasound contrast agent, Optison®, which is administered via the IV route. Optison is a suspension of microspheres of HSA with octafluoropropane. Heat treatment and sonication of appropriately diluted human albumin, in the presence of octafluoropropane gas, is used to manufacture microspheres in Optison injection. The protein in the microsphere shell makes up approximately 5% to 7% (wt/wt) of the total protein in the liquid. The microspheres have a mean diameter range of 2.0 to 4.5 μm with 93% of the microspheres being less than 10 μm.

Sucrose acetate isobutyrate (SAIB) is a high-viscosity liquid system, which converts into free flowing liquid when mixed with 10% to 15% ethanol (104). Upon SC or IM injection, the matrix rapidly converts to water insoluble semisolid, which is capable of delivering proteins and small molecules for a prolonged period. SAIB is biocompatible and biodegrades to natural metabolites.

Several other biodegradable, biocompatible, injectable polymers being investigated for drug delivery systems include polyvinyl alcohol, block copolymer of PLA-PEG, polycyanoacrylate, polyanhydrides, cellulose, chitosan, alginate, collagen, modified HSA, albumin, starches, dextrans, hyaluronic acid and its derivatives, and hydroxyapatite (105). It is impossible to cover all the aspects in the field of excipient development, control, and usage in a single chapter, but it is clear that many of the new drug modalities like delivery of genes, immunomodulators, RNAi, anti-sense, aptamers, and other novel therapeutic agents will invariably require new excipients to be successful (106).

REFERENCES

1. Robertson MI. Regulatory issues with excipients. Int J Pharm 1999; 187:273–276.
2. <1059> Excipient performance, proposed new USP general information chapter. Pharmacopoeial Forum 2007; 33(6):1311–1321.
3. Moreton C. Functionality and performance of excipients in a quality-by-design world: Part1. Am Pharm Rev 2009; 12(1):40–44.

4. Industry is developing a more profound understanding of excipients. Gold Sheet September, 2009.
5. British Pharmacopoeia. Parenteral Preparations. Vol. 2. London: Stationary Office, 1999:1575.
6. European Pharmacopoeia. Parenteral Preparations. 6th ed. Strasbourg: Council of Europe, 2009:1765.
7. Uchiyama M. Regulatory status of excipients in Japan. Drug Inf J 1999; 33:27–32.
8. Boylan JC, DeLuca PP. Formulation of small volume parenterals. In: Avis KE, Lieberman HA, Lachman L. eds. Pharmaceutical Dosage Forms: Parenteral Medications. Vol. 1, 2nd ed. New York: Marcel Dekker, Inc, 1992:173–248.
9. Matthews B. Excipients used in products approved through the EU centralised procedure. Regul Aff J 2002; 13(12):1036–1044.
10. Matthews B. Excipients for non-oral routes of administration. Regul Aff J 2002; 13(11):897–908.
11. Nema S, Washkuhn RJ, Brendel RJ. Excipients and their use in injectable products. PDA J Pharm Sci Technol 1997, 51(4):166–171.
12. Powell MF, Nguyen T, Baloian L. Compendium of excipients for parenteral formulations. PDA J Pharm Sci Technol 1998; 52(5):236–311.
13. Strickley RG. Parenteral formulations of small molecules therapeutics marketed in the United States (1999)—Part I. PDA J Pharm Sci Technol 1999; 53(6):324–349.
14. Strickley RG. Parenteral formulations of small molecules therapeutics marketed in the United States (1999)—Part II. PDA J Pharm Sci Technol 2000; 54(1):69–96.
15. Strickley RG. Parenteral formulations of small molecules therapeutics marketed in the United States (1999)—Part III. PDA J Pharm Sci Technol 2000; 54(2):152–169.
16. Wang YJ, Hanson MA. Parenteral formulations of proteins and peptides: stability and stabilizers. PDA J Parenter Sci Technol 1988; 42(suppl):S4–S26.
17. Wang YJ, Kowal RR. Review of excipients and pH's for parenteral products used in United States. J Parenter Sci Technol 1980; 34(6):452.
18. Mosby's GenRx. 8th ed. St. Louis: Mosby-Year Book, Inc, 1998.
19. Physician's Desk Reference. 63rd ed. Toronto: Thomson Corporation, 2009.
20. FDA. Inactive Ingredient Guide. Division of Drug Information Resources, FDA, CDER; 2009.
21. Rowe RC, Sheskey PJ, Quinn ME. Handbook of Pharmaceutical Excipients. 6th ed. London: The Pharmaceutical Press, 2009.
22. Trissel LA. Handbook on Injectable Drugs. 10th ed. Bethesda: American Society of Health-System Pharmacists, Inc, 1998.
23. EMEA. Note for Guidance on Quality of Water for Pharmaceutical Use. CPMP/QWP/158/01 revision—EMEA/CVMP/115/01 revision; May 2002.
24. EMEA. Reflection Paper on Water for Injection Prepared by Reverse Osmosis. EMEA/CHMP/CVMP/QWP/28271/2008; March 5, 2008.
25. Sweetana S, Akers MJ. Solubility principles and practices for parenteral drug dosage form development. PDA J Parenter Sci Technol 1996; 50(5):330–342.
26. Yalkowsky SH, Roseman TJ. Solubilization of drugs by cosolvents. In: Techniques of Solubilization of Drugs. New York: Marcel Dekker, Inc, 1981:91–134.
27. Hancock BC, York P, Rowe RC. The use of solubility parameters in pharmaceutical dosage form design. Int J Pharm 1997; 148:1–21.
28. Rubino JT, Yalkowsky SH. Cosolvency and cosolvent polarity. Pharm Res 1987; 4(3):220–230.
29. Brazeau GA, Cooper B, Svetic KA, et al. Current perspectives on pain upon injection of drugs. J Pharm Sci 1998; 87(6):667–677.
30. Brazeau GA, Fung H. Use of an in-vivo model for the assessment of muscle damage from intramuscular injections: in-vitro-in-vivo correlation and predictability with mixed solvent systems. Pharm Res 1989; 6(9):766–771.
31. Reed KW, Yalkowsky S. Lysis of human red blood cells in the presence of various cosolvents. J Parenter Sci Technol 1985; 39(2):64.
32. Yalkowsky SH, Krzyzaniak JF, Ward GH. Formulation-related problems associated with intravenous drug delivery. J Pharm Sci 1998; 87(7):787–796.
33. Mottu F, Laurent A, Rufenacht DA, et al. Organic solvents for pharmaceutical parenterals and embolic liquids: a review of toxicity data. PDA J Pharm Sci Technol 2000; 54(6):456–465.
34. Johnson DM, Gu LC. Autoxidation and antioxidants. In: Swarbick J, Boylan JC, eds. Encyclopedia of Pharmaceutical Technology. Vol. 1. New York: Marcel Dekker, Inc, 1988:415–449.
35. Herman AC, Boone TC, Lu HS. Characterization, formulation, and stability of neupogen (filgrastim), a recombinant human granulocyte-colony stimulating factor. In: Pearlman R, Wang YJ, eds. Formulation, Characterization, and Stability of Protein Drugs: Case Histories. Vol. 9. New York: Plenum Press, 1996:325.
36. Frank DW, Gray JE, Weaver RN. Cyclodextrin nephrosis in the rats. Am J Pathol 1976; 83(2): 367–382.

37. Stella VJ, Rajewski RA. Cyclodextrins: their future in drug formulation and delivery. Pharm Res 1997; 14(5):556–567.
38. Thompson DO. Cyclodextrins-enabling excipients: their present and future use in pharmaceuticals. Crit Rev Ther Drug Carrier Syst 1997; 14(1):1–104.
39. Loftsson T, Johannesson HR. The influence of cyclodextrins on the stability of cephalothin and aztreonam in aqueous solutions. Die Pharmazie 1994; 49:292–293.
40. Lehner SJ, Muller BW, Seydel JK. Effect of hydroxylpropyl-beta-cyclodextrin on the antimicrobial action of preservatives. J Pharm Pharmacol 1994; 46:186–191.
41. Fatouros A, Osterberg T, Mikaelsson M. Recombinant factor VIII SQ—Influence of oxygen, metal ions, pH and ionic strength on its stability in aqueous solution. Int J Pharm 1997; 155:121–131.
42. Stadtman ER. Metal ion catalyzed oxidation of proteins: biochemical mechanism and biological consequences. Free Radic Biol Med 1990; 9(4):315–325.
43. Enever RP, Li Wan Po A, Shotton E. Factors influencing decomposition rate of amitriptyline hydrochloride in aqueous solution. J Pharm Sci 1977; 66(8):1087–1089.
44. Munson JW, Hussain A, Bilous R. Precautionary note for use of bisulfite in pharmaceutical formulations. J Pharm Sci 1977; 66(12):1775–1776.
45. Akers MJ. Antioxidants in pharmaceutical products. J Parenter Sci Technol 1982; 36(5):222–228.
46. CPMP. Note for Guidance on Inclusion of Antioxidants and Antimicrobial Preservatives in Medicinal Products. CPMP; January 1998.
47. EMEA. Guideline on Excipients in the Dossier for Application for Marketing Authorization of a Medical Product. EMEA/CHMP/QWP/396951/2006; June 19, 2007.
48. Dabbah R. The use of preservatives in compendial articles. Pharmacopeial Forum 1996; 22(4):2696.
49. Martindale: The Extra Pharmacopoeia. 31st ed. London: Royal Pharmaceutical Society, 1996:1128.
50. British Pharmaceutical Codex. London: Royal Pharmaceutical Society, 1973:100.
51. Oishi S. Effects of propyl paraben on the male reproductive system. Food Chem Toxicol 2002; 40(12):1807–1813.
52. Oishi S. Effects of butyl paraben on the male reproductive system in mice. Arch Toxicol 2002; 76:423–429.
53. Oishi S. Lack of spermatotoxic effects of methyl and ethyl esters of p-hydroxybenzoic acid in rats. Food Chem Toxicol 2004; 42:1845–1849.
54. Hoberman AM, Schreur DK, Leazer T, et al. Lack of effect of butylparaben and methylparabben on the reproductive system in male rats. Birth Defects Res B Dev Reprod Toxicol 2008; 83:123–133.
55. Dabbah R. Harmonization of microbiological methods—a status report. Pharmacopeial Forum 1997; 23(6):5334–5344.
56. European Pharmacopoeia. 3rd ed. Strasbourg: Council of Europe, 1997:286.
57. USP <51> Antimicrobial Effectiveness Testing. 32nd ed. Rockville: U.S. Pharmacopeial Convention, Inc, 2009.
58. CPMP. Note for Guidance on Maximum Shelf life for Sterile Products for Human Use After First Opening or Following Reconstitution. CPMP, July 1998.
59. Lam XM, Costantino HR, Overcashier DE, et al. Replacing succinate with glycolate buffer improves the stability of lyophilized interferon-gamma. Int J Pharma 1996; 142:85–95.
60. Piedmonte DM, Summers C, McAuley A, et al. Sorbitol crystallization can lead to protein aggregation in frozen protein formulations. Pharm Res 2007; 24(1):136–146.
61. Pikal MJ. The correlation of structural relaxation time with pharmaceutical stability. Presented at: The Freeze-drying of Pharmaceuticals and Biologicals Conference, Brownsville, VT, September 23–26, 1998.
62. Arakawa T, Kita Y, Carpenter JF. Protein-solvent interactions in pharmaceutical formulations. Pharm Res 1991; 8(3):285–291.
63. Miller DP, Anderson RE, de Pablo JJ. Stabilization of lactate dehydrogenase following freeze-thawing and vacuum-drying in the presence of trehalose and borate. Pharm. Res 1998; 15(8):1215–1221.
64. Carpenter JF, Crowe JH. Modes of stabilization of a protein by organic solutes during desiccation. Cryobiology 1998; 25:459–470.
65. Baumann TJ, Smythe MA, Kaufmann K, et al. Dissolution times of adriamycin and adriamycin RDF. Am J Hosp Pharm 1988; 45:1668.
66. Jain NK, Jain S, Singhai AK. Enhanced solubilization and formulation of an aqueous injection of piroxicam. Pharmazie 1997; 52(12):942–946.
67. Meyer JD, Manning MC. Hydrophobic ion pairing: altering the solubility properties of biomolecules. Pharm Res 1998; 15(2):188–192.
68. Wang YJ, Dahl TC, Leesman GD, et al. Optimization of autoclave cycles and selection of formulation for parenteral product. Part II: Effect of counter-ion on pH and stability of diatrizoic acid at autoclave temperatures. J Parenter Sci Technol 1984; 38(2):72–77.

69. Japanese Pharmaceutical Excipients Directory. Tokyo: Yakuji Nippo, Ltd, 2004.
70. Excipients in Pharmaceutical Dosage Forms: The Challenge of the 21st Century Conference Proceedings. Nice, France, May 14–15, 1998.
71. Benzyl alcohol. Pharmacopeial Forum 1995; 21(5):1240.
72. USP <1078> Good Manufacturing Practices for Bulk Pharmaceutical Excipients. 32nd ed. Rockville: U.S. Pharmacopeial Convention, Inc, 2009.
73. European Commission. Excipients in the Label and Package Leaflet of Medicinal Products for Human Use. Vol. 3B. European Commission, July 2003.
74. CPMP. Note for Guidance on the Use of Bovine Serum in the Manufacture of Human Biological Medicinal Products. CPMP/BWP/1793/02; June 18, 2003.
75. CPMP. Note for Guidance on Development Pharmaceutics. Committee for Proprietary Medicinal Products. CPMP; July 1998.
76. Akers MJ. Excipient-drug interactions in parenteral formulations. J Pharm Sci 2002; 91(11):2283–2300.
77. USP <1> Injections. 32nd ed. Rockville: U.S. Pharmacopeial Convention, Inc, 2009.
78. Paul WL. Excipient intake and heavy metals limits. Pharmacopeial Forum 1995; 21:(6):1629.
79. Aluminum in large and small volume parenterals used in total parenteral nutrition. Fed Regist 1998; 63(2):176–185.
80. FDA. Guideline on Validation of the Limulus Amebocyte Lysate Test as an End-Product Test for Human and Animal Parenteral Drugs. Biological Products and Medical Devices, FDA, December 1987.
81. Opalchenova GA. Comparison of the microbial limit tests in the British, European, and United States pharmacopeias and recommendation for harmonization. Pharmacopeial Forum 1994; 20 (4):7872–7877.
82. Taylor P. Regulation for excipients is brewing across the Atlantic. Pharm Technol 2009; 33(2):86–87.
83. Morbidity and Mortality Weekly Report, U.S. Department of Health and Human Services, 1996; 45:649–650.
84. FDA. Pharmaceutical Components at Risk for Melamine Contamination. U.S. Department of Health and Human Services. August 2009.
85. FDA. Interim Inactive Ingredient Policy. OGD, FDA, 1994.
86. USP <1074> Excipient Biological Safety Evaluation Guidelines. 32nd ed. Rockville: U.S. Pharmacopeial Convention, Inc, 2009.
87. FDA. Nonclinical Studies for the Safety Evaluation of Pharmaceutical Excipients. In: Guidance for Industry. FDA, May 2005.
88. Matthews B. BSE/TSE risks associated with active pharmaceutical ingredients and starting materials: the situation in Europe and the global implications for healthcare manufacturers. PDA J Pharm Sci Technol 2001; 55(5):295–328.
89. CPMP. Note for Guidance on Minimizing the Risk of Transmitting Animal Spongiform Encephalopathy Agents via Medicinal Products. CPMP; April 21, 1999.
90. FDA. The Sourcing and Processing of Gelatin to Reduce the Potential Risk Posed by Bovine Spongiform Encephalopathy (BSE) in FDA-Regulated Products for Human Use, Guidance for Industry. U.S. Dept. of Health and Human Services, FDA, September 1997.
91. EMEA. Gelatin for use in Pharmaceuticals: Explanatory Note. EMEA/CPMP/4306/00/v0.2; December 13, 2000.
92. EMEA. First Cases of BSE in USA and Canada: Risk Assessment of Ruminant Materials Originating from USA and Canada. EMEA/CHMP/BWP/27/04, July 21, 2004.
93. EMEA. CHMP Position Statement on Creutzfeldt-Jakob Disease and Plasma-Derived and Urine Derived Medicinal Products. EMEA/CPMP/BWP/2879/02/rev 1; June 23, 2004.
94. Chuang VTG, Kragh-Hansen U, Otagiri M. Pharmaceutical strategies utilizing recombinant human serum albumin. Pharm Res 2002; 19(5):569–577.
95. CPMP. Position State on Polysorbate 80. CPMP/BWP/1952/98; October 22, 1998.
96. Jain R, Shah NH, Malick AW, et al. Controlled drug delivery by biodegradable poly(ester) devices: different preparative approaches. Drug Dev Ind Pharm 1998; 24(8):703–727.
97. Middleton JC, Tipton AJ. Synthetic biodegradable polymers as medical devices. Med Plast Biomater 1998; 5(2).
98. Pettit DK, Lawter JR, Huang WJ, et al. Characterization of poly(glycolide-co-D,L-lactide)/poly(D,L-lactide) microspheres for controlled release of GM-CSF. Pharm Res 1997; 14(10):1422–1430.
99. Katre NV, Asherman J, Schaefer H. Multivesicular liposome (depofoamtm) technology for the sustained delivery of insulin-like growth factor-I. J Pharm Sci 1998; 87(11):1341–1346.
100. Wang P, Johnston TP. Sustained-release interleukin-2 following intramuscular injection in rats. Int J Pharm 1995; 113(1):73–81.

101. Moghimi SM. Mechanisms regulating body distribution of nanospheres conditioned with pluronic and tetronic block co-polymers. Adv Drug Deliv Rev 1995; 16:183–193.
102. Zheng JY, Bosch HW. Sterile filtration of nanocrystal™ drug formulations. Drug Dev Ind Pharm 1997; 23(11):1087–1093.
103. Knepp VM, Muchnik A, Oldmark S, et al. Stability of non-aqueous suspension formulations of plasma derived factor IX and recombinant human alpha interferon at elevated temperatures. Pharm Res 1998; 15(7):1090–1095.
104. Sullivan SA, Gilley RM, Gibson JW, et al. Delivery of taxol and other antineoplastic agents from a novel system based on sucrose acetate isobutyrate. Pharm Res 1997; 14:291.
105. Gombotz WR, Pettit DK. Biodegradable polymers for proteins and peptide drug delivery. Bioconjug Chem 1995; 6:332–351.
106. Apte SR, Ogwu SO. A review and classification of emerging excipients in parenteral medications. Pharm Technol 2003; 27(3):46–60.

8 | Techniques to evaluate damage and pain on injection

Gayle A. Brazeau, Jessica Klapa, and Pramod Gupta

BACKGROUND AND OPTIMIZING PARENTERAL FORMULATIONS

Injectable products have and will continue to be an important aspect in the medication management of patients for cancer, acute cardiovascular disease, infection, central nervous system disorders, and traumatic injuries. It is often also important for the development of parenteral product formulations for existing oral drugs for use in institutional, long-term, and home health care settings, given many of the chronic conditions may necessitate treatment in these types of facilities. The formulation of injectables can be a challenging project, given the complexity of the formulations from the perspective of optimizing the formulation requirements for the product, the physiological constraints associated with administration of the product, and the therapeutic characteristics of the drug (Fig. 1).

With respect to a given formulation, pharmaceutical scientists must consider the specific therapeutic requirements such as the indication or use of the drug, the optimal route of drug administration for the treated condition or disease, the targeted patient population(s) for the condition or disease treatment, the type of product (viz., immediate vs. sustained release), and the pharmacokinetic and pharmacodynamic profile of the drug. These therapeutic considerations must be balanced with the formulation requirements in optimizing the type of dosage form (e.g., solution, suspension, emulsion, or the newer and innovative drug delivery systems), solubility, stability, compatibility, injection volume, and viscosity. Finally, the formulation and therapeutic requirements must be optimized in considering the physiological constraints associated with parenteral administration, such as the route and site of injection, specifically the injection volume, injection speed, frequency of injections, and the local site reactions, namely the tissue damage on injection and pain on injection.

While the tools and methodological approaches are readily available for optimizing and understanding the elements with respect to the formulation requirements and characterizing the therapeutic requirements, one specific area that is often difficult to characterize during formulation development is the evaluation of the potential for causing tissue damage and/or pain on injection. The goal of this chapter is to provide pharmaceutical scientists with a general overview of available in vitro and in vivo methods in animals to screen drugs, excipients, and formulations for their potential to cause tissue damage and pain. While this chapter will provide a general discussion and summary of these topics, readers are encouraged to review the specific references for additional details. Furthermore, the characterization and determination of the extent of tissue damage and/or pain associated with a parenteral formulation is an ideal example of the need for professional collaboration between pharmaceutical scientists, pharmacologists, toxicologists, and neuroscientists, given the complexity of the physiological, biological, and biochemical interactions between the formulation and the site of injection.

Definitions and Relationship Between Tissue Damage and Pain on Injection

It is critical to understand the key definitions with respect to tissue damage and/or pain associated with injectables. Tissue damage can be defined as a *formulation*-induced reversible or irreversible change in the anatomy, biochemistry, or physiology at the injection site. *Formulation* in this specific definition can range from a single drug to one or more excipient(s) to final product composed of the drug and other excipients or a delivery system. The specific type of tissue damage includes hemolysis or phlebitis associated with intravenous administration and myotoxicity associated with intramuscular administration. For subcutaneous injections, the damage could be associated with those structures associated with this injection space such as the skin or skeletal muscle. The evaluation of the extent of tissue damage on intramuscular or intravenous injection is relatively easy to evaluate, given the availability of a

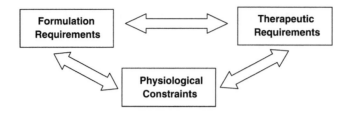

Figure 1 Optimization considerations in the formulation of injectable products.

Formulations requirements	Physiological constraints	Therapeutic requirements

• Dosage form type
• Solubility
• Stability and compatibility
• Injection volume
• Viscosity
	• Route/site injection
	  ○ Injection volume
	  ○ Injection speed
	  ○ Frequency of injection
	• Local site reactions
	  ○ Tissue damage
	  ○ Pain
		• Therapeutic indication and use
		  ○ Administration route
		  ○ Patient population
		• Formulation release profile
		• Pharmacokinetic and pharmacodynamic profile

Table 1 In Vitro and In Vivo Markers to Evaluate Tissue Damage

In vitro markers	In vivo tissue markers

• Hemoglobin—erythrocytes
• Cytosolic cellular components
  ○ Creatine kinase
  ○ Lactate dehydrogenase
  ○ Potassium
• Histological evaluation
  ○ Extracellular membrane disruptions
  ○ Intracellular membrane disruptions
  ○ Changes in intracellular organelles
	• Release of proteins/cytosolic components
	  ○ Creatine kinase—specifically MM isozyme
	  ○ Lactate dehydrogenase
	  ○ Myoglobin
	  ○ Aldolase
	  ○ Carbonic anhydrase III
	  ○ Myloperoxidase—indicative of neutrophils
	  ○ N-acetyl-β-glucosaminidase—indicative of monocytes
	  ○ Potassium
	• Blinded histological examination
	  ○ Lesion size
	  ○ Severity
	  ○ Presence of necrosis/degeneration
	  ○ Presence of inflammatory cells
	  ○ Edema
	  ○ Hemorrhage

wide array of biochemical and histological markers associated with these specific sites. A listing of the various available in vitro and in vivo markers for evaluating tissue damage is provided in Table 1.

Pain on injection is an unpleasant sensation associated with the injection of a *formulation* (as defined in the above paragraph). Pain on injection is often acute in nature as it is limited to the normal time for healing or the time necessary for neutralization of the initiating or causative factors. Evaluating the potential of a *formulation* to cause pain has been found to be more difficult to quantify experimentally as this process is associated with the activation of pain receptors, nociceptors, at the injection site. The sensation of pain is mediated in the periphery by multiple sets of specialized afferents called nociceptors. A brief overview is discussed in the following text as an introduction to this topic, but for additional and more specific information about acute versus chronic pain, the reviews by Brazeau, Schmelz, Dussor, and Mense are useful in this regard (1–4).

There are three different relationships linking tissue damage with pain on injection. The most likely relationship is the *formulation* causes tissue damage, and this damage results in the release of intracellular molecules that activate nociceptors, resulting in pain as suggested by outward behavioral indicators such as licking the injection site or guarding/minimizing the use of the limb. Alternatively, a formulation could result in the direct activation of nociceptors and produce pain without any specific tissue damage. A third potential relationship is tissue damage associated with the *formulation*, but the *formulation* itself may inhibit the nociceptive pathways. This later relationship may be the hardest to screen *formulations* unless specific markers of tissue damage and approaches are included in the evaluation. An easy way to consider the relationships and considerations between tissue damage in muscle with pain following intramuscular injection is provided in Figure 2.

Why the Importance of In Vitro and In Vivo Animal Studies to Evaluate Tissue Damage and/or Pain on Injection?

It might be questioned why it is necessary for the utilization of in vitro and in vivo animal methods to evaluate and screen *formulations* for tissue damage and/or pain on injection. Ideally and initially, it is advantageous and cost effective to identify any potential tissue damage and/or pain on injection of a given formulation prior to the clinical trials. However, in vitro methods can provide formulators with the opportunity to screen various excipients, evaluate different formulation compositions and delivery systems, as well as evaluate the mechanisms of acute tissue damage to optimize the initial selection of a formulation. In vivo studies not only provide the opportunity to further confirm the in vitro results but can also allow investigators to look at the effect of blood flow, the immune system, and the intact pain system as shown in Figure 3. As such, formulators are encouraged to consider both in vitro and in vivo *studies* to thoroughly optimize injectable formulations prior to commencing any clinical studies.

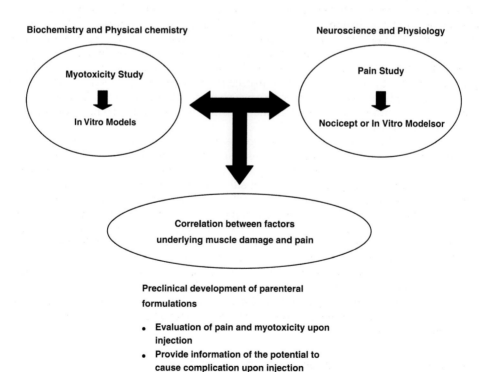

Figure 2   Link between myotoxicity and pain on intramuscular injection.

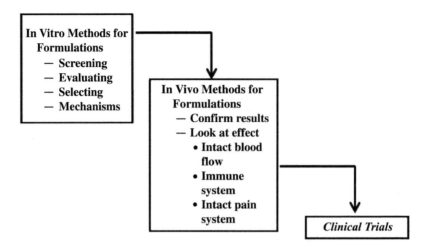

Figure 3   Importance of in vitro and in vivo methods in optimizing formulations.

General Overview on the Mechanisms of Tissue Damage

It is important to define key terms when considering tissue damage or pain on injection. An irritant is the molecule that can be linked to the source of irritation, either pain or tissue damage. Alternatively, a vesicant is a highly reactive molecule that combines with DNA, proteins, or other cell components resulting in cellular alterations that can be reversible or irreversible. It is essential to know and characterize the chemistry of molecules in a parenteral formulation as this can provide insight as to whether the structural elements may be likely to react with cellular components at the injection site. The knowledge of the structural elements provides a key as to whether the excipients or the therapeutic agent in the parenteral formulation has the potential to be an irritant or a vesicant. This highlights the importance of systematically screening all the components in a formulation or to avoid the use of specific agent if there is a potential for tissue damage/pain based on the chemical structure, the literature, or previous experimental findings. One needs to consider all parenteral formulations from the pathological perspective, specifically whether a given injectable component can result in inflammation, soreness, or irritability of a cell, tissue, or organ system, and from the physiological perspective, whether this compound results in an elicitation of an activity or response in an organ or tissue that could result in a pathological alteration.

It becomes critical for investigators to become familiar with the various types of mechanisms that could result in damage to the tissues at the site of injection. Consultation with toxicologists can provide important insight into identifying the potential mechanisms responsible for tissue damage at the injection site. For example, in skeletal muscle, there are several mechanisms that can be initially considered when evaluating formulations for their potential to cause tissue damage. These potential mechanisms by which a molecule could cause muscle damage include (*i*) a disruption of the sarcolemma (the muscle membrane), which could disrupt intracellular homeostasis; (*ii*) a disruption or alteration in the mechanisms responsible for maintaining intracellular calcium homeostasis as this is essential to muscle functioning, and increased cytosolic calcium is associated with tissue damage; (*iii*) an interference in mitochondrial functioning thus disrupting homeostatic processes; (*iv*) an increased oxidative stress leading to formulation of reactive molecules, thus disrupting cellular functioning; and (*v*) dramatic changes in intracellular or extracellular pH or tonicity, which can result in cellular distress (5–13).

General Overview on the Mechanisms for Pain on Injection

Pain on injection involves the activation of nociceptors at the injection sites (1). Three types of nociceptors seem to be involved primarily with pain on injection and involve chemical, thermal, or mechanical sensitivity. This includes the acid-sensing ion channels that are activated by protons and have a preference for sodium, the heat-gated vanilloid receptors (VR-1 capsaicin) that are activated by heat (> 45°) and capsaicin and nonselective to cations, and the mechanosensitive or stretch-activated channels that are responsive to membrane stress and mechanical forces when cells are exposed to either hypo-osmotic or hyperosmotic fluids.

CONSIDERATIONS IN MODEL SELECTION FOR TISSUE DAMAGE

The selection of the in vitro or in vivo model for evaluating the potential of a drug, excipient, or formulation to cause tissue damage on injection requires the investigator to be knowledgeable of the particular aspects of these particular methodologies. These aspects include the advantages and disadvantages of the model; the parameters utilized to evaluate the tissue damage; the key experimental assumptions; important experimental cautions, limitation, and the requirements or approaches for data analysis. An investigator who neglects to take these aspects into consideration in utilizing these approaches may end up with experimental results that may not be that useful for screening, evaluation, and selection of parenteral formulations that are not associated with tissue damage on injection.

IN VITRO METHODS FOR EVALUATING TISSUE DAMAGE

In vitro methods can play a critical role in the selection of excipients or the development and comparison of various parenteral formulations. These methods, in general, can be easily developed and implemented in any laboratory setting and can provide an approach for the establishment of a database related to specific excipients and formulations useful for future studies, given the experimental assumptions and limitations are taken into consideration.

Red Blood Cell Hemolysis Methods

The utilization of red blood cell hemolysis with the release of hemoglobin as a marker for evaluating formulation-induced irritation continues to be an important approach in developing and optimizing injectables, particularly those intended for intravenous injection. Two types of experimental systems have been implemented, and involved either a static evaluation or flow through dynamic evaluation of the acute interaction between the test formulation and red blood cells as reported by Yalkowsky and coworkers (14–21) and Obeng and Cadwallader (22). Yalkowsky and his team have contributed significantly to the use of red blood cell hemolysis as an indicator of tissue damage (14,21). In a static evaluation of the interaction of a formulation with red blood cells, there are several key issues to be addressed. This includes limiting the sources of the red blood cells and ensuring adequate and consistent time for the interaction of the formulation with the red blood cells as this will minimize the variability. Furthermore, it is critical to keep the ratio of the test vehicle to the red blood cells constant, to incorporate in the study design the appropriate negative or positive controls, and to incorporate during the hemoglobin quantification an extraction method that avoids possible changes in hemoglobin absorption maxima by the test solution through the use of a standard matrix for the spectrophotometric analysis. Additional considerations for the dynamic flow through system include ensuring there is a consistent flow through the system to allow adequate mixing and interaction with between the test solution and the red blood cells (22). One advantage of the dynamic flow through system is it enables the investigator to vary the injection speed to look at dilutional effects and the impact on this interaction between the formulation and red blood cells.

Cell Culture Methods

The use of muscle cell cultures can be an important tool for evaluating tissue damage on injection. Two muscle cell lines have been found to be particularly useful in looking at parenteral induced tissue damage, specifically for intramuscular injectables. These cell lines are the rat L6 myoblasts and mouse C2C12 myoblasts, and both are available commercially (23–29). Cell culture methods can be easily adopted in the laboratory and are advantageous, given this is a relatively rapid approach to evaluate the acute effect of the test compound. Cell culture methods to evaluate tissue damage can employ the release of intracellular components (often cytosolic enzymes such as creatine kinase or lactate dehydrogenase) into the medium, the concentration of intracellular components remaining in the cells after removal of the medium or an assessment of cell viability or cell death caused by the treatment. If the measure of tissue damage is the release of cytosolic enzymes into the medium, the investigator must also conduct the requisite preliminary studies showing the presence of the treatment does not interfere with the activity of the specific enzyme. It is critical to evaluate the extent to which a treatment formulation may reduce the number of cells as they may be lifted from the plate during the experiment, particularly if one is analyzing the release or retention of intracellular components. As such, it becomes critical for the investigators to always normalize their experimental findings for cell number, protein, DNA, or other markers useful to characterize the cell population. Furthermore, it is critical to include in the experimental design the appropriate negative and positive control treatments as the benchmark for evaluating the extent of tissue damage.

A limitation of any cell culture approach is that the investigators must be cognizant of the specific passage number for the cell line. Secondly, experimental results can be confounded by complications associated with formulations that are not isotonic, as this could result in cell swelling and lysis associated with hypotonic solutions and cell shrinkage associated with hypertonic solutions. Cell passage number and tonicity can impact upon the concentrations of intracellular components often utilized as parameters for evaluating tissue damage, thus confounding experimental results.

Another key issue associated with muscle cell culture methods is whether to utilize either myoblast (immature muscle cells) or to differentiate the cells into mature muscle cells (myotubules) as this can impact on the concentration of intracellular components used as markers in the screening process. Both the L6 and C2C12 cell lines can be differentiated into myotubules as judged by increases in cytosolic enzymes and morphological changes. Figure 4 shows the difference in L6 and C2C12 in growth medium (2% fetal bovine serum in Dulbecco's

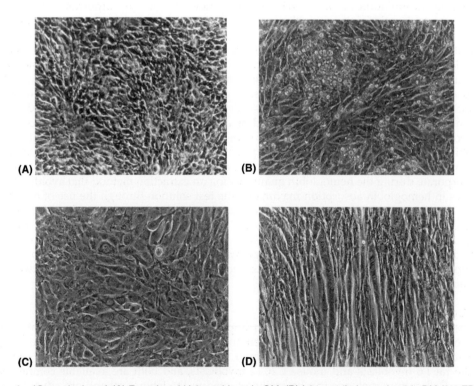

Figure 4 (*See color insert*) (**A**) Four-day-old L6 myoblasts in GM. (**B**) L6 myotubules at day 6 in DM (2% FBS in DMEM) during fusing process. (**C**) Four-day-old C2C12 in GM. (**D**) C2C12 myotubules at day 6 in DM (10% HS in DMEM). *Abbreviations*: GM, growth medium; DM, differentiation medium; FBS, fetal bovine serum; HS, horse serum.

Table 2 Creatine Kinase Activity During Differentiation in
L6 and C2C12 Cell Lines

| Cell line | Creatine kinase activity (U/L)[a] | |
	Myoblasts	Myotubules
L6	132.2 ± 19.8	73.6 ± 11.6[b]
C2C12	2905 ± 46	3599 ± 308[c]

[a]Means ± SEM. Data was obtained from the appropriate
initial cell density of 6×10^6 cells per sample.
[b]Significantly lower than that of myoblasts ($p < 0.05$).
[c]Significantly higher than that of myoblasts ($p < 0.05$).

Modified Essential Medium (DMEM)) compared with these cell lines in differentiation medium (10% horse serum in DMEM). There is a pronounced morphological alteration during differentiation process as shown by the change formation of myotubules in panels B and D. Futhermore, there can be a pronounced difference between the two cell lines in the creatine kinase activity in myoblasts versus myotubules as shown in Table 2.

Tissue Reactivity Model

Silva and colleagues reported a tissue reactivity model that can be useful to look at biocompatibility or toxicity of biomaterials, parenteral formulations, or delivery systems (30). In this experimental system, L-929 cells are grown to near confluent monolayers followed by removal of the culture medium that is replaced with agar-containing medium and neutral red vital stain (marker of cell viability). Following solidification of the agar, the treatment is placed on the cells with control treatments (on filter paper), and the cells are then incubated for 24 hours at 37°C. The culture can be evaluated microscopically around the treatments, and toxicity is measured by the loss of the vital stain. The investigator is able to evaluate the biological reactivity (cellular degeneration, lysis, malformation, and sloughing) by calculating a zone index (ZI) with a range of reactivity of the treatment ranging from 0 with no detectable zone around the sample to 5, which involves the entire dish (as the numerator) and a lysis index ranging from none to severe (80% of the zone affected) as the denominator for the controls and treatments. It is critical to include the appropriate positive and negative control treatments in this system (30).

Isolated Skeletal Muscle Systems

Rodent isolated muscles can also be useful in screening *formulations* for their potential to cause tissue damage for both intramuscular and subcutaneous injectables (31,32). This method involves direct administration of small volume (15 μL) of the treatment into either the extensor digitorum longus (EDL) muscle or the soleus (SOL) muscle. These two muscles are utilized because (*i*) they can be easily isolated and removed via the tendon connections from the rear legs without directly touching or damaging the respective muscle, (*ii*) the treatments can be injected easily into the body of the muscle belly using a small gas chromatographic syringe, and (*iii*) the muscles can be saved at the end of the experiment for possible histological evaluation. It is recommended to utilize both muscles in these studies as the EDL and SOL muscles can provide an indication of potential tissue damage to fast twitch glycolytic muscles or slow twitch oxidative muscles, respectively, and most human skeletal muscle is primarily composed of mixtures of these two muscle fiber types. The experimental design can involve using the two EDL and SOL muscles for one specific treatment, thus enabling duplicates for each animal for both muscles. Alternatively, the experimental design can utilize one EDL or SOL as the treated muscle, while the contralateral muscles could be used as the control (no treatment or solvent control).

In general, this experimental protocol involves male Sprague Dawley or one consistent strain of rats, six weeks old, 150 to 200 g, that are humanely sacrificed using cervical dislocation after being anesthetized. After carefully isolating and dissecting out the EDL or SOL muscles (their weight is ~200 mg for an adult rat), the treatment being investigated is injected lengthwise into the belly of the muscle. An optimal injection volume is 15 µL as this causes a small welt on the muscle and the investigator can visualize whether the treatment has leaked out of the muscle. Larger volumes (25 µL) are associated with more difficulties in the treatment leaking out of the muscle, while smaller volumes (5 µL) may not be sufficient to elucidate a response in the skeletal muscle. Once the muscle has been injected, it is suspended in the incubation vessel via placing the muscle into a small basket (a long narrow circumference teflon tube with holes to prevent the muscle from floating or being disrupted by the aeration process) and placed into 8 to 10 mL of balanced salt solution through which is being bubbled a carbogen (95% O_2–5% CO_2) at 37°C. One such balanced salt solution that has been utilized is composed of 116 mM sodium chloride, 5.4 mM potassium chloride, 5.6 mM dextrose, and 262 mM sodium bicarbonate adjusted to pH 7.4. This solution does not contain calcium, which has been shown to exacerbate skeletal muscle damage (33).

The extent of tissue damage can be measured by the release of cytosolic enzymes into the incubation medium over a specific period of time. The most useful markers of tissue damage are the release of enzymes such as creatine kinase or lactate dehydrogenase that can be easily quantified using available spectrophotometric kinetic assays. The most useful approach has been to measure the activity of the released enzymes at 30-minute intervals from the time the muscle was injected with the formulation. This is easily facilitated by draining the incubation medium at 30-minute intervals and replacing the incubation vessel with fresh balanced salt solution. Experiences with these isolated muscles indicate that tissue viability is maintained for 90 to 120 minutes as noted by a dramatic increase in enzyme release after 90 or 120 minutes. As such, tissue damage is quantified by the cumulative release of the enzyme (as measured by activity) over the experimental period. One caution, though it may be minor, given the injection volume of the test formulation (15 µL), the muscle size (~200 mg), and the incubation medium (8–10 mL) in this study design, is to always consider whether the treatment has the potential to interfere with enzyme activity or with the measurement of this enzyme activity. This issue can easily be addressed through simple preliminary studies looking at enzyme activity in the absence or presence of the treatment.

One overall advantage of this experimental system is it involves direct injection into the muscle tissue like an intramuscular administration. In addition, it can also provide a basis for evaluating subcutaneous injectables for their potential to cause tissue damage since this injection site is often adjacent to muscle tissue that may become damaged. Additional advantages of this type of experimental system for screening *formulations* for their potential to cause tissue damage is that the process is relatively rapid, uses a minimal amount of the test formulation, can easily be learned by new investigators, and is reproducible over time and location with minimum variability as measured by coefficient of variation in the experimental results of 10% to 20%. This system has also been shown to correlate well with in vivo results in animals and clinical trials (32,34,35).

A limitation in this experimental system is that it measures only the acute toxicity to the muscle tissue caused by either a direct effect on the muscle membrane (sarcolemma) or rapid biochemical changes as a result of the injected formulation. It is also critical to include the appropriate negative and positive control formulations in the study design as a basis for evaluating the magnitude of the tissue damage caused by a given formulation. Useful negative controls (those formulations that do not cause tissue damage) can include an uninjected muscle, a needle puncture alone with no vehicle, normal saline, and 5% dextrose, while positive controls (those formulations that have been shown to cause tissue damage) can include directly slicing or damaging the muscle, slicing the muscle in half, or other formulations such as surfactants at higher than normally used concentrations, solvents such as propylene glycol at 40% vol/vol or higher or available parenteral formulations that have been shown to cause muscle damage (32). Cautions, as stated earlier, in utilizing this experimental system or in selecting the appropriate positive or negative controls must take into account the viability of the isolated muscle and ensuring there is no interference in the measurement of the released enzyme or other cytosolic component being evaluated as the marker for tissue damage (36).

IN VIVO METHODS FOR EVALUATING TISSUE DAMAGE
Infusion-Related Thrombosis
Several studies have reported infusion-related thrombosis; however, this seems to be attributed most frequently to device used in the drug administration. Examples include continuous infusion of interleukin-2 and total parenteral nutrition (37,38).

Rabbit Model
The rabbit lesion model is a generally accepted method for the prediction of muscle damage following intramuscular administration of drugs (39,40). This is because the damaged area is readily visible and can be quantified using histological approaches. If the damaged area is sufficiently large, it is generally considered a lesion. This model has been extensively referenced since 1949 (41) and remains the "gold standard" for predicting formulation tolerability in humans (42). Being more sensitive to intramuscular inflammation than humans, rabbits serve as a good and gentle animal model to screen formulations that might be intolerable in humans.

The typical method of rabbit lesion assessment involves injecting groups of animals with 1 mL of test and reference articles, approximately 0.6 cm deep into the sacrospinalis muscle using 23-guage sterile needles. The animals are euthanized over period of time postdosing, for example, 1, 2, 3, 6, 12 through 24 days, and lesions monitored for hemorrhage, lesion volume, and histology. Protocols that benefit with creatine kinase measurements in blood involve sample and testing blood samples through 72 hours postinjection. The results are typically converted into area under the creatine kinase curve and used to compare formulations versus control treatments or other treatments.

Given inherent variability with animals, typically a group of six animals are advised per treatment, which allows good differentiation between different formulations and drug concentration effects (43). Key advantages of the method include its broad acceptability, opportunity for testing multiple treatments per animal to permit crossover comparison, correlation of test data with historical or published data, and ability to monitor the lesion size. The disadvantages of the method are somewhat inherent to animal models, like relatively more expensive over the in vitro methods and need for training in handling of animals, including dosing of test articles and blood sampling from the ear vein for creatine kinase measurements.

Rodent Model
While the rabbit model has certainly been useful in evaluating tissue damage associated with parenteral injections, the cost, time, and difficulties associated with the use of this animal model may limit the enthusiasm of such an approach. Alternatively, a rat model can be useful for evaluating injectable formulations, given the reduced costs, easier experimental design, and time considerations (35,41,43). The rodent model has been shown to be useful and complements the findings observed with the isolated muscle model (35,44–46).

In this specific experimental design, the rodent is canulated via the jugular vein and allowed to recover for at least 12 hours prior to initiating the study. In previous studies, this 12-hour period is sufficient to allow the serum creatine kinase levels (the marker of tissue damage) to return to baseline following the surgery for the placement of the jugular cannula. The treatment (200–500 µL) can then be injected either into the gastrocnenius muscle (one of the two main muscles of the calf with the SOL adjacent to the gastrocnemius) or the gluteus medius muscle (in the pelvic area on the dorsal side). The advantage of using a rat versus the rabbit is the duration of the experiment one needs to utilize in characterizing the serum creatine kinase levels. Experiences with measuring serum creatine kinase levels at specific times as a marker of tissue damage may only require up to 72 hours and in most cases will only require 24 hours (as compared to an average 7–10 days in the rabbit). This would enable the investigator to easily design crossover studies as based on the patency of the jugular cannula. An additional advantage of the rodent system in studies have shown that peak creatine kinase levels occur at two hours after injection and is independent of the magnitude of the tissue damage caused by the formulation (in rabbit studies peak creatine kinase levels varied as to the severity of the tissue damage, with the most damaging formulations peaking at an earlier time compared with less damaging formulations) (35,39). This would enable investigators to evaluate serum creatine kinase levels as a second measure of tissue damage. Additional advantages are the ease of working with rodents versus rabbits, given their size differences and costs in housing and caring between the two species. A limitation of using rats is the volume of and number of blood samples that can be taken daily or during the course of the study.

IN VIVO METHODS FOR EVALUATING PAIN ON INJECTION

Pain on injection can occur following local administration of drug (e.g., subcutaneous and intramuscular) as well as on infusion, like in the case of intravenous or intra-arterial injection. Often model selection for assessing pain on injection is dependent on the route, frequency, and duration of injection. Three animal models have been successfully applied for assessing pain on injection for both local as well as infused drug administrations.

Rabbit and Rat Vein Models

In these models, the respective animal is used for infusing test and comparator samples intravenously, and often the results are assessed visually in terms of local site reaction and the associated changes. Implicitly, an article with least local visual change on infusion is considered the least painful to the animal.

Rabbit ear vein has been used to assess pain on injection. In this model, groups of three to five animals receive a fixed drug concentration (e.g., 1–10 mg/mL) and a set total dose (e.g., 1–10 mg) over a predetermined infusion rate through their marginal ear vein (e.g., 1 mL/min). Use of set dosing parameters allows comparison of results among different formulation groups as well as with negative control like saline or dextrose. Following dosing, each animal is examined carefully at the site of injection, for up to 24 or 48 hours, for swelling, bruises, and/or discoloration of the injection site and surrounding tissue. Generally, no change on injection is indication of a relatively well-tolerated formulation. This model has been used for assessment of numerous drugs known to be painful on injection such as clarithromycin (47). The model assumes that pain on injection will translate into visual change at and around the site of injection.

The rabbit vein irritation test has been shown to be effective in that a comparison of a lactobionate solution of macrolide antibiotic clarithromycin with its emulsion formulation and a dextrose control demonstrated the negative control (dextrose) to cause no local changes. The solution formulation of drug caused flushing of blood at the site of injection in all three animals in this test group, with bruises lasting through 24 hours after dosing. However, an emulsion formulation intended to reduce pain on injection caused no local change in two of the three animals in this test group immediately on injection, one of which demonstrated no change through 24 hours after dosing and the other showed limited bruising during this period. The third animal in this group exhibited some bruises immediately on injection that lasted through the 24 hours observation period. The emulsion formulation was deemed to be more tolerable than the solution formula based on correlation of the results from this animal model with those from other models (48).

The rat tail vein test complements and provides data comparable with the rabbit ear vein model. Because of smaller size, typically six animals are used per test group in this model. The infusion rate is kept low, for example, 0.3 mL/min, and the results can be compared against controls after one or multiple dosing. Again, in a study comparing an emulsion formulation of macrolide clarithromycin against its solution form, the negative control dextrose demonstrated purple, pink, and red spots near the area of injection in five out of the six animals in this group. Upon administration of the drug lactobionate solution, all six animals in this group demonstrated pink, red, and purple area covering large portion of the tail around the injection site. As a comparison, the emulsion formulation indicated limited spots near the site of injection in fewer animals (48). The similarity in results in these two animal models is generally believed to be a good predictor of similar manifestations in human clinical trials.

Conscious Rat Model

Subjectivity and lack of good correlation in assessment of pain on injection based on visual scores and patient response led Marcek et al. (49) to investigate the response to the intravenous injection of test articles in rats restrained in a tube that, in turn, is connected with a data acquisition system. The model is based on the premise that the rapid onset of vocalization and struggle in the restraint tube following injection, and the disappearance of these signs on completion of injection, is indicative of pain caused by the product injected to the animal. The authors have validated this model using isotonic and hypertonic formulations, increasing concentrations of pain-inducing chemicals that demonstrated good correlation with the results, as well as with range of marketed products known to cause pain on injection. Although the model has been shown to discern formulations with differing pain-inducing abilities on injection, it is somewhat complex to step up specifically for rapid screening in preclinical drug development.

Rat Paw-Lick Model
This model is based on the theory that if a substance is injected in the paw of a rat, the frequency of paw licks by the animal are proportional to the pain at the site of injection. Implicitly, a formulation that does not cause pain on injection, for example, saline or dextrose, would not stimulate the animal for paw licks; however, a more painful chemical injected in the same area would trigger the animal to lick its paw. The model was initially developed for testing local pain on injection, for example, subcutaneous injection (50). The authors demonstrated good correlation between concentrations of pain-inducing drugs cefoxitin and cefazolin and paw licks over a 12-minute interval after injection. In addition, the authors were able to prove local anesthetic effect, and hence reduction in paw licks, after injection of these drugs that also contained lidocaine. The model also seems to correlate reasonably well with creatine kinase levels on injection, a marker of pain/irritation following injection (51).

A good correlation has been noted between normal and extreme pH of injectable samples with rat paw licks, as well as between cosolvent concentration in treatment and rat paw licks (52). Finally, a good correlation has been demonstrated between pain-causing formulation of macrolide clarithromycin and its less-painful emulsion formula with paw licks, and the results corroborate well with other models like rabbit ear vein and rat tail vein results.

A major limitation of this model is local drug administration and small injection volume. Although these limitations may not play a role in testing samples intended for local injection, the model may identify pain through paw licks for formulations that may not cause the same physiological response on intravenous injection due to dilution.

**Now and the Future Use of Molecular Genetic Methods
in Evaluating Tissue Damage and Pain on Injection**
With continued advancements in the area of molecular and genomic technologies, parenteral formulators will have the opportunity to employ screening techniques to identify specific biomarkers for tissue damage or pain on injection and whether these biomarkers are upregulated or downregulated with given excipients, drugs, or formulations. The availability of specific quantitative polymerase chain reaction (QPCR) gene-array systems to evaluate specific biochemical pathways will enable investigators to employ experimental systems ranging from cells culture methods and animal studies to even simple initial clinical trials to rapidly screen formulations for their potential to activate cellular entities thought to be associated with tissue damage and/or pain on injection. In clinical studies, this could be crucial if there is concern that repeated injections may result in the development of tissue damage. Furthermore, these experimental methods would enable investigators to identify and quantify specific biomarkers as a measure for tissue damage or pain for a given formulation or classes of compounds.

CONCLUSIONS
While not all injectables may be associated with damage and/or pain on injection during preclinical and clinical trials, when this occurs in an injectable product it can be a challenge to the subsequent optimization of the final formulation and to the acceptance by clinician and patients. It becomes necessary, therefore, for pharmaceutical scientists to be aware of the available experimental approaches, both in vitro and in vivo, to screen and evaluate excipients, drugs, or various formulations for their potential to cause tissue damage and/or pain early on in the development of injectables. The available literature can provide important insight into

the types of excipients, drugs, or formulations, which have been associated with these adverse effects during injection. Careful design of the parenteral formulation based on early screening and evaluation studies for any potential tissue damage and/or pain on injection can result in the savings of time and financial resources during subsequent studies in the development and approval processes. Furthermore, the development of an in-house database related to the potential for chemicals to cause damage and/or pain using existing experimental methods will enable the rational design of future formulations intended for parenteral administration.

ACKNOWLEDGMENTS

The authors would like to thank Dr Daniel Brazeau and Ms Kellyn Wendt for their assistance in reading this chapter.

REFERENCES

1. Brazeau GA, Cooper B, Svetic KA, et al. Current perspectives on pain upon on injection. J Pharm Sci 1998; 87(6):667–677.
2. Schmelz M. Translating nociceptive processing into human pain models. Exp Brain Res 2009; 196(1): 173–178.
3. Dussor G, Koerber HR, Oaklander AL, et al. Nucleotide signaling and cutaneous mechanisms of pain transduction. Brain Res Rev 2009; 60(1):24–35.
4. Mense S. Muscle pain: mechanisms and clinical significance. Dtsch Arztebl Int 2008; 105(12):214–219.
5. Adachi J, Asano M, Ueno Y, et al. Alcoholic muscle disease and biomembrane perturbations. J Nutr Biochem 2003; 14(11):616–625.
6. Authier FJ, Chariot P, Gherardi RK, et al. Skeletal muscle involvement in human immunodeficiency virus (HIV)-infected patients in the era of highly active antiretroviral therapy (HAART). Muscle Nerve 2005; 32(3):247–260.
7. Masini A, Scotti C, Calligaro A, et al. Zidovidine-induced experimental myopathy: dual mechanisms of mitochondria damage. J Neurolog Sci 1999; 166(2):131–140.
8. Moylan JS, Reid MB. Oxidative stress, chronic disease, and muscle wasting. Muscle Nerve 2007; 35(4): 411–429.
9. Napaporn J, Thomas M, Svetic KA, et al. Assessment of the myotoxicity of pharmaceutical buffers using an in vitro muscle model: effect of pH, capacity, tonicity and buffer type. Pharm Dev Technol 2000; 5(1):123–130.
10. Brazeau GA, Chu A. Solvent, dependent influences on skeletal muscle sarcoplasmic reticulum calcium uptake and release. Toxicol Appl Pharmacol 1994; 125(1):142–148.
11. Brazeau GA, Fung HL. Mechanisms of creatine kinase release from isolated rat skeletal muscles damaged by propylene glycol and ethanol. J Pharm Sci 1990; 79(5):393–397.
12. Persky AM, Green PS, Stubley L, et al. Protective effect of estrogens against oxidative muscle damage to heart and skeletal muscle in vivo and in vitro. Proc Soc Exp Biol Med 2000; 223(1):59–66.
13. McArdle A, Jackson MJ. Intracellular mechanisms involved in skeletal muscle damage. In: Salmons S, ed. Muscle Damage. Oxford: Oxford University Press, 1997:90–106.
14. Reed KW, Yalkowsky SH. Lysis of human red blood cells in the presence of various cosolvents. J Parenter Sci Technol 1985; 39(2):64–69.
15. Reed KW, Yalkowsky SH. Lysis of human red blood cells in the presence of various cosolvents. II. The effect of differing NaCl concentrations. J Parenter Sci Technol 1986; 40(3):88–94.
16. Reed KW, Yalkowsky SH. Lysis of human red blood cells in the presence of various cosolvents. III. The relationship between hemolytic potential and structure. J Parenter Sci Technol 1987; 41(1):37–39.
17. Ward GH, Yalkowsky SH. The role of the effective concentration in interpreting hemolysis data. J Parenter Sci Technol 1992; 46(5):161–162.
18. Krzyzaniak JF, Raymond DM, Yalkowsky SH. Lysis of human red blood cells 1: Effect of contact time on water induced hemolysis. PDA J Pharm Sci Technol 1996; 50(4):223–226.
19. Krzyzaniak JF, Alvarez Núñez FA, Raymond DM, et al. Lysis of human red blood cells. 4. Comparison of in vitro and in vivo hemolysis data. J Pharm Sci 1997; 86(11):1215–1217.
20. Krzyzaniak JF, Yalkowsky SH. Lysis of human red blood cells. 3: Effect of contact time on surfactant-induced hemolysis. PDA J Pharm Sci Technol 1998; 52(2):66–69.
21. Yalkowsky SH, Krzyzaniak JF, Ward GH. Formulation-related problems with intravenous drug delivery. J Pharm Sci 1998; 87(7):787–796.
22. Obeng EK, Cadwallader DE. In vitro dynamic method for evaluating the hemolytic potential of intravenous solutions. J Parenter Sci Technol 1989; 43(4):167–173.
23. Williams PD, Masters BG, Evans LD, et al. An in vitro model for assessing muscle irritation due to parenteral antibiotics. Fundam Appl Toxicol 1987; 9(1):10–17.

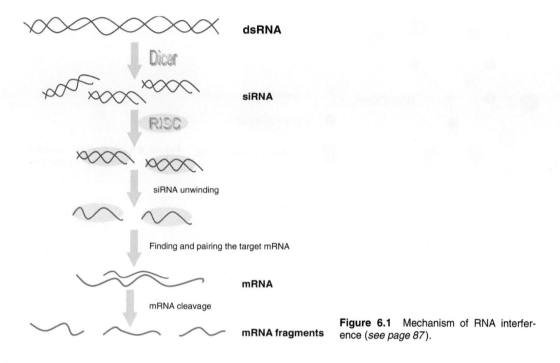

Figure 6.1 Mechanism of RNA interference (*see page 87*).

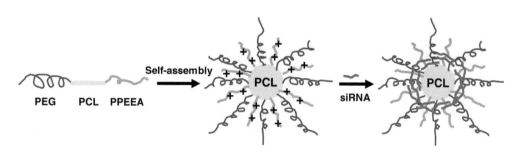

Figure 6.5 Self-assembling of cationic micellar nanoparticles and loading of siRNA. *Abbreviation*: siRNA, small interfering RNA. *Source*: Adapted from Ref. 73 (*see page 95*).

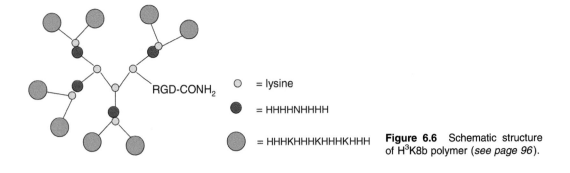

RGD-CONH₂

○ = lysine

● = HHHHNHHHH

⬤ = HHHKHHHKHHHKHHH

Figure 6.6 Schematic structure of H³K8b polymer (*see page 96*).

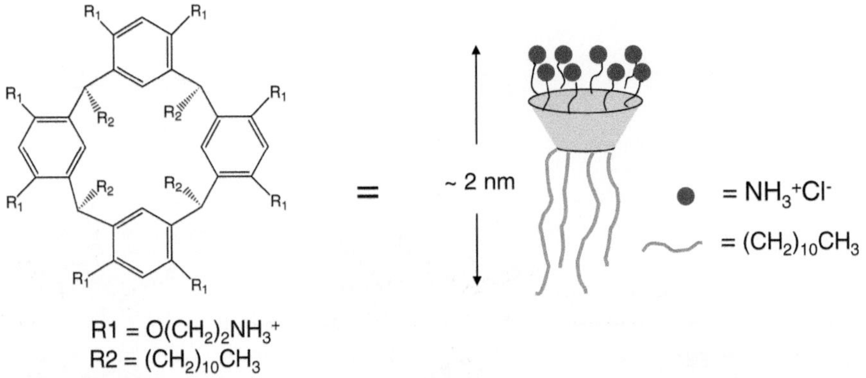

R₁
R₁
R₁
R₂   R₂
R₁
R₂   R₂
R₁
R₁   R₁
R₁   R₁

= ~ 2 nm

● = NH₃⁺Cl⁻

∿ = (CH₂)₁₀CH₃

R1 = O(CH₂)₂NH₃⁺
R2 = (CH₂)₁₀CH₃

Figure 6.8 Cone-shaped structure of macrocyclic octaamine. *Source*: Adapted from Ref. 88 (*see page 98*).

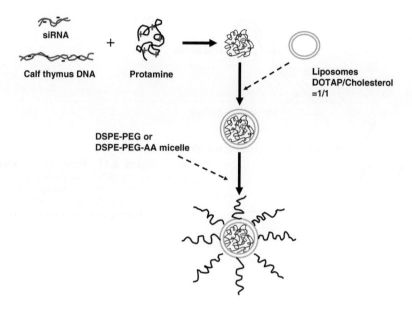

Figure 6.9 Preparation of PEGylated LPD. *Abbreviation*: LPD, liposome-polycation-DNA. *Source*: Adapted from Ref. 90 (*see page 99*).

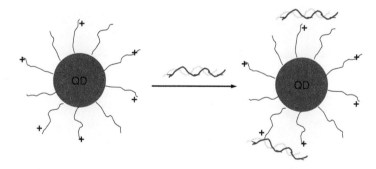

Figure 6.10 Adsorption of siRNA onto surface-modified QDs. *Abbreviations*: siRNA, small interfering RNA; QDs, quantum dots (*see page 102*).

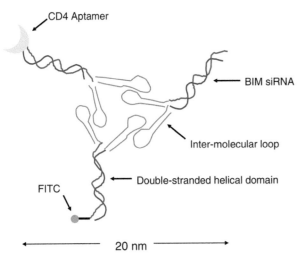

Figure 6.11 Schematic structure of engineering pRNA nanoparticle containing siRNA, aptamer, and fluorescent label. *Abbreviations*: pRNA, packing RNA; siRNA, small interfering RNA (*see page 103*).

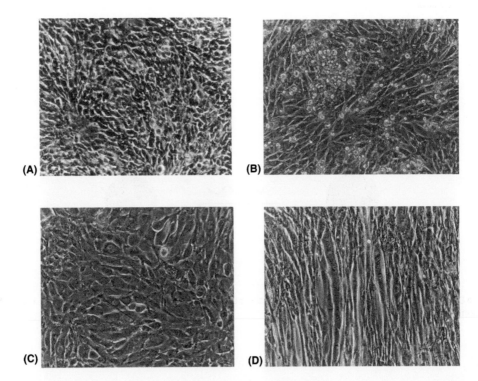

Figure 8.4 **(A)** Four-day-old L6 myoblasts in GM. **(B)** L6 myotubules at day 6 in DM (2% FBS in DMEM) during fusing process. **(C)** Four-day-old C2C12 in GM. **(D)** C2C12 myotubules at day 6 in DM (10% HS in DMEM). *Abbreviations*: GM, growth medium; DM, differentiation medium; FBS, fetal bovine serum; HS, horse serum. (*see page 140*).

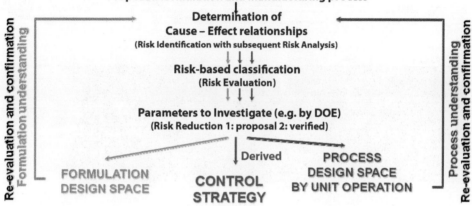

Figure 13.2 Process flow diagram describing the approach to developing process understanding and building quality into formulation and manufacturing process design (*see page 249*).

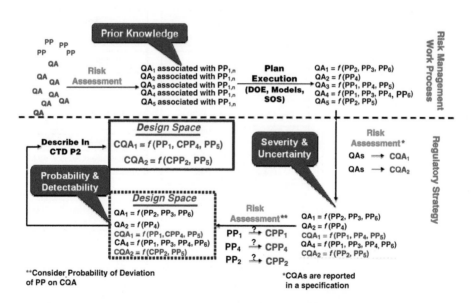

Figure 13.5 Schematic example of the quality risk management process. *Abbreviations*: PP, process parameters; CQA, critical quality attribute; CTD, Common Technical Document. *Source*: From Refs. 22,23,26 (*see page 253*).

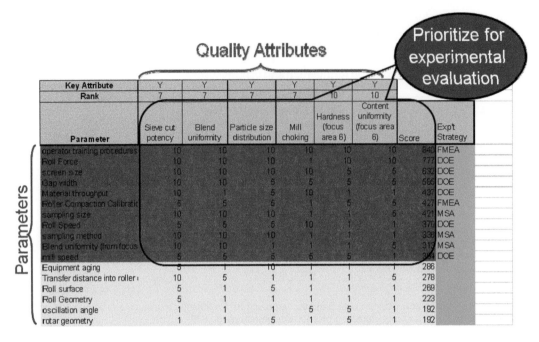

Key Attribute	Y	Y	Y	Y	Y	Y		
Rank	7	7	7	7	10	10		
Parameter	Sieve cut potency	Blend uniformity	Particle size distribution	Mill choking	Hardness (focus area 6)	Content uniformity (focus area 6)	Score	Expt Strategy
operator training procedures	10	10	10	10	10	10	840	FMEA
Roll Force	10	10	10	1	10	10	777	DOE
screen size	10	10	10	10	5	5	632	DOE
Gap width	10	10	5	5	5	5	565	DOE
Material throughput	10	1	5	10	1	1	437	DOE
Roller Compaction Calibration	5	5	5	1	5	5	427	FMEA
sampling size	10	10	10	1	1	5	421	MSA
Roll Speed	5	5	5	10	1	1	379	DOE
sampling method	10	10	10	1	1	1	336	MSA
Blend uniformity (from focus	10	10	1	1	1	5	313	MSA
mill speed	5	5	5	5	5	1	314	DOE
Equipment aging	5	1	10	1	1	1	286	
Transfer distance into roller	10	5	1	1	1	5	278	
Roll surface	5	1	5	1	1	1	269	
Roll Geometry	5	1	1	1	1	1	223	
oscillation angle	1	1	1	5	5	1	192	
rotar geometry	1	1	5	1	5	1	192	

Figure 13.6 Cause-and-effect matrix for distinguishing important quality attributes and process parameters for subsequent evaluation (*see page 254*).

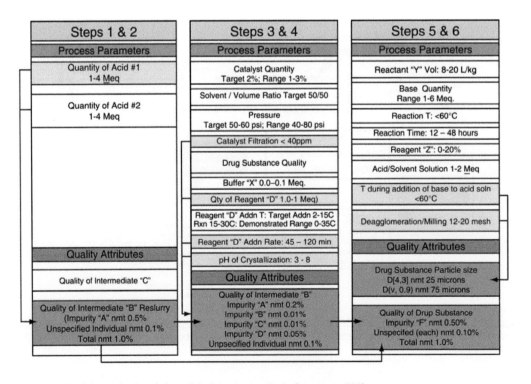

Figure 13.10 Schematic description of design space criteria (*see page 260*).

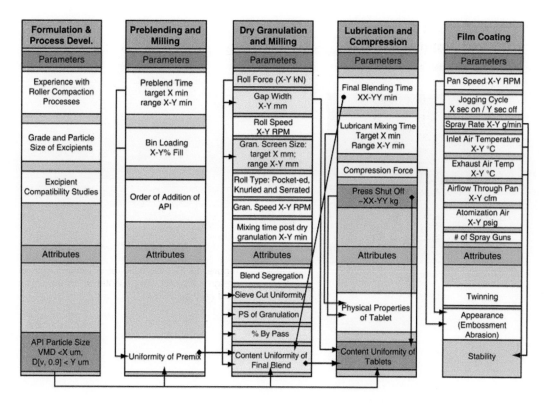

Figure 13.11 Design space for a drug product manufacturing process (*see page 261*).

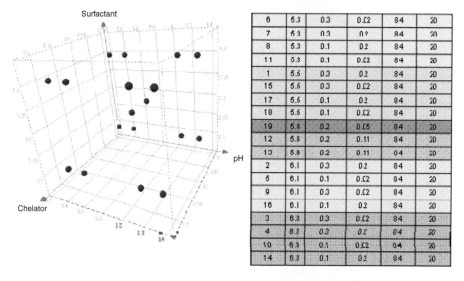

Figure 13.12 Example of multifactorial design to determine optimum concentrations of formulation parameters for a biologic (*see page 262*).

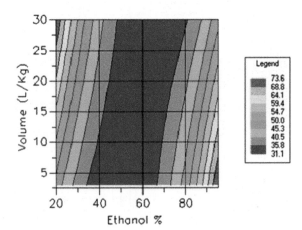

API Yield Relative to Total Volume X% EtOH

Yield (%) Contour Plot

Legend

73.6
68.8
64.1
59.4
54.7
50.0
45.3
40.5
35.8
31.1

Figure 13.14 Example of a contour plot of design space for drug substance crystallization yield (*see page 264*).

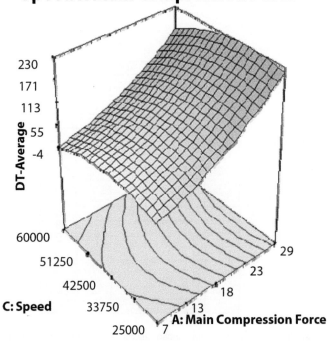

Tablet Disintegration Time Relative to Speed X Main Compression Force

Figure 13.15 Contour plot describing design space for tablet disintegration time (*see page 264*).

24. Laska DA, Williams PD, White SL, et al. In vitro correlation of ultrastructural morphology and creatine phosphokinase release in L6 skeletal muscle cells after exposure to parenteral antibiotics. In Vitro Cell Dev Biol 1990; 26(4):393–398.

25. Laska DA, Williams PD, Reboulet JT, et al. The L6 muscle cell line as a tool to evaluate parenteral products for irritation. J Parenter Sci Technol 1991; 45(2):77–82.

26. Horn JW, Jensen CB, White SL, et al. In vitro and in vivo ultrastructural changes induced by macrolide antibiotic LY281389. Fundam Appl Toxicol 1996; 32(2):205–216.

27. Lomonte B, Angulo Y, Rufini S, et al. Comparative study of the cytolytic activity of myotoxic phospholipases A2 on mouse endothelial (tEnd) and skeletal muscle (C2C12) cells in vitro. Toxicon 1999; 37(1):145–158.

28. da Costa KA, Badea M, Fischer LM, et al. Elevated serum creatine phosphokinase in choline-deficient humans: mechanistic studies in C2C12 mouse myoblasts. Am J Clin Nutr 2004; 80(1):163–170.

29. McClung JM, Judge AR, Talbert EE, et al. Calpain-1 is required for hydrogen peroxide-induced myotube atrophy. Am J Physiol Cell Physiol 2009; 296(2):C363–C371.

30. Silva VV, Lameiras FS, Lobato ZI. Biological reactivity of zirconia-hydroxyapatite composites. J Biomed Mater Res Appl Biomater 2002; 63(5):583–590.

31. Brazeau GA, Fung HL. An in vitro model to evaluate muscle damage following intramuscular injections. Pharm Res 1989; 6(2):167–170.

32. Brazeau GA, Fung HL. Use of an in vitro model for the assessment of muscle damage from intramuscular injections: in vitro-in vivo correlation and predictability with mixed solvent systems. Pharm Res 1989; 6(9):766–771.

33. Brazeau GA, Fung HL. Mechanisms of creatine kinase release from isolated rat skeletal muscles damaged by propylene glycol and ethanol. J. Pharm Sci 1990; 79(5):393–397.

34. Brazeau GA, Sciame M, Al-Suwayeh SA, et al. Evaluation of PLGA microsphere size effect on myotoxicity using the isolated rodent skeletal muscle model. Pharm Dev 1996; 1(3):279–283.

35. Al-Suwayeh, Tebbett IR, Wielbo D, et al. In vitro in vivo myotoxicity of intramuscular liposomal formulations. Pharm Res 1996; 13(9):1384–1388.

36. Brazeau GA, Fung HL. In vitro assay interferences of creatine kinase activity. Biochem J 1989; 257(2): 619–621.

37. Eastman ME, Khorsand M, Maki DG, et al. Central venous device-related infection and thrombosis in patients treated with moderate dose continuous-infusion interleukin-2. Cancer 2001; 91(4):806–814.

38. Andrew M, Marzinotto V, Pencharz P, et al. A cross-sectional study of catheter-related thrombosis in children receiving total parenteral nutrition at home. J Pediatr 1995; 126(3):358–363.

39. Brazeau GA, Fung HL. The effect of organic cosolvent-induced skeletal muscle damage on the bioavailability of intramuscular ^{14}C-diazepam. J Pharm Sci 1990; 79(9):773–777.

40. Sutton SC, Evans LA, Rinaldi MT, et al. Predicting injection site muscle damage I: evaluation of immediate release parenteral formulation in animal models. Pharm Res 1996; 13(10):1507–1513.

41. Nelson AA, Price CW, Welch H. Muscle irritation following the injection of various penicillin preparations in rabbits. J Am Pharm Assoc 1949; 38(5):237–239.

42. Gray JE. Appraisal of the intramuscular irritation tests in rabbits. Fund Appl Toxicol 1981; 1:290–292.

43. Sutton SC. Lesion and edema models. In: Gupta PK, Brazeau GA, eds. Injectable Drug Development. Denver: Interpharm 1999:91–117.

44. Kranz H, Brazeau GA, Napaporn J, et al. Myotoxicity studies of injectable biodegradable in-situ forming drug delivery systems. Int. J Pharm 2001; 212(1):11–18.

45. Rungseevijitprapa W, Brazeau GA, Simpkins JW, et al. Myotoxicity studies of O/W-in situ forming microparticle systems. Eur J Pharm Biopharm 2008; 69(1):126–133.

46. Kranz H, Yilmaz E, Brazeau GA. In vitro and in vivo drug release from a novel *in situ* forming drug delivery system. Pharm Res 2008; 25(6):1347–1354.

47. LingHao Q, Wei L. Formulation and evaluation of less-painful clarithromycin lipid microspheres. Arch Pharm Res 2007; 30(10):1336–1343.

48. Lovell MW, Johnson HW, Hui HW, et al. Less painful emulsion formulations for intravenous administration of clarithromycin. Int J Pharm 1994; 109:45–57.

49. Marcek JM, Seaman WJ, Weaver RJ. A novel approach for the determination of the pain producing potential of intravenously injected substances in the conscious rats. Pharm Res 1992; 9(2):182–186.

50. Celozzi E, Lotti VJ, Stapley EO, et al. An animal model for assessing pain-on-injection of antibiotics. J Pharmacol Methods 1980; 4(4):285–289.

51. Gupta PK. Rat-paw lick model. In: Gupta PK, Brazeau GA, eds. Injectable Drug Development. Denver: Interpharm 1999:119–130.

52. Gupta PL, Patel P, Hahn KR. Evaluation of pain and irritation following local administration of parenteral formulations using the rat paw-lick model. J Pharm Sci Technol 1994; 48(3):159–166.

9 | Parenteral product specifications and stability

Michael Bergren

INTRODUCTION

Specifications and stability of parenteral products are set in the broader context of drug product development of small molecules, biologics, and devices. The specifications for a finished pharmaceutical product are an accepted list of requirements that a product must meet before it is released into distribution. Typically, these requirements are laboratory tests and associated acceptance criteria. International Conference on Harmonisation (ICH) Q6A (1) provides a suitable working definition of specifications. For purposes of this chapter, the definition should be broadened to extend beyond "new" drug substances and products.

> A specification is defined as a list of tests, references to analytical procedures, and appropriate acceptance criteria, which are numerical limits, ranges, or other criteria for the tests described. It establishes the set of criteria to which a new drug substance or new drug product should conform to be considered acceptable for its intended use. "Conformance to specifications" means that the drug substance and/or drug product, when tested according to the listed analytical procedures, will meet the listed acceptance criteria. Specifications are critical quality standards that are proposed and justified by the manufacturer and approved by regulatory authorities as conditions of approval.

The concept of specifications is central to globally accepted principles of Good Manufacturing Practice (GMP). Specifications are product-specific and must ensure, in the words of U.S. GMP regulations, that "drug products conform to appropriate standards of identity, strength, quality, and purity" (2). Further, existing U.S. GMP regulations state "For each batch of drug product, there shall be appropriate laboratory determination of satisfactory conformance to final specifications for the drug product." Specifications must also be developed for investigational products used in human clinical trials. In contrast to marketed products, specifications for clinical trial materials generally reflect the more limited understanding of the product and assume greater commensurate restrictions on its use in a carefully monitored clinical setting.

The topic of specifications for the broad category of parenteral medications shares elements common to all drug products, but it includes many additional complexities unique to parenterals, particularly within the broad scope of modern parenterals.

1. Because parenteral administration is fundamentally invasive, specifications always include requirements for sterility, limits on byproducts of microbes (e.g., endotoxin), and limits on particulates, particularly in the case of intravenous administration.
2. A large number of parenteral products are not sufficiently stable to be marketed as solutions. These products must be reconstituted prior to injection, in many cases from a lyophilized solid or sterile powder. Specifications on these products must incorporate an understanding of factors that influence the solid-state stability of lyophiles, frequently incorporating partially or completely amorphous drugs.
3. Biopharmaceuticals are a large and rapidly growing category of complex substances that are delivered almost exclusively via the parenteral route. Specifications for biopharmaceuticals generally reflect a qualitative difference in both our capacity for analytical characterization of these molecules and the robustness of these products.
4. Increasingly, parenteral medications are being developed to achieve either sustained or targeted delivery. Many of these products require specifications based on chemical or functional tests to help ensure consistent drug release or targeting. While there is some overlap with considerations applied to other categories of modified release drugs, the physiology, chemistry, and requirements are usually quite different.

5. Parenterals are commonly in intimate contact with product packaging and may be delivered via admixtures or through a variety of devices. Specifications may need to include formulation quality attributes affected by packaging or administration sets.

The topic of biopharmaceuticals will benefit from a brief explanation of terminology in this introductory section. For purposes of this chapter, drug substances will be classified in one of two broad categories: synthetic/semisynthetic drug substances or biopharmaceutical drug substances. This is a somewhat arbitrary division that is largely drawn to recognize two quite different situations from a regulatory and specifications perspective (3). While these categories appear to relate to the origin of the drug substance, they are equally associated with our capability to achieve complete molecular characterization of the drug using modern analytical methods. A prototypical synthetic drug substance is a single chemical entity produced largely or entirely by organic chemical synthesis, with an impurity profile and a degradation profile that can be almost completely known and measured with available analytical methods. For such a substance, a chemical assay, or a collection of chemical assays, can be used with confidence to assure both biological potency and drug safety. Indeed, in many cases the terms "assay" and "potency" are used interchangeably. From a product and process design perspective, as well as a regulatory and specifications perspective, there are clear benefits to achieving this detailed level of understanding. In contrast, a prototypical biopharmaceutical drug substance is the isolated high molecular weight product of a biological organism. It is less well understood on a molecular level, sometimes much less well understood. Existing analytical tools fail to completely measure the molecular attributes that contribute to the biological activities—desired or undesired—of the prototypical biopharmaceutical drug substance, and an array of physicochemical analyses need to be supplemented by biological or biochemical assays to assess both potency and safety with confidence. The prototypical biopharmaceutical drug substance is more susceptible to changes resulting from stresses encountered in processing, formulation, or storage. Clearly, these prototypical examples are constructs that represent two extremes, but the categories they represent require significant differences in approach to specifications development. These differences are reflected in the regulatory guidances covering specifications and stability, and they will be emphasized periodically throughout this chapter.

Specifications are a broad topic that can be discussed in different contexts, but the focus of this chapter will be specifications in the context of parenteral product stability. Consequently, the final product specifications will be discussed with an emphasis on the relationship between specifications and product shelf life. Other aspects of specifications—including specifications on components, drug substance, raw materials, or in-process specifications—will not be specifically addressed. Even with this restriction, it is not possible to cover all areas in depth in a single chapter, in particular for areas where approaches to setting specifications are still evolving, such as biopharmaceuticals and controlled release parenterals.

DEVELOPMENT OF SPECIFICATIONS

Specifications gain acceptance as an outcome of the regulatory process associated with product registration. This application process is a formal dialog between the product manufacturer, or applicant, and the appropriate regulatory authority. Acceptability of final product specifications is predicated on the following elements of the application.

- Product definition
  The rationale for specifications is predicated on a well-defined product—including formulation, process, and packaging—manufactured according to current GMP standards. Although specifications on components will not be discussed in this chapter (note—other chapters address this topic), it should be noted that there are significant and stringent global requirements for specifications on raw materials, especially water, and packaging used in parenteral products.
- Product performance
  The product development process is expected to result in clear expectations for product performance and a corresponding understanding of attributes critical to

product performance. Specifications must be designed to ensure that marketed product performs safely and effectively, consistent with the performance in clinical studies that provided basis for approval.

- Regulatory expectations
  Proposed specifications are considered within the context of current regulatory expectations for the category and region in which the product will be marketed. Because of the intrinsically invasive nature of parenteral therapy, a number of specific requirements are integral to specifications for parenteral dosage forms.
- Stability
  Stability of the product must be understood and supported by confirmatory data. Storage conditions should be established that ensure the product continues to meet specifications throughout its shelf life or expiration dating period. Product attributes, and related test results, are expected to change over time, but specifications should be designed to ensure that products meet requirements throughout shelf life.
- Test procedures
  Test procedures and sampling procedures are integral to product specifications. Test procedures must be validated, and validation is specific to the formulation and process. Likewise, acceptance criteria are specific to the test procedure; changes in procedures may require revalidation of the method and the associated acceptance criteria may need to be revised, even if the product is unchanged.

From a manufacturing perspective, the central outcomes of drug product development are the final product specifications and the manufacturing process. Specifications created for products in early stages of clinical investigation are designed to ensure the safety of the product in the clinical setting, and to provide reasonable assurance of the integrity of conclusions derived from the outcome of the study. (In these early stages, risks are concomitantly reduced because clinical exposure is also limited, and there is significantly greater control over clinical setting.) Because knowledge is limited, specifications with quantitative acceptance criteria are commonly fewer in number and acceptance criteria may be less restrictive, with the exception of specifications for impurities. As formulation, and clinical, and process experience with the drug develops, specific product attributes may be identified that are critical to product quality (4). Identification of these attributes provides a framework for defining the experimental studies and data required to establish manufacturing process and specifications.

As a useful example, consider the case where release rate of drug from the formulation is critical for achievement of effective blood levels, and one of the important determinants of release rate is suspected to be particle size. Specific developmental studies—including combinations of clinical, nonclinical, and in vitro studies—may be designed to assess the dependence of blood levels on particle size, including interactions with other formulation factors. In addition, data collected from processing studies provide an assessment of the capability of the process to manufacture drug within a targeted range of particle size. In the end, the limits on particle size (i.e., acceptance criteria for the specification) may be established on the basis of requirements imposed by drug release, which has an established relationship to safe and therapeutic blood levels. However, if the manufacturing process is typically capable of producing particles in a much tighter range, the limits may instead be chosen to reflect the process capabilities. A third outcome is that developmental studies could be used to justify the absence of a specification for particle size, in the event that the process is shown to be consistently capable of producing material in a range where no meaningful variation in drug release could be demonstrated. The rational development of specifications based on an understanding of critical quality attributes is fundamental to concepts of Quality by Design, which are elaborated in a subsequent chapter.

REGULATORY EXPECTATIONS

Many long-standing requirements for parenterals have tests and acceptance criteria that are thoroughly vetted and well documented in regulations and compendia. Although requirements have long been similar across market regions, they were sufficiently distinct to create great complexity in specifications for a "global" pharmaceutical product. Multiple tests were occasionally required for the same attribute to ensure regional regulatory approvals.

During the last two decades however, members of the Pharmacopoeial Discussion Group (PDG) have harmonized several important tests in the regional compendia in traditional major pharmaceutical markets—the United States, Europe, and Japan. This progress is reflected in significantly improved consistency among the compendia in these three regions, including *The United State Pharmacopoeia* (USP) (5), the *European Pharmacopoeia* (Ph. Eur.) (6), and the *Japanese Pharmacopoeia* (JP) (7). (For clarity and brevity, in subsequent text, references will be made to titles of specific compendial chapters without additional parenthetic references to the bibliography. All such references should be understood to refer to the editions, through indicated supplements.) All three compendia have specific umbrella discussions of require-ments for parenterals: USP <1> *Injections*, Ph. Eur. *Parenteral Preparations*, and JP *General Rules for Preparations 11. Injections*, and many of the test requirements in these chapters provide references to chapters that have been partially or entirely harmonized.

In a parallel manner, regulatory expectations have been increasingly harmonized through the initiatives of regulatory and industry groups in these three major market regions. Their collaboration across a broad range of topics, under the banner of the *International Conference on Harmonisation of Technical Requirements for Registration of Pharmaceuticals for Human Use*, has resulted in agreements commonly referred to as ICH guidelines (8). As an integral part of the ICH process, the ICH guidelines are adopted by regulators in participating regions. In some cases these guidelines are new, but commonly they displace prior guidance documents that were not harmonized. A parallel process for harmonization efforts for veterinary drugs have occurred under the auspices of VICH (9), which in many cases has resulted in quality guidelines analogous to those from ICH. The ICH harmonization process is woven together with the compendial harmonization efforts through a process described in ICH Q4B: Evaluation and Recommendation of Pharmacopoeial Texts for Use in the ICH Regions (10). Through this process, the ICH Expert Working Group recommends harmonized compendial text for adoption by regulators as interchangeable across ICH regions. The harmonized tests, along with considerations for implementation in each region, are published in annexes to the ICH Q4B guidance.

Although the ICH process centered on regulatory process in regions associated with major pharmaceutical research, from the beginning it involved observers from other bodies representing non-ICH regions, particularly the World Health Organization (WHO). As a consequence, ICH guidelines have had significant influence on regulatory processes and expectations well beyond Europe, the United States, and Japan. Guidelines have provided reasonable frameworks for regional harmonization initiatives, such as ASEAN Harmonization effort (11). Not infrequently however, regional initiatives have made changes to the guidelines to meet regional requirements. As a consequence, for example, the initial ICH guideline (12) on stability in Climate Zones III and IV was withdrawn because of concerns in Zone IV countries that stability conditions were insufficiently stressful to reflect climatic conditions prevalent in some locations. WHO and the regional initiatives are filling the gap through introduction of regionally harmonized guidelines (13). Hence, despite the value of global efforts toward regulatory harmonization, it is still important to understand regional regulatory expectations. Subsequent discussion will draw largely on the ICH guidelines, which adequately capture central global themes for parenterals specifications and stability, but regional details may differ, particularly outside the ICH regions.

Shelf Life Specifications and Release Specifications

Specifications are universally understood to be requirements that the product is expected to meet throughout its shelf life, when it is stored and dispensed according to the instructions provided in its labeling. Few products are timeless however. When products age, results of some tests will change over time. Stability studies are conducted to confirm that these changes occur in a reproducible manner that is characteristic of the product design. For those tests that change with time, test results at the time the product is released into distribution must be sufficiently within the acceptance criteria to ensure the product remains within the acceptance criteria throughout its shelf life. In effect, the product must be released according to a narrower set of acceptance criteria. In some regulatory regions, the European Union in particular, "release specifications" must be submitted for regulatory approval in addition to "shelf life specifications." Whether release specifications are submitted for approval or not, Quality Assurance groups in most manufacturing organizations necessarily decide on product release by utilizing a set of criteria narrower than the shelf life specification. These criteria are normally designed to ensure the product meets specifications when tested throughout shelf life, accounting for changes that result from both product stability as well as measurement variability. Generally, and throughout the remainder of this chapter, "specifications" will refer to shelf life specifications. Where requirements refer to release specifications, they will specifically be noted as such.

Individual Specification Requirements

This section provides brief introduction to elements of common parenterals specifications as presented in ICH Q6A guideline (1) and major regional compendia. ICH recognized the distinctive challenges posed by biopharmaceutical products and consequently issued ICH Q6B (14), the scope of which covers "proteins and polypeptides, their derivatives, and products of which they are components (e.g., conjugates). These proteins and polypeptides are produced from recombinant or nonrecombinant cell-culture expression systems." ICH Q6B provides additional guidance for this category of biopharmaceutical products but refers to compendial testing requirements for many standard parenteral tests.

Volume of Injection

For liquid parenteral products, fill volume must be sufficient for withdrawal of the specified volume of injection from the container using the recommended configuration of needle and syringe for injection. Some overfill is generally required, the magnitude of which depends on both container volume and product viscosity. Specific requirements for volume of injection, both for single-dose and multiple-dose products, are found in USP *<1> Injections*, which is harmonized with JP 6.05 *Test for Extractable Volume of Parenteral Preparations* and Ph. Eur. 2.9.17. *Test for Extractable Volume of Parenteral Preparations.* The methods are considered interchangeable within the current compendia, as indicated in ICH Q4B Annex2 (15). The volume of injection may be determined as part of in-process testing, but the limit should be justified on the basis of the actual volume requirement for administration. Normally, volume of injection would not be evaluated as a stability-related attribute unless the product was packaged in a semipermeable container.

Description or Appearance

Description is a standard requirement that provides valuable qualitative information on the visual appearance of a product relative to a standard description. It typically includes an assessment of color, clarity of solution, physical integrity of lyophilized cake, homogeneity of dispersion, or visual indication of presence of foreign matter. Attributes such as physical separation of emulsions or dispersions or cake collapse, may be readily detected by appearance testing. As a supplement to the qualitative appearance test, specific quantitative compendial tests allow measurement of color or turbidity relative to a set of standards (e.g., Ph. Eur. 2.2.2. *Degree of coloration of liquids*; 2.2.1. *Clarity and degree of opalescence of liquids*; USP *<851> Spectrophotometry and light-scattering*). These tests may provide particularly appropriate tools for establishing thresholds of acceptability for appearance, or for allowing more quantitative trending of color and clarity on stability. Appearance is a vital requirement for stability testing.

Visible Particles

Solutions for parenteral administration are expected to meet compendial requirements. USP31 *<1> Injections* states that solutions should be "essentially free from visible particulates." Ph. Eur. 6 *Parenterals* requires that solutions for injection "are clear and practically free from particles," and JP XV requires that "Injections must be clear and free from readily detectable foreign insoluble matter." Products are expected to meet these criteria at the time of release on the basis of established and qualified visual or machine-based inspection methods of each unit. Extensive history and many of the details associated with the requirement for visual inspection are provided in references (16). The limit on visible particles applies to all small volume parenterals, infusions, and reconstituted solutions. Where the character of the product does not allow for 100% inspection—either because of packaging or because the product must be reconstituted—the product must be suitably sampled for visual inspection. In some cases, the product may need to be transferred to an alternate vessel for inspection.

Parenteral products should typically be examined for visible particles as part of developmental and registration stability program for parenterals to ensure that instability related to decomposition or incompatibility does not give rise to the appearance and growth of visible particles. Although the product will typically be examined for subvisible particulates, some common mechanisms for particle nucleation and growth may result in visible particles without exceeding the limits for subvisible particulate. In general, visible particles should not appear on stability, but because of the probabilistic nature of the inspection process, visible particles may occasionally be evident in product on stability, even though product met inspection criteria at time of release and even in the absence of stability-related particle formation. Therefore, when particles are observed, it is particularly helpful to identify and characterize the particle to assess whether it is a foreign contaminant from the manufacturing process (extrinsic) or whether it is formulation-related (intrinsic). Trends in the occurrence of intrinsic particles may signify important stability-related changes in the product.

Subvisible Particulate
The requirement for subvisible particulates in parenteral solutions is harmonized among the three major compendia (USP *<788> Particulate Matter in Injections*; Ph. Eur. *2.9.19. Particulate Contamination: Sub-Visible Particles*; and JP *6.07 Insoluble Particulate Matter Test for Injections*). This requirement is uniformly applicable to both large- and small-volume injectables. However, the limits may be relaxed for injectables that are strictly administered via the subcutaneous or intramuscular route or for powders that are reconstituted prior to injection. (There is, however, a possibility that this exclusion may be dropped.) The test currently can be conducted via light obscuration (LO) or membrane microscopy (MM) methods. Although LO is frequently preferred, the microscopic method has advantages for formulations where optical or flow characteristics preclude the use of LO method. Additionally, if the results of the LO test exceed the limits, the procedure prescribes a second stage of testing using the MM method to assess whether the sample meets the requirement. Second stage testing may be required, for example, in the testing of prefilled syringes, where silicone oil droplets may contribute to high LO counts, but the oil droplets are filtered during the MM procedure. Applicable current limits for compendial articles are listed in Table 1.

Until demonstrated otherwise, the quantity of subvisible particulate should be considered a stability-related attribute. During product development, the MM procedure may be particularly useful for tracking and trending types of particles on stability, allowing the user to

Table 1 Subvisible Particulate Matter Acceptance Criteria for Parenterals

Nominal volume of container	Test attribute (μm)	Acceptance criteria	
		Light obscuration	Membrane microscopy
>100 mL[a]	Count ≥10	NMT 25/mL	NMT 12/mL
	Count ≥ 25	NMT 3/mL	NMT 2/mL
≤100 mL[a]	Count ≥10	NMT 6000/container	NMT 3000/container
	Count ≥ 25	NMT 600/container	NMT 300/container

[a]In the JP, containers with a nominal volume of 100 mL have the same requirement as containers with nominal volume of >100 mL.
Abbreviations: JP, Japanese Pharmacopoeia; NMT, not more than.

identify particles and the possible root cause associated with specific stability-related trends. Subvisible particulates, in the range of 1 to 10 µm, are increasingly of interest for therapeutic protein products because particulates containing aggregated proteins may have greater potential for stimulating undesired immune responses (17).

Sterility/CCI
Sterility is an absolute, universal requirement, specifically required by statute for essentially all types of parenteral products—human and veterinary drugs, biologics, and devices. Although sterility testing is commonly required, it is widely acknowledged that sterility testing of the final packaged product is a relatively weak tool for demonstrating acceptable levels of sterility assurance, which depends on establishing, maintaining, and monitoring appropriate process and environmental controls throughout the manufacturing process. Sterility testing only provides a final confirmatory evaluation of sterility for a validated sterile manufacturing process. Sterility test methods are in final stages of harmonization among three major compendia (USP <71> *Sterility Tests*; Ph. Eur. *2.6.1 Sterility*; JP *4.06 Sterility Test*). The harmonized texts are referenced in ICH Q4B Annex 8 (18), which provides references to the compendial editions/supplements containing versions of the test method that will be considered interchangeable by regulators across ICH regions. The methods are written to apply to a broad variety of parenteral products, and they include requirements for assessing the test suitability. Under some conditions, a terminally sterilized product may qualify for parametric release (19), in which case sterility testing is not part of the product release specification.

Sterility must be maintained throughout product shelf life, and therefore sterility testing is an essential requirement of specifications for stability testing, even if sterility testing is not required for product release in a parametric release environment. The critical factor in maintaining the sterility of the product is usually the integrity of container/closure system. A recent guidance from the FDA (20) allows for container/closure integrity (CCI) testing in lieu of sterility testing in stability protocols. The guidance does not specify or recommend a specific method for CCI testing, but requires that the validated method, and its relationship to microbial integrity, be discussed in the application or supplement. A growing variety of approaches to CCI testing (16) may offer useful alternatives to sterility testing on stability protocols. In contrast to sterility testing, many of these methods have the advantage of being nondestructive.

Endotoxin or Pyrogen Testing
Limits on bacterial endotoxin or pyrogen are a standard requirement for parenterals, and "endotoxins/pyrogen" is listed as a recommended test for parenterals in ICH Q6A. Bacterial endotoxin is an impurity introduced during product manufacture. The sterility of parenteral products ensures that bacterial endotoxin content will not increase over time, therefore endotoxin is not a common test requirement for stability.

Measurements of endotoxin are based on response to test reagents prepared from *Limulus* amoebocyte lysate (LAL). The response of the reagent is standardized relative to Reference Standard Endotoxin, the strength of which is expressed in endotoxin units, or EU (1 EU = 1 IU or international unit). Compendial methods for endotoxin are largely harmonized (Ph. Eur. *2.6.14 Bacterial Endotoxins*; USP <85> *Bacterial Endotoxins Test*; JP *4.01 Bacterial Endotoxins Test*). Results for endotoxin, as measured by these procedures, are expressed in EU/unit of dose—typically EU/mg or EU/mL. The general acceptance criteria for compendial tests are based on calculation of K/M, where K is a threshold dose for pyrogen response in EU/kg body weight, and M is the maximum recommended dose per kg body weight in a one-hour period. The threshold pyrogen dose is 5 EU/kg for most parenterals, but 0.2 EU/kg for parenterals administered intrathecally. Specific compendial monographs may contain alternative acceptance criteria, and alternative criteria for new drugs may be proposed on the basis of development experience and route of administration. Endotoxin specifications are also commonly required for medical devices, where the endotoxin limit is commonly expressed as 0.5 EU/mL of extract solution obtained by rinsing the device.

If the endotoxin method cannot be suitably validated, animal-based testing for pyrogens may provide an acceptable alternative. Each of the three compendia include pyrogen tests, but the tests have not been harmonized and therefore differ somewhat with respect to factors such as number of animals tested and acceptance criteria for temperature deviations (Ph. Eur. *2.6.8 Pyrogens*; USP *<151> Pyrogen Test*; and JP *4.04 Pyrogen Test*).

The displacement of the animal-based pyrogen test with the LAL test for bacterial endotoxin is an interesting example of widespread industry and regulatory adoption of, and transition to, improved methodology. Two recent books (21,22) review test methods, history, and test applications.

Uniformity of Dosage Units

Individual dosage units are required to be uniform with respect to drug content (ICH Q6A, ICH Q6B) based on standards established in compendia. The test and acceptance criteria for dose uniformity are being harmonized among compendia in the ICH regions (Ph. Eur. *2.9.4 Uniformity of Dosage Units*; USP *<905> Uniformity of Dosage Units*; JP *6.02 Uniformity of Dosage Units*). Specific regional considerations for acceptability and publication dates for harmonized texts are given in Annex 6 of ICH Q4B (23). As written, the dose uniformity test applies to a broad variety of dosage forms, including many single-dose parenterals. Dose uniformity can always be tested by measuring content uniformity, using an assay method for the active ingredient, or it may be assessed by the simpler alternative of mass variation in special cases. With one exception, content uniformity approach is required for single-dose injectable suspensions, emulsions, gels, and many solids containing additional ingredients. Mass variation can be employed for solid powder fills where the active is the sole ingredient, and for lyophilized products where the product was filled as a true solution and subsequently lyophilized. Content uniformity testing is always required for single-dose products containing multiple active substances, and the limits apply to each active. Dose uniformity is not a requirement for single-dose solutions because these meet requirements for both assay and extractable volume.

Identification

Identification is a requirement common to specifications for all drug products. As stated in ICH Q6A, the identification test should be able to establish the identity of the drug substance in the drug product and should discriminate among drugs closely related in structure. Multiple test methods and acceptance criteria may be applied in the event that a single method fails to show "sufficient" specificity. High-performance liquid chromatography (HPLC) retention times alone are regarded insufficient to establish identity, but the combination of retention time with distinctive UV spectrum or mass spectrum is usually considered adequate. An infrared spectral match of the extracted active ingredient is broadly regarded as a good example of a single identification method that is sufficient for lower molecular weight drugs.

ICH Q6B also requires identification testing in product specifications, although it is referred to as an "Identity" test in the Q6B guideline. Identity tests for biopharmaceuticals may be significantly more challenging because the molecular weight and molecular diversity of the drug may be greater, and it may be necessary to differentiate between materials that are more similar in structure. Alternative biological, immunochemical, or biochemical test methods may be utilized.

Identification tests are qualitative in nature, and method validation requirements (24) only include demonstration of specificity. Identity is not, therefore, a stability-indicating test, and the identity of the product is not typically reconfirmed as part of stability-testing protocols. Clearly, however, the identification test must retain suitable specificity to ensure that it can be used to identify the product throughout its shelf life.

Assay or Potency

On the basis of ICH Q6A, ICH Q6B, and broadly by statute in most regions, drug products require testing to assess content of the active substance (i.e., "strength" of the drug product as required by regulation). For many drugs, and particularly for synthetic drug substances, drug content is commonly and readily measured by precise analytical assay of quantity per dose, or of drug concentration per unit volume administered. Assays are expected to be specific—meaning they can differentiate the active drug from closely related substances including components of the formulation matrix and potential drug impurities. In addition, assays normally should be stability-indicating, meaning they can analyze the drug accurately in the presence of its degradation products. HPLC, which has been developed in a wide variety of formats, is the methodology most commonly utilized for assay. Less specific methods, such as titrations, may still be utilized, if justified, but under these circumstances, supplemental analysis of degradation products on stability is required. Low-specificity methods are not commonly utilized for synthetic drugs in new drug applications, but they are common to many pharmacopoeial monographs.

Establishing an appropriate acceptance range for assay in the case of small-molecule pharmaceuticals requires an understanding of manufacturing variation, analytical method variation, and product stability. While there are no rigorous requirements for limits on assay, a range of 90% to 110% of labeled content or concentration is typical for small-molecule pharmaceuticals. Values outside of this range generally require some justification. A maximum range of 95% to 105% of label claim at time of release is required by European Union guidance on release specifications (25), unless special circumstances justify a broader range. This release specification is consistent with common manufacturing and analytical variation, as well as the broader acceptance range of the shelf life specification. Frequently, these ranges can be reduced on the basis of manufacturing experience. If analytical methods employed for content uniformity determinations are adequate, the assay value can typically be taken as an average of the content uniformity measurements.

For biopharmaceuticals, ICH Q6B recognizes significant challenges that complicate this assay of the active ingredient. Biopharmaceuticals—by virtue of their size, complexity, and sometimes molecular diversity—may be very difficult to analyze in the presence of substances that are chemically similar but not pharmaceutically active. Further, biopharmaceuticals typically have complex higher-order structures that depend on noncovalent interactions to achieve and retain molecular conformations, or shapes, associated with activity. Analyzing the population of active molecular species in properly folded conformations is challenging, and in many cases is only achieved by a combination of tools, including physicochemical assays, immunochemical assays, and biological assays. Physicochemical alterations are frequently monitored by a combination of methods, including chromatography, electrophoresis, and a variety of spectroscopic and spectrometric methods. Biological potency assays, either in vitro or in vivo, are typically required to ensure therapeutic efficacy throughout shelf life, and to provide some ability to select and interpret the physicochemical measurements. Specific acceptance criteria are unique to each product, but the same general principles apply: quantitative ranges must account for variation in assay, manufacture, and the maximum extent of change anticipated on stability. In addition, biopharmaceuticals typically require a specification for quantity, or total concentration, in the final product. The method may be nonspecific, such as an assay for total protein concentration.

Degradation Products

Impurities appearing in the product as a result of chemical changes in the drug over time are classified as "degradation products." These substances may result from a variety of reaction pathways: unimolecular decomposition, interaction with light or radiation, or from specific chemical reactions with components of the formulation or packaging system. They are distinct from "process impurities," which are substances produced as byproducts of synthesis of the drug substance, although some degradation products (DgPs) may also be process impurities. ICH Q3B (26) establishes expectations for specifications on DgPs in new drug products for small-molecule pharmaceuticals. The principles and language of this guidance have broader application, so the guidance is worth summarizing here because it provides a sound set of principles for the case where modern analytical technology is brought to bear on specifications development process for the category of drugs that are presently capable of relatively thorough chemical characterization with these tools. Biopharmaceutical molecules, however, may degrade in ways that cannot be adequately or comprehensively characterized by current analytical methods.

On the basis of ICH Q3B, specifications for substances related to the drug in the drug product are required for DgPs only. Process impurities in a new drug product are assumed to be limited by specifications on the drug substance using principles of ICH Q3A (27), and the process impurities need not be monitored in the drug product unless they are also DgPs. To establish suitable specifications, a chemical understanding of degradation pathways (i.e., a degradation profile) must be acquired from scientific stability studies conducted with highly specific analytical methods during drug development. Final analytical methods should be capable of selectively quantifying individual DgPs and should exclude process impurities from the analysis. The exclusion of a related substance from the list of DgPs requires justification—on the basis of their chemical structure and established chemistry and/or data from scientific stability studies, including studies commonly referred to as "stress-stability studies" (28). Because DgPs are usually close structural relatives of the drug, the quantity of a DgP is expressed as a weight percentage relative to the drug at its label content.

With respect to specifications, DgPs are classified in three ways in the guidance.

- Identified vs. unidentified
  Identified DgPs have known molecular structure. Unidentified DgPs are specific substances with undetermined molecular structure. They are recognized by some characteristic behavior in analytical systems, such as chromatographic retention time.
- Qualified vs. unqualified
  Qualified DgPs have established biological safety profiles that support their safe administration under a dosing protocol when levels are below an established "qualification level." Unqualified DgPs are those for which a qualification level has not been established on the basis of safety data.
- Specified vs. unspecified
  If an individual specification (i.e., procedure and acceptance criterion) is associated with the DgP, it is considered a "specified" DgP. An "unspecified" DgP is subject to a general acceptance limit, which is applied to all individual DgPs that are not specified DgPs.

In addition to the requirement for individual degradation products, a specification must be established for total degradation products, which is defined as the sum of all reported DgPs.

ICH Q3B established clear guidelines for specific levels above which DgPs should be (i) detected and reported, (ii) identified, and (iii) qualified. These levels are derived in a manner that is dose-dependent, because exposure to impurities decreases with decreasing dose. The guidelines therefore provide a framework for establishing acceptance criteria during drug product development based on stability studies, analytical method characteristics, manufacturing experience, and safety studies. For some reconstitutable products, use-period stability studies may reveal increases in DgPs that need to be factored in to provide acceptable ranges of DgPs throughout product shelf life.

The fundamental concepts for biopharmaceutical products—as discussed in ICH Q6B—are similar, although the terms are further refined to capture distinctions less commonly encountered for small molecules. Further, precise guidance on thresholds for identification and qualification is not provided. Impurities in molecules derived from biotechnological synthetic pathways are classified in ICH Q6B as "process-related impurities" and "product-related impurities." *Process-related impurities* typically cover a broad range of cell-derived and process-derived constituents that must be evaluated and monitored, typically in the drug substance. Process-related impurities are structurally unrelated to *product-related impurities*, which are regarded as "molecular variants of the desired product which do not have properties comparable to those of the desired product with respect to activity, efficacy, and safety." The guidance further differentiates "product-related substances" as molecular variants that are comparable to the desired product in terms of activity, efficacy, and safety, and therefore are not considered impurities. ICH Q6B requires that specific tests and acceptance criteria be developed to monitor product-related impurities when they increase during manufacturing or storage of the drug product, but there is no need to establish specifications for the product when evidence indicates that these impurities do not change during manufacture or storage. Clearly, a great deal of effort may be required during development to develop methods for product-related impurities (e.g., truncated forms, isomers, posttranslational modifications, deamidated forms) and to assess whether they can be appropriately classified as product-related substances.

Residual Solvents

ICH Q3C (29) set guidelines for a broad variety of residual solvents in drug products based on safety considerations for permissible daily exposure. Solvents are divided into three classes on the basis of risk:

> Class 1: Solvents to be avoided
> Class 2: Solvents to be limited
> Class 3: Solvents with low toxic potential

Limits are provided and monitoring for specific solvent is strongly recommended for class 1 and class 2 solvents. The guideline recommends a general limit of 50 mg/day for class 3 solvents, and nonspecific test methods, such as loss-on-drying, are considered suitable for monitoring. The residual solvent requirement applies to drug products, but it is commonly met by limiting the concentration of residual solvents in all drug substances and excipients. To simplify, a broadly acceptable set of concentration limits are provided in ICH Q3C on the basis of a total daily product dose of 10 g/day. Alternatively, limits for individual substances can be calculated on the basis of their specific contributions to the daily intake (option 2).

ICH Q3C requires validation of analytical methods for residual solvents, but does not provide specific methods. Broadly applicable general test methods are provided in compendia (USP <467> *Residual Solvents*; Ph. Eur. *2.4.24 Identification and Control of Residual Solvents*) on the basis of methods originating in Ph. Eur. Both USP and Ph. Eur. adopt the ICH Q3C limits for compendial articles, unless otherwise stated in specific monographs, and the limits have thus become legal requirements for drug products covered by compendial monographs in these regions. Residual solvents are not commonly considered a stability-related attribute.

Leachables/Extractables

Leachables are a category of impurities that originate in packaging and migrate into a pharmaceutical product under the normal range of storage conditions. Leachables are specifically excluded from consideration in the ICH Q3B guidance, and there is currently relatively little prescriptive policy guidance that lists specific regulatory expectations for monitoring of leachables in injectable products. To a large extent, many concerns associated with leachables are addressed by development studies that demonstrate safety of packaging extractables—substances extracted from packaging using forcing conditions, typically combinations of solvents and temperature. Both the existing FDA packaging guidance (30) and ICH Q6A raise concern for the influence of package chemistry on safety and compatibility of parenterals. These guidances focus on demonstrating acceptable levels of packaging extractables in product through development studies, including stability studies, if needed. An EMEA guidance addresses specific requirements for studies on extractables in plastic materials (31), including a requirement for migration studies or stability studies on leachables to confirm the extractables from packaging do not migrate into product to significant extent.

Leachables in general are an active topic for guidance development. Detailed guidances are available for inhalers and nasal products, and regulatory expectations for leachables in ophthalmics are also under discussion. Parenterals are typically packaged in complex materials, including glass, polymers, and elastomers (32). They have a documented history of product/package interactions, including various examples where small amounts of leached substances—including metal ions, silicones, benzothiazoles, formaldehyde, and a variety of organics—have altered product appearance, safety, or stability (33,34). Large volume parenterals represent a particular case where low concentrations may still give rise to a large total exposure because of the volume administered, and analytical methods with very low quantitation limits may be needed. During development studies, leachables should be considered as a candidate for potential specification development unless levels are low enough to justify these substances are not a concern. The rationale needs to be revisited as a consequence of packaging changes. Their exclusion from the specifications should be based on sound rationale developed from packaging extractable studies and confirmed through migration studies or through leachable analysis conducted during the formal stability program.

Water Content

ICH Q6A recommends a specification on water content for sterile solids for reconstitution as well as nonaqueous parenteral products. Water may be assessed by a variety of methods, the most common being loss-on-drying or Karl Fischer titration (Ph. Eur. *2.05.12 Water Semi-micro Determination, 2.05.32 Water Micro-determination*; USP *<921> Water Determination*; JP *2.41 Loss on Drying Test, 2.48 Water Determination Karl Fischer Method*), but a variety of other methods exist, including near-infrared spectroscopy, which can be utilized as a nondestructive probe of water content. The potential influence of water on solid-state chemical and physical stability of solids, particularly amorphous solids and lyophilized powders, has been extensively documented (35–37). Likewise, water may play a critical role in stability of nonaqueous parenterals, including physical stability of some nonaqueous suspensions (38). Water content may change significantly on stability, depending on permeability of packaging and water sorption characteristics of components, such as stoppers.

pH

ICH Q6A recommends a specification on pH for parenteral products where applicable. Changes in pH may alter product characteristics including pain on injection or product stability. pH is commonly measured potentiometrically with glass electrodes using compendial methods (Ph. Eur. *2.02.03 Potentiometric Determination of pH*; USP *<791> pH*; JP *2.54 pH Determination*). Acceptance criteria for pH should be developed on the basis of batch data and product development studies that address the influence of pH on key quality attributes. Stability attributes, such as rates of appearance of degradation products, may be strongly sensitive to pH (39). Developmental kinetic studies with supporting data may be needed to ensure the acceptance range for pH is consistent with the acceptance range for the degradation product throughout shelf life. Changes in pH should be monitored through the stability program.

Preservative Effectiveness and Antimicrobial Preservative Content

With few exceptions, multidose parenterals incorporate a chemical preservative (40), and suitable levels of the preservative must be maintained throughout shelf life. Minimum effective levels of preservative are typically established during product development based on results of pharmacopoeial preservative effectiveness testing (Ph. Eur. *5.1.3 Efficacy of Antimicrobial Preservation*; USP *<51> Antimicrobial Effectiveness Testing*) across a range of preservative concentrations. Although preservative effectiveness testing directly assesses the resistance of the formulation to microbial growth, it has some distinct disadvantages. Results of the test are difficult to trend quantitatively, and the test requires extended times, 14 to 28 days, for culturing of microbes. Therefore, in the registered product specifications, antimicrobial preservatives are commonly assayed using a stability-indicating method such as HPLC, which provides a suitable measure of preservative effectiveness as long as the formulation parameters (e.g., pH) critical to preservative action are properly controlled and understood. The rationale for acceptance of preservative content is captured in ICH Q6A guidance, which also recommends testing of preservative effectiveness through development and scale-up, including stability studies.

Acceptance criteria for preservative levels throughout shelf life should be clearly supported by the relationship established between preservative levels and preservative effectiveness during product development and confirmed in the new drug product stability studies. In many cases, for preservatives in common use, there may be considerable decrease in preservative levels throughout shelf life or usage duration of the product, in part because preservatives are frequently sorbed into plastics or elastomers. The European Union guidance (25) on product release specifications stipulates that preservative content should be 90% to 110% at time of release, unless suitable justification is provided for a broader specification. Since levels of preservatives also need to be justified (41), the stability of the preservative system needs to be understood well enough to allow reasonable initial preservative levels to be established.

Antioxidant/Chelating Agent

Antioxidants and/or chelating agents are sometimes utilized in parenteral formulations to enhance product stability, and development of specifications for these components is aligned with recommendations of ICH Q6A. Specific identification and test procedures are required for release specifications and stability specifications (41). Depending on the stability of the antioxidant and its specific role in the formulation, stability testing may not be required, but this should be justified.

Functionality Testing

ICH Q6A recommends that parenteral formulations packaged with delivery devices, for example, prefilled syringes, include testing to ensure specific functional characteristics of the delivery system (e.g., syringability, extrusion force, glide force, or break force), have test procedures and acceptance criteria. Depending on requirements for the device in question, functionality may be required for the drug/device combination on stability to ensure that delivery characteristics are consistent through shelf life. Syringability should also be properly regarded as a formulation attribute of injectable products that reflects the ease with which the product flows through the needle, including the force required to deliver it. This characteristic is largely related to rheology of the formulation, but it may be more easily measured by practical subjective assessment of ejection force through an appropriate gauge needle. In some cases, it may be appropriate to include a test for syringability, or a rheological test, in the product specification, particularly during development and primary stability studies.

Osmolarity or Osmolality

Osmolarity or osmolality specifications are expected when the tonicity of a parenteral product is declared in its labeling. Osmolarity is commonly a consistent function of composition and is not likely to change on stability unless there is significant breakdown of product or loss of water. Osmolarity is typically determined based on standard compendial methods (Ph. Eur. *2.2.35 Osmolality*; USP *<785> Osmolality and Osmolarity*; JP *2.47 Osmolarity Determination*). Data from development and validations studies may allow reduction of testing for osmolarity in the marketed product.

Particle Size Distribution

Key performance attributes, including drug release rate and rheology, may depend on the particle size distribution of injectable dispersions, such as suspensions, emulsions, and liposomes. ICH Q6A recommends consideration of testing for particles size distributions with quantitative acceptance criteria. If release rate is a primary concern, developmental data should be considered when determining the need for either a dissolution procedure or a particle size distribution procedure. A variety of methods exist for measuring particle size distribution of suspended particles. The most commonly used methods are laser light diffraction, electrozone particle counting, and optical microscopy (42,43). For nanoparticle distributions, more suitable tools include dynamic light scattering, field-flow fractionation, and analytical ultracentrifugation. None of these methods are chemically specific, so particle size distribution measurements on products are difficult at best in cases where multiple species are present that must be chemically differentiated.

Particle size distribution specifications commonly include assessments of various percentage points on a cumulative distribution curve representing the fraction of particles below a given size range. ICH Q6A recommends specifications based on some measure of the mean of the distribution, as well as an upper and/or lower percentage point of the distribution. It would be common to utilize the median, or 50th percentile (a.k.a. D_{50}), as well as the 90th percentile D_{90} to measure the large particle tail of the distribution and the 10th percentile to measure the small particle tail of the distribution. The volume-weighted mean diameter would also be a common measure of mean particle size and a suitable replacement for D_{50}. There are numerous complexities associated with particle size measurements, including selection of methods, sampling, and sample preparation. It is important to recognize that instrumental methods for particle size differ fundamentally in their response to particles of differing sizes and shapes. Hence, acceptance criteria developed for the method and suspension have little meaning when the method is changed or if the process is changed in ways that alter particle shapes. Validation of particle size methods includes assessment of precision and robustness, but not accuracy.

ICHQ6A recommends "acceptance criteria should be set based on the observed ranges of variation, and should take into account the dissolution profiles of the batches that showed acceptable performance in vivo and the intended use of the product." While it would be uncommon to include acceptance criteria for particle size in specifications for early clinical batches, the measurement of particle sizes at this stage would be provide important data for subsequent justification of acceptance criteria. In some cases, specific clinical studies may need to be designed to establish the limits of an acceptable range of particle sizes. Particle size is a stability-related attribute, and changes in particle size on stability should be assessed. The particle size distribution may change through dissolution and regrowth, or through aggregation/agglomeration.

Redispersibility

ICH Q6A recommends a test for resuspendability of injectable suspensions that settle on storage (produce sediment). Shaking is considered an appropriate procedure, if properly controlled, and the time required to achieve resuspension provides a measurement of redispersibility. If adequate redispersibility is demonstrated on the basis of product development and stability studies, elimination of this attribute from the specifications may be proposed. Redispersibility is a stability-related attribute: significant changes in redispersibility occasionally occur because of caking of suspension solids, which may only become evident over time.

Reconstitution Time

For parenteral products that require reconstitution, acceptance criteria for reconstitution time should be provided on the basis of ICH Q6A guidance. The choice of diluent should be justified. Data generated during product development and process validation may be sufficient to justify skip lot testing or elimination of this attribute from the specification for rapidly dissolving products. Reconstitution time can be affected by physical changes in powders or solid cakes, so reconstitution time should probably be assessed on stability unless there is evidence that the solids dissolve sufficiently rapidly that such changes are not likely to affect reconstitution time.

Other Specifications

A wide variety of other specifications may be appropriate or expected depending on product attributes or specific product type. Examples include testing unique to liposomal formulations, lipid emulsions, or microparticulate controlled release formulations. The interested reader should consult current literature and guidance sources, including compendia.

Biopharmaceuticals include numerous requirements for process-related impurities like residual DNA, residual host cell proteins, Protein A, mycoplasma, antibiotics, and viruses.

Similar to residual solvent testing, where specifications for the drug substance eliminate the need to test for solvents in the drug product, testing for these process-related impurities is rarely done as part of drug product specifications but is commonly managed at the drug substance stage.

One major topic—in vitro release testing for parenteral suspensions or parenteral controlled release dispersions—represents a persistent area of interest, which is not discussed extensively in current guidance or compendial monograph testing (44). (Interestingly, ICH Q6A indicates that under some circumstances particle size testing may be performed in lieu of dissolution testing, but it does not mention dissolution testing as a requirement.) A variety of approaches have been developed for in vitro release testing of parenteral dispersions, partially in an effort to develop a test procedure suited to specifications and capable of measuring an attribute correlated to in vivo release rate. The monograph on injectable dispersed systems by Burgess is recommended (45). Some guidelines regarding preferred test systems and methods for setting acceptance criteria have been published by professional associations (46).

DRUG PRODUCT STABILITY

The investigation of stability for a new parenteral product progresses through the same stages as all new drug products that ultimately gain approval on the basis of successful regulatory applications. In the end, wherever appropriate, these applications follow established guidelines for submission and approval of data to support the marketed product shelf life. Most of the remainder of this chapter will summarize the content of these guidelines, but first it must be stressed that registration stability studies are only intended to confirm an understanding of product stability. The fundamental stability characteristics of the drug must be elucidated during the formulation design stage. Formulation choices at the design stage provide the opportunity to eliminate, or at least minimize, formulation factors that lead to poor stability and to optimize formulation properties that protect the drug from stress factors in the surrounding environment, sometimes including the environment in vivo. In a limited sense, "poor stability" may be interpreted as a shelf life shorter than desired. In the broadest sense, poor stability should equally be understood to include inconsistency. Consistent stability performance requires an adequate level of control over factors that affect stability, and the importance of achieving an understanding of the comprehensive array of factors that yield consistent batch-to-batch stability cannot be overemphasized.

Stability and Formulation Development

Successful formulation development requires expertise in analytical, bioanalytical, and physical organic chemistry to design studies that elucidate the major chemical degradation pathways. For biopharmaceuticals, an array of tools may be needed to explore the consequences of stresses imposed by time and temperature as well as processing and packaging. Assessments for biopharmaceuticals may be chemical, physical, and biological in an effort to not only identify changes but to correlate changes in physical and chemical measurements with changes in activity and overall biological response. From a stability perspective, package-related liabilities should also be identified early, because parenterals belong to a relatively limited category of pharmaceutical products that are typically in intimate contact with packaging. These packaging studies include some characterization of packaging extractables and the potential effect they may have on stability. In addition to chemistry, physical attributes of the parenteral product may become stability-limiting, particularly in dispersed systems, and it is not uncommon to find that chemical stability is predicated on physical stability or physical consistency. Finally, largely because of considerations related to stability, successful formulation development goes hand-in-hand with successful analytical method development. There is extensive literature on chemical and physical stability of drugs, which should be consulted for design of studies as well as specific reaction chemistries (see Refs. 29 and 40 for good introductions).

Registration stability studies or formal stability studies are lengthy and relatively expensive, and are usually undertaken immediately prior to registration. To reduce the risk of failure, the fundamental science needed to support the understanding of the product should be acquired before initiating formal stability studies for product registration. Ideally, at the initiation of formal stability, there should be a thorough knowledge of product attributes (specifications) that are most likely to restrict shelf life, and their kinetic behavior. In most cases, this level of stability understanding develops concurrently with clinical studies and nonclinical safety studies that provide some of the basis for setting acceptance limits used in product specification development.

Stability Requirements for Product Registration

Currently, regulatory expectations for stability studies of new parenteral products are best reflected in ICH guidelines. Stability requirements for product registration were an important topic in ICH guidelines related to quality, because disparities in regional requirements frequently resulted in excessive stability testing. The resulting guidelines reduced the number of conditions, test points, and approaches to data evaluation that were needed to establish shelf lives for a global product, and distilled most of the sound principles developed across multiple regions into a common and rational framework.

ICH Q1A:	Stability Testing of New Drug Substance and Products (47)
ICH Q1B:	Photostability Testing of New Drug Substances and Products (48)
ICH Q1C:	Stability Testing of New Dosage Forms (49)
ICH Q1D:	Bracketing and Matrixing Designs for Stability Testing of New Drug Substances and Products (50)
ICH Q1E:	Evaluation of Stability Data (51)
ICH Q1F:	Stability Data Package for Registration Applications in Climatic Zones III and IV (withdrawn) (12)
ICH Q5C:	Stability Testing of Biotechnological/Biological Products (52)

ICH Q1A is the parent guideline that establishes a number of key definitions. Most of the terms are in common use, but they have less precise meanings. The precision of some of the ICH definitions is sufficiently important for the current discussion to repeat the definitions here. In a few cases the definitions have been abbreviated but not otherwise altered.

Formal Stability Studies and Shelf Life

Formal stability studies are "Long term and accelerated (and intermediate) studies undertaken on primary and/or commitment batches according to a prescribed stability protocol to establish or confirm the retest period of a drug substance or the shelf life of a drug product." The results of these studies should provide sufficient confirmatory data to allow the marketing authority to approve the product shelf life. The studies are conducted based on written protocols, with samples stored in chambers under controlled conditions, and pulled for testing at preselected times. The testing is regulated under GMPs. Test methods must be validated and associated equipment and laboratories must be qualified to do the testing. Laboratory records must be maintained for inspection, and in many cases these will be reviewed prior to product approval. The *shelf life*, or *expiration dating period*, is "the time period during which a drug product is expected to remain within the approved shelf life specification, provided that it is stored under the conditions defined on the container label."

Batch Requirements for Formal Stability

On the basis of ICH Q1A, formal stability studies require a minimum of three *primary batches* of drug product. Two of the three primary batches should be at least pilot scale batches, and the third may be smaller, although laboratory batches are not acceptable. A *pilot scale batch* is "a batch of a drug substance or drug product manufactured by a procedure fully representative of and simulating that to be applied to a full production scale batch," and typical pilot scale batches are at least 1/10 of production scale. *Commitment batches* are "Production batches of a drug substance or drug product for which the stability studies are initiated or completed post approval through a commitment made in the registration application." ICH Q1A requires the evaluation of stability from the first three production batches on the basis of the approved protocol and specifications as a postapproval commitment.

Multiple batches are utilized in formal stability to provide some understanding of consistency in stability performance across factors that vary between batches. Therefore, different batches of drug substance should be utilized where possible. Preferably, although not specifically stated in ICH Q1A, different batches of key excipients should be used as well. The formulation of primary batches must be the same as requested for production batches for the product, and the process must at least simulate the process used for production batches. Samples for the formal stability study must be packaged in the same package (i.e., primary container/closure and secondary packaging) that will be used for marketed product. Primary batches should meet the same specifications requested for the marketed product, and they should be manufactured to provide the labeled amount of drug.

One factor that is not well addressed in current guidance is container orientation on stability. The only ICH guidance that addresses container orientation is ICH Q5C for biopharmaceutical products, which recommends "stability studies should include samples maintained in the inverted or horizontal position (i.e., in contact with the closure), as well as in the upright position, to determine the effects of the closure on product quality." The withdrawn 1998 FDA draft stability guidance (53) was more specific on the issue of orientation, requiring that solutions, dispersed systems, and semisolids be stored on stability in both horizontal/inverted and upright orientations until the effect of storage orientations could be assessed and stability could be restricted to the most stressful orientation. The draft guidance for metered dose inhaler products takes a similar approach (54). It is reasonable to anticipate the need for some data to demonstrate the influence of container orientation on storage stability of the product, in accord with ICH Q5C.

Stability Storage Conditions and Test Intervals

In considering the appropriate storage conditions for stability studies, ICH Q1A exploits concept of climate zones, which provides different conditions to account for regional variations in two major stability factors: temperature and relative humidity. (The influence of remaining environmental factors is accounted for in special studies, such as photostability testing as discussed in ICH Q1B.) Pharmaceutical markets are divided among four major climate zones, which differ with respect to ambient annual averages in these two variables, and therefore require long-term stability under conditions that are sufficiently suitable to reflect the upper extremes of average ambient storage in each area. Zahn provides a recent review (55), including data that cover most global regions. The ICH regions are largely part of Climate Zones I and II, corresponding to temperate and subtropical or Mediterranean climates.

ICH Q1A recommendations appropriate for Zones I and II are summarized in Table 2, on the basis of the storage condition specified for the product. There are two cases where

Table 2   ICH Stability Storage Conditions for Zones I and II

Case;	Temperature/relative humidity[a] and duration of study		
Recommended storage conditions	Long term	Intermediate	Accelerated
General case; Room temperature storage	25°C/60% RH through shelf life	30°/65% RH 12 mo	40°/75% RH 6 mo
Aqueous product/semipermeable package; Room temperature storage	25°/40% RH through shelf life	30°/65% RH 12 mo	40°/ NMT 25% RH 6 mo
General case; Refrigerator storage	5°C ± 3°C through shelf life	None	25°/40% RH 6 mo
General case; Freezer storage	−20°C ± 5°C through shelf life	None	None

[a]Where tolerances are not specifically indicated in the table, they are ±2°C for temperature and ±5% RH for relative humidity.
Abbreviation: NMT, not more than.

packaging plays a role. For products in impermeable packaging (e.g., sealed glass ampoules), relative humidity is not significant and storage conditions do not need to provide for controlled relative humidity. For aqueous products in semipermeable packaging (e.g., large-volume parenterals in low-density polyethylene bags, ophthalmic solutions in flexible polyolefin containers) intended to be stored under ambient conditions, water loss can be a significant problem. Clearly for these products, high water activity inside of the package is not a stressful condition, but water loss in a dry environment may be a significant limitation for stability. Storage at alternate lower relative humidities is required.

Product testing is recommended at minimum test intervals of three months through the first year, six months through the second year, and annually thereafter. This recommendation is clearly a general minimum requirement. Where specific kinetic information is available to guide selection of test points, points may be added at suitable locations to improve the analysis. Not all testing is required at all intervals or for all batches, as noted in the discussions on specifications, particularly for the microbiological tests. In general, testing at the intermediate condition is not needed if the product does not undergo "significant change" at the accelerated condition, as discussed in the data evaluation section of ICH Q1A and ICH Q1E, and in the subsequent section on Stability Data Evaluation. Recommendations for stability data at time of regulatory submission are a minimum of 12 months long-term data and 6 months of accelerated or intermediate data.

Climate Zones III and IV cover hotter regions of the globe. Zone III climates are drier, and Zone IV climates are more humid. Temperature requirements for room temperature storage in these zones are 30°C ± 2°C. Relative humidity requirements for testing in Zone IV are split between 65% and 75% for the general case, because some Zone IV countries (specifically those that declare themselves to be Zone IVb) have conditions where products are consistently subjected to higher relative humidities. Where global stability products are considered, this leaves manufacturers with a range of options, depending on specific regions and willingness to consider different expiration periods for the various regions.

Label storage statements differ by region. Examples of regional label storage statement recommendations are provided in USP 31, an EMEA guidance (56), and a WHO guideline (57).

In-Use Stability

Parenterals include groups of reconstitutable or multidose products with specific requirements for in-use stability testing. ICH Q1A requires in-use stability testing for products through the label use-period after reconstitution or dilution. This in-use stability testing should be conducted as part of the formal stability protocol at both initial and final (or end of shelf life) time points for long-term stability of primary batches. In addition, data should be collected at 12 months or the intermediate time point immediately prior to submission. The type of testing that might be conducted through the use-period on a reconstituted product was probably best covered in the now-withdrawn 1998 FDA draft stability guidance (53). It included appearance, clarity, color, pH, assay, preservative (if present), degradation products/aggregates, sterility, pyrogenicity, and particulate matter.

An EMEA guidance (58) recommends testing to be completed prior to submission to ensure that multidose products retain quality throughout their label use period, when product is taken from the multidose container on the basis of label instructions under normal environmental conditions. At least one of these protocols should be conducted on a batch near the end of shelf life.

Analysis of Stability Data

Basic principles of data analysis for formal stability studies are presented in the ICH Q1A parent guidance. These principles are further elaborated in the ICH Q1E guidance, particularly for instances where the requested shelf life at time of submission for a new drug product requires extrapolation beyond existing long-term data. Data for submission are expected to be properly tabulated and reviewed, with results and trends for each attribute individually discussed, as appropriate. The shelf life for the product should not exceed the shelf life predicted for any individual attribute.

The combination of accelerated, intermediate, and long-term stability data are first evaluated for significant change. The following definition of significant change is adapted directly from ICH Q1A:

- A 5% change in assay from its initial value, or failure to meet the acceptance criteria for potency when using biological or immunological procedures.
- Any degradation product's exceeding its acceptance criterion.
- Failure to meet the acceptance criteria for appearance, physical attributes, and functionality test (e.g., color, phase separation, resuspendability, caking, hardness, dose delivery per actuation); however, some changes in physical attributes (e.g., softening of suppositories, melting of creams) may be expected under accelerated conditions.
- Failure to meet the acceptance criterion for pH.
- Failure to meet the acceptance criteria for dissolution for 12 dosage units.
- Loss of water from semipermeable containers greater than 3% after three months at 40°C/NMT 25%RH, unless justified for small containers (1 mL or less).

ICH Q1A provides a decision tree to help evaluate the data and assess whether, and to what extent, extrapolation can be used to justify a shelf life request. As an example, consider a product to be stored at ambient conditions. The outcomes of the ICH decisions tree are tabulated in Table 3, which shows the greater the extent of change with time and temperature, the less extrapolation can be used to justify shelf life beyond that allowed by existing long-term data. It also shows the advantage of analyzing data statistically, where possible, to provide justification for extrapolation beyond that which could be justified in the absence of statistical analysis. A similar table can readily be constructed for refrigerated storage, where in general the extent of allowed extrapolation is shorter.

For biologics, generally extrapolation of not more than six months beyond the real-time stability data is granted due to potential concerns of non-Arrhenius stability behavior of drug product.

Statistical Analysis of Stability Data
The ICH Q1E guidance provides a suitable framework for statistical trend analysis of quantitative data, where the data can be modeled using linear regression. Simpler data, such as discrete pass/fail responses, or more complex data, such as dissolution profile data, can also be trended statistically (59), but these are beyond the scope of current guidance. Because the extent of degradation of the drug in most drug products is relatively small throughout shelf life, most stability data can be trended accurately with simple zero-order kinetics, even if the fundamental model kinetics are first order or more complex orders. (In this case, the models are commonly referred to as pseudo-zero-order.) Carstensen (60) has shown the relative difference between first-order and zero-order expressions is quite small as long as the extent of

Table 3   Summary of Allowed Extrapolation Based on (no) Changes in Quality Attributes

Significant change at 6 mo?	Little change over time or little variability	Statistical analysis	Allowed extrapolation[a]
No—Accel.	Yes—both accelerated and long term	Not required	2X but NMT +12 mo
	No—accelerated or long term	Yes	2X but NMT +12 mo
		No	1.5X but NMT + 6 mo
Yes—Accel. No—Inter.	Not relevant	Yes	1.5X but NMT + 6 mo
		No	NMT + 3 mo
Yes—Accel. Yes—Inter.	Not relevant	Not relevant	None

[a]Times and multipliers are referenced to the time established by current long-term data. Extrapolations must be supported by statistical analysis or supportive data.

change is below about 15%. The appearance of primary degradation products also commonly follows pseudo-zero-order kinetics, unless the products themselves degrade, in which case more complex models may need to be considered (39). Changes in other components, such as antioxidants or preservatives, may cover much more significant fractions of the initial content, so alternate kinetic models may also be required in these cases. For simplicity, the overview of statistical treatment presented here will focus on zero-order kinetic models, but many of the concepts can be extended to alternate kinetic models via variable transformations or nonlinear regression modeling.

Linear regression, as utilized for analysis of zero-order trends in stability data for a single batch, is a straightforward application of linear regression as described in multiple introductory texts on applied statistics (61,62). To briefly summarize, consider the assay values as the dependent variable, y_i, which is modeled as a simple linear function of time, t_i, for stability time points $i = 1, \ldots n$. The stability data at each time point are limited in precision by variability in measurements and sampling. Although they follow a general zero-order trend, deviations of individual points about the line are common. The linear model is given in equation (1), where the least squares slope β and intercept α are the values that minimize the sum-of-square errors $\sum \varepsilon_j^2$ in the model expression over all time points j. The errors, ε, are assumed to be normally distributed, with mean of zero, and standard deviation of σ.

$$y_j = \alpha + \beta t_j + \varepsilon_j \quad \text{(Model 1)} \tag{1}$$

To exemplify, consider the stability data for assay collected for a one of three primary stability batches (Batch A). Assay values collected at 0, 3, 6, 9, 12, and 18 months are plotted in Figure 1. In Figure 1, a horizontal line is drawn through the lower specification limit (LSL) of 90% of label. The solid trend line obtained by standard linear regression computer program is shown in Figure 1. Where the trend line crosses the LSL just below 30 months, it provides a "median" estimate of the shelf life for this batch. However, because there is error in the fitted data, there is inferential uncertainty in the shelf life. The standard error of the data about the line in this example is 0.78%, a value that is very consistent with the precision of analytical methods for assay. The "true" shelf life for the batch has equal likelihood of being longer or shorter than 30 months, because the variability in the data limits the degree of certainty in the estimation of the shelf life. The greater the variation of data about the line, the greater the degree of uncertainty. Clearly, since products are expected to meet specifications throughout

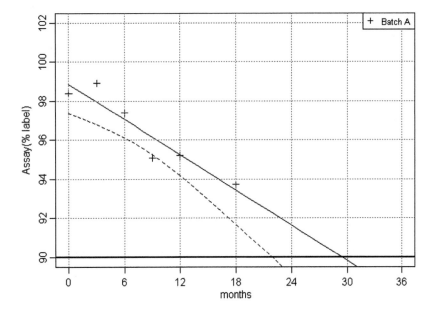

Figure 1 Linear regression analysis of assay results for 18 months of stability from a single batch. The dashed line is the lower 95% two-sided confidence interval for the regression line.

shelf life, the regression line estimate of shelf-life is a poor choice because it is very likely (50% probability) to overestimate the true shelf life.

The accepted resolution to this dilemma, as recommended in ICH Q1E, is to calculate a confidence limit about the regression line. Assuming the errors are normally distributed, confidence limits can be calculated on the basis of standard formulas. These limits estimate the region within which the true trend line for the batch is expected to lie at a chosen level of confidence. The dashed line in Figure 1 represents the lower 95% two-sided confidence limit for the regression line. In this case, the true trend line has a 97.5% likelihood of being above the region shown by this curve. If the tentative shelf life is conservatively assigned at not-more-than (NMT) 22 months, the point where the 95% confidence limit intersects the LSL, there is only a small likelihood that the shelf life will be an overestimate and the batch is likely to remain within specifications through shelf life. Note, however, that there is considerable difference between the regression line estimate of NMT 29 months and this accepted shelf life estimate of NMT 22 months. This difference is the penalty of uncertainty. The ability of the statistical tools to account for uncertainty in conservatively estimating shelf lives is the major reason why ICH Q1E allows more extended extrapolation in cases where the data are analyzed statistically.

Typically, the formal stability studies will include results from three batches. ICH Q1E requires that the shelf life of the product cannot be longer than the shortest estimated shelf life of these batches, using the intersection point of the 95% confidence interval with the lower or upper specification limits. With appropriate statistical software, data from multiple batches can be analyzed together using the model given by equation (2).

$$y_{ij} = \alpha_i + \beta_i t_{ij} + \varepsilon_{ij} \qquad \text{(Model 2)} \qquad (2)$$

In this model, the "i" subscript denotes the batch. Each batch has an independent slope and intercept. Although equation (2) may seem equivalent to three instances of equation (1), the regression using equation (2) assumes a single pooled estimate of the residual standard error. In some cases, this pooled estimate may extend the shelf life estimate somewhat by reducing the estimated standard deviation associated with the batch that has the shortest estimated crossing point. As an example, consider Figure 2, where data from two additional batches have been added to the data from Batch A. Regression lines, and the lower 95%

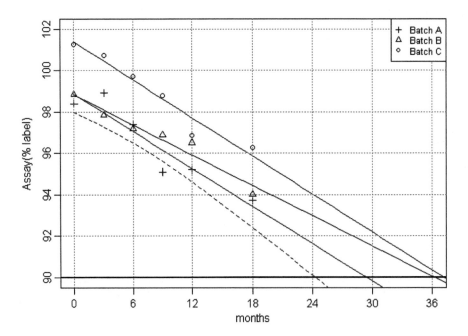

Figure 2   Linear regression of assay results for 18 months of stability from three batches. The model was pooled for residuals. The dashed line is the lower 95% two-sided confidence interval for the regression line of Batch A.

confidence interval for Batch A have been plotted based on analysis of Model 2. Batch A is still the batch with the shortest estimated shelf life. The regression line associated with Batch A has not changed, but the confidence limit has moved closer to the line, and it crosses the lower specification at a point just beyond 24 months, an increase of greater than two months beyond the confidence limit in Figure 1. The additional data included in the model of equation (2) resulted in an improved estimate of the residual standard error (0.59%), an increase in the residual degrees of freedom, and a shrinkage of the confidence limit about the line. An increase in the estimated shelf life is common, but by no means universal. The residual standard error may increase when the estimate is improved, but if it results in a more accurate assessment of actual measurement precision, it should also result in a better conservative estimate of shelf life.

Additional improvements in shelf life estimation can be achieved through batch pooling, as discussed in ICH Q1E. Where batch data pooling can be justified, based on a set of appropriate decision criteria, more extended shelf lives can frequently be justified, and in most cases more realistic shelf life estimates can be obtained. The decision criteria are based on ANCOVA (analysis of covariance) comparison of regression models (60), with time as a covariate and batches (or strengths, package, orientation, etc.) as a factor. As a typical example of how this procedure is applied to batch pooling, consider the first regression model to be that given by equation (2), where slope and intercept are allowed to vary by batch. The second regression model is the pooled slope model, given by equation (3), where intercepts are allowed to vary by batch but the slope is equal for all batches.

$$y_{ij} = \alpha_i + \beta t_{ij} + \varepsilon_{ij} \qquad \text{(Model 3)} \qquad\qquad (3)$$

Regression analysis of the same data shown in Figure 2, with the model equation given in equation (3), provided the results shown in Figure 3. In this case, equality of slopes has decreased the slope of the regression line for Batch A, so it intersects with the LSL at a point

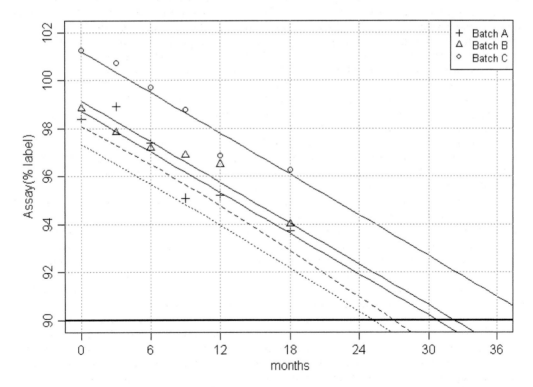

Figure 3　Linear regression of assay results for 18 months of stability from three batches. The model allowed pooling for estimation of a common slope. The dashed line is the lower 95% two-sided confidence interval for the regression line of Batch A. The dotted line is the lower 95% two-sided prediction interval for the regression line of Batch A.

just beyond 30 months. The 95% confidence limit for Batch A, given by the dashed line, now intersects the LSL at 27 months. (The additional dotted line will be discussed further below.) For this example, pooling has increased the estimated shelf life from 22 to 27 months, a substantial improvement if allowable.

The question arises, "How can we justify pooling of slopes within the guidelines of ICH Q1E?" Intuitively, if the slopes visibly differ to a greater extent than the distribution of points about each line, the difference in slopes would seem to be meaningful, and pooling of the slopes would be expected to obscure real differences between batches. However, if differences in the slopes over the length of the line segment appear small relative to the distribution of points about the line, it would seem unlikely that the differences would contribute much to the overall fit of the model. In this case, the slopes could be pooled without loss of valuable information regarding batch-to-batch differences in stability behavior. This intuitive assessment can be readily translated into a test statistic when the data are distributed normally about the regression lines. Let $SSE_x = \Sigma \varepsilon^2$ for model x. Also, let the residual degrees of freedom for model x be given by DF_x, which is equal to the number of data points minus the number of fitted parameters (independent slopes and intercepts). The test statistic for comparison of model 2 and model 3 is given by the expression.

$$F = \frac{(SSE_3 - SSE_2)/(DF_3 - DF_2)}{SSE_2/DF_2}$$

Both the numerator and denominator are mean square errors, which are random variables with chi-square distributions because the data are normally distributed. Their ratio, F, is therefore distributed as the F-distribution with $DF_3 - DF_2$ and DF_2 degrees of freedom, and values of the ratio can be compared with percentage points of standard F-distribution to assess the statistical significance of model improvements as additional terms are added. In the case of the examples given above, $DF_3 = 14$, $DF_2 = 12$, and the value of F computed from the sum of square error terms is $F = 0.7332$. This value corresponds to a probability point on the $F_{2,14}$ distribution of $p = 0.50$, which suggests that any difference between the two models is not statistically significant. Under these circumstances, pooling is accepted. ICH Q1E recommends batches not be pooled when $p < 0.25$, which is a relatively conservative choice that helps ensure any true batch-to-batch differences are retained in the analysis. A similar analysis can be completed for pooling of intercepts. In this case, for the data shown, $p < 0.0001$, and the intercepts were not poolable. Where factors do not involve cross-batch comparisons (e.g., package sizes, orientations), ICH Q1E recommends that batches pooling be disallowed only when the significance level is $p < 0.05$.

Within the limitations of the linear regression models, the ANCOVA analysis is versatile enough to account for multiple factors and sequential comparisons of models. This type of evaluation can readily be conducted with a number of standard statistical packages, including SAS (63) and R, an open-source software package (64).

Two caveats should to be kept in mind. (*i*) Future batches may show variation that is not evident in the primary batches. For example, if a batch similar to those in Figure 3 had an assay intercept of 96%, it likely would not meet acceptance criteria through a 24-month shelf life. Possible variations in future batches need to be considered when release criteria are established, and the relationship between expiration period, shelf life specifications, and release specifications must be understood. (*ii*) For batches that will be tested on stability as part of a commitment, or as part of an ongoing annual stability program, the risk of out-of-specification (OOS) test events may need to be assessed and managed. The 95% confidence interval reflects confidence in the estimation of linear model parameters, and this confidence interval shrinks as the number of points included in the analysis increases. However, when the distribution of future individual test measurements is considered, it is affected by both the inferential uncertainty associated with the model parameters and the measurement uncertainty associated with the precision of individual measurements. This contribution does not diminish. The 95% prediction interval represents both of these contributions and delineates the range within which 95% of future measurements are predicted to lie. An example of the lower two-sided 95% prediction interval is shown as the dotted line in Figure 3. When expiration periods are set, it may be useful to consider whether the prediction intervals show significant likelihood of OOS prior to the end of shelf life. Approaches may need to be developed to manage this risk.

Matrixing and Bracketing
Products that are available in multiple container volumes or strengths may benefit from either bracketing or matrixing (ICH Q1D). These practices allow for some overall reduction in the stability testing workload. Bracketing is the practice of testing stability for products at the extremes of strength or container size, and using the results to support the expiration period for the product at intermediate strength or container size. For example, results from a solution formulated at 0.5% and 2% wt/vol concentrations could be used to support a 1.0% concentration as long as the formulations did not differ significantly with respect to type and concentrations of excipients. If the high and low strengths differ in stability, the expiration period of the intermediate strength is constrained to the shorter of the two expiration periods. Some care is obviously required for multiphase products like suspensions, where changes in strength may be associated with change in the distribution of drug among phases that differ in stability.

Matrixing is the practice of testing multiple combinations of factors in a design that allows for reduction of testing at some time points. In statistics, matrix designs are commonly referred to as fractional factorial designs. Matrix designs commonly include combinations of factors such as strength, container size, container type, and even minor changes in formulation. Where a full factorial design would require testing of all combinations at all time points, a matrixed design allows for a subset of the samples to be tested at many of the time points. While the stability workload can be reduced, there is a risk that the expiration period that can be supported in a matrixed design will be shorter than that given by the full design, particularly in cases where measurement variability is large. Depending on the product and testing schedule, stability testing may be reduced by as much as 30% to 50% or more, a substantial decrease for designs that include multiple factors. Matrixing may be applied to bracketed designs. For additional discussion, see Chow (59) and references therein.

Special Considerations in Stability of Biopharmaceutical Products
Recommendations for stability of new biopharmaceutical products are captured in ICH Q5C. Basic principles are similar to those for ICH Q1A, but specific recommendations differ from those of ICH Q1A, reflecting both increased challenges of biopharmaceuticals, as well as the more limited capacity to capture comprehensive product quality attributes through a collection of precise analytical measurements. Briefly, the major points are as follows:

- Data from three primary batches on formal stability studies of at least six months duration are expected at time of filing. There are no prescribed stability storage conditions. Each product is considered unique with respect to definition of appropriate storage conditions.
- Lots may be pilot scale, but process and final package should be the same as batches to be manufactured.
- The shelf life request is granted on the basis of real time data (no extrapolations) at the label storage condition. Accelerated stability studies are recommended to help characterize the degradation profile and support excursions.
- In all cases, some measurement of biological potency of stability samples is required.

Photochemical Stability
ICH Q1B guidance describes a standard confirmatory photostability stress testing scheme for assessing susceptibility of both drug substance and a new drug product to light. The light sources have well-defined visible and near UV spectral characteristics typical of filtered daylight. The product is exposed to a minimum of 1.2 million lux hours and an integrated near ultraviolet energy of not less than 200 watt hours/m^2. It is subsequently analyzed for changes in appearance, color, clarity, assay, and degradants. Products should be exposed in a

transparent container, in the immediate package, and in the market pack to assess whether the product is photolabile, and whether the packaging is sufficiently protective to prevent the product from degradation that could affect its shelf life. If the product is not sufficiently stable after exposure to light, changes in packaging or special labeling may be required.

Stability Studies to Support Excursions

Although pharmaceutical products may generally be stored in controlled environments prior to dispensing, they may be shipped under conditions where they are subjected to short-term temperature excursions or rapid temperature changes. The increasing proportion of biopharmaceutical products, which are usually more sensitive to these excursions, has drawn considerable attention to shipping and distribution practices and the types of stability information required to support them. ICH Q1A accelerated stability studies provide useful information regarding exposure to higher temperatures, but at least two types of additional short-term studies are commonly run. Temperature excursion studies expose samples to high- or low-temperature excursions (including $-20°C$ exposure), typically for intervals of two days. Temperature cycling studies cycle the product through drastic changes in temperature over short intervals. For instance, where drug product labeled for room temperature storage might be exposed to subfreezing temperatures, thermal cycling would expose the product to three successive cycles of $-20°C$ for two days, followed by $40°C$ for two days. At the end of the cycle, the product would be examined for appearance, assay, and degradation products, and physical attributes including precipitation, aggregation, or phase segregation. Designs for cycling studies are somewhat specific to both the product and the stresses present in the intended distribution network. Specific designs are not currently outlined in regulatory guidances, except for the Metered Dose Inhaler guidance and the withdrawn 1998 stability guidance, but they are discussed in the Parenteral Drug Association Technical Report 39 (65).

REFERENCES

1. ICH Q6A: Specifications: Test Procedures and Acceptance Criteria for New Drug Substances and New Drug Products: Chemical Substances, 1999.
2. US Code of Federal Regulations, 21CFR 211.160(b): Current Good Manufacturing Practice for Finished Pharmaceuticals, Subpart I—Laboratory Controls.
3. Shah VP, Crommelin DJA. Regulatory issues and drug product approval for biopharmaceuticals. In: Crommelin DJA, Sindelar RD, Meibohm B, eds. Pharmaceutical Biotechnology: Fundamentals and Applications. New York: Informa Healthcare, 2008:447–453.
4. ICH Q8(R1): Pharmaceutical Development, 2008.
5. USP 31—NF 26, through 2nd Supplement. Rockville, MD: United States Pharmacopeial Convention, 2008.
6. European Pharmacopoeia 6.0–6.2. European Directorate for the Quality of Medicines and Healthcare, 2008.
7. The Japanese Pharmacopeia, 15th ed. MHLW, 2006.
8. International Conference on Harmonisation. Available at: http://www.ich.org. Accessed December 2008.
9. International Cooperation on Harmonisation of Technical Requirements for Registration of Veterinary Products. Available at: http://www.vichsec.org/. Accessed December 2008.
10. ICH Q4B: Evaluation and Recommendation of Pharmacopoeial Texts for Use in the ICH Regions, 2007.
11. Awang DCMZBC. ASEAN Initiatives toward pharmaceutical regulatory harmonization. Drug Info J 2003; 37:55–58.
12. ICH Q1F: Stability Data Package for Registration Applications in Climatic Zones III and IV.
13. Molzon JA. Current International Harmonization Efforts: Opportunities and Challenges. AAPS Workshop: Pharmaceutical Stability Testing to Support Global Markets, 2007.
14. ICH Q6B: Specifications: Test Procedures and Acceptance Criteria for Biotechnological/Biological Products, 1999.
15. ICH Q4B Annex 2: Evaluation and Recommendation of Pharmacopoeial Texts for Use in the ICH Regions on Test for Extractable Volume of Parenteral Preparations General Chapter, 2008.
16. Akers MJ, Larrimore DS, Guazzo DM. Particulate matter testing. In: Parenteral Quality Control: Sterility, Pyrogen, Particulate, and Package Integrity Testing. 3rd ed. New York: Informa Healthcare, 2007:197–280.

17. Carpenter JF, Randolph TW, Jiskoot W, et al. Overlooking subvisible particles in therapeutic protein products: gaps that may compromise product quality. J Pharm Sci 2009; 98:1201–1205.
18. ICH Q4B Annex 8: Evaluation and Recommendation of Pharmacopoeial Texts for Use in the ICH Region on Sterility Test General Chapter, 2008.
19. US FDA Draft Guidance for Industry: Submission of Documentation in Applications for Parametric Release of Human and Veterinary Drug Products Terminally Sterilized By Moist Heat Processes, 2008.
20. US FDA Guidance for Industry: Container and Closure System Integrity Testing in Lieu of Sterility Testing as a Component of the Stability Protocol for Sterile Products, 2008.
21. Akers MJ, Larrimore DS, Guazzo DM. Parenteral Quality Control: Sterility, Pyrogen, Particulate, and Package Integrity Testing. 3rd ed. New York: Informa Healthcare, 2007:119–196.
22. Williams K, ed. Endotoxins, Pyrogens, LAL Testing, and Depyrogenation. 3rd ed. New York: Informa Health Care, 2007.
23. ICH Q4B Annex 6: Evaluation and Recommendation of Pharmacopoeial Texts for Use in the ICH Regions on Uniformity of Dosage Units General Chapter, 2008.
24. ICH Q2(R1): Validation of Analytical Procedures: Text and Methodology, 2005.
25. EMEA Directive 75/318/EEC as amended: Specifications and Control Tests on the Finished Product, 1992.
26. ICH Q3B(R2): Impurities in New Drug Products, 2006.
27. ICH Q3A(R2): Impurities in New Drug Substances, 2006.
28. Baertschi SW. Pharmaceutical stress testing: predicting drug degradation. Boca Raton: Taylor & Francis, 2005.
29. ICH Q3B(R2): Impurities: Guideline for Residual Solvents, 1997.
30. US FDA Guidance for Industry: Container Closure Systems for Packaging Human Drugs and Biologics Chemistry, Manufacturing, and Controls Documentation, 1999.
31. EMEA CPMP/QWP/4359/03: Guideline on Plastic Immediate Packaging Materials, 2005.
32. Jenke D. Extractable substances from plastic materials used in solution contact applications: an updated review. PDA J Pharm Sci Technol 2006; 60:191–207.
33. Akala EO. Effect of packaging on the stability of drugs and drug products. In: Gad SC, ed. Pharmaceutical Manufacturing Handbook: Regulations and Quality. Hoboken, N.J: John Wiley & Sons, 2008.
34. Jenke D. Evaluation of the chemical compatibility of plastic contact materials and pharmaceutical products; safety considerations related to extractables and leachables. J Pharm Sci 2007; 96:2566–2581.
35. Yoshioka S, Aso Y. Correlations between molecular mobility and chemical stability during storage of amorphous pharmaceuticals. J Pharm Sci 2007; 96:960–981.
36. Hageman MJ. Water sorption and solid-state stability of proteins. Pharm Biotechnol 1992; 2:273–309.
37. Shamblin SL. The role of water in physical transformations in freeze-dried products. Biotechnol Pharm Aspects 2004; 2:229–270.
38. Medlicott NJ, Waldron NA, Foster TP. Sustained release veterinary parenteral products. Adv Drug Deliv Rev 2004; 56:1345–1365.
39. Yoshioka S, Stella VJ. Stability of drugs and dosage forms. New York, NY: Kluwer Academic/Plenum Publishers, 2000.
40. Meyer BK, Ni A, Hu B, et al. Antimicrobial preservative use in parenteral products: past and present. J Pharm Sci 2007; 96:3155–3167.
41. EMEA CPMP/CVMP/QWP/115/95: Note for Guidance on Inclusion of Antioxidants and Antimicrobial Preservatives in Medicinal Products, 1997.
42. Allen T. Powder sampling and particle size determination. Amsterdam, Boston: Elsevier, 2003.
43. Shekunov BY, Chattopadhyay P, Tong HHY, et al. Particle size analysis in pharmaceutics: principles, methods and applications. Pharm Res 2007; 24:203–227.
44. Burgess DJ, Crommelin DJA, Hussain AS, et al. Assuring quality and performance of sustained and controlled release parenterals. Eur J Pharm Sci 2004; 21:679–690.
45. Burgess DJ. Injectable Dispersed Systems: Formulation, Processing, and Performance. Boca Raton: Taylor & Francis, 2005.
46. Siewert M, Dressman J, Brown CK, et al. FIP/AAPS guidelines to dissolution/in vitro release testing of novel/special dosage forms. AAPS PharmSciTech 2003; 4:E7.
47. ICH Q1A(R2): Stability Testing of New Drug Substances and Products, 2003.
48. ICH Q1B: Stability Testing: Photostability Testing of New Drug Substances and Products, 1996.
49. ICH Q1C: Stability Testing for New Dosage Forms, 1996.
50. ICH Q1D: Bracketing and Matrixing Designs for Stability Testing of New Drug Substances and Products, 2002.
51. ICH Q1E: Evaluation for Stability Data, 2003.
52. ICH Q5C: Quality of Biotechnological Products: Stability Testing of Biotechnological/Biological Products, 1995.

53. US FDA Draft Guidance for Industry (withdrawn): Stability Testing of Drug Substances and Drug Products, 1998.
54. US FDA Draft Guidance for Industry: Metered Dose Inhaler (Mdi) and Dry Powder Inhaler (Dpi) Drug Products Chemistry, Manufacturing, and Controls Documentation, 1998.
55. Zahn M. Global stability practices. In: Huynh-Ba K, ed. Handbook of Stability Testing in Pharmaceutical Development : Regulations, Methodologies, and Best Practices. New York: Springer, 2008.
56. EMEA CPMP/QWP/609/96/Rev 2: Guideline on Declaration of Storage Conditions: A: in the Product Information of Medicinal Products B: for Active Substances, Annex to Note for Guidance on Stability Testing of New Drug Substances and Products, Annex to Note for Guidance on Stability Testing of Existing Active Substances and Related Finished Products, 2003.
57. World Health Organization WHO Technical Report Series, No. 908, Annex 9: Guide to Good Storage Practices for Pharmaceuticals, 2003.
58. EMEA CPMP/QWP/2934/99: Note for Guidance on in-Use Stability Testing of Human Medicinal Products, 2001.
59. Chow SC. Statistical Design and Analysis of Stability Studies. Boca Raton: Chapman & Hall/CRC, 2007.
60. Rhodes CT, Carstensen JT. Drug Stability: Principles and Practices. New York: Marcel Dekker, 2000.
61. Bolton S. Pharmaceutical Statistics: Practical and Clinical Applications. New York: M. Dekker, 1997.
62. Sen AK, Srivastava MS. Regression Analysis: Theory, Methods, and Applications. New York: Springer, 1997.
63. SAS/STAT software. Available at: http://www.sas.com/. Accessed December 2008.
64. R Project for Statistical Computing. Available at: http://www.r-project.org/. Accessed December 2008.
65. Technical Report # 39. Cold chain guidance for medicinal products: maintaining the quality of temperature-sensitive medicinal products through the transportation environment. Parenteral Drug Association. PDA J Pharm Sci Technol/PDA 2005; 59:1–12.

10 | The management of extractables and leachables in pharmaceutical products

Edward J. Smith and Diane M. Paskiet

INTRODUCTION: ORIGIN AND IMPORTANCE OF EXTRACTABLES AND LEACHABLES

In this chapter the important elements of extractables and leachables, with which those in the pharmaceutical industry should be familiar, are discussed. On completion of this chapter, the reader should be able to

- Understand the extractables/leachables expectations in guidelines, guidances, United States Pharmacopoeia (USP), International Conference on Harmonization (ICH), U.S. Code of Federal Regulations (CFR), and other regulatory documents
  What is expected or required?
- Identify sources of extractables from packaging and process materials such as plastic, glass, and rubber
  Where do E&L (extractables and leachables) originate?
- Know what specific extractables/leachables information must be present in the chemistry, manufacturing, and controls documentation (CMC) sections of applications for the various types of drug products (injectables, oral, etc.)
  What information must be contained in new drug applications?
- Design and execute an extractables/leachables study
  What are the elements of an E&L study?

Packaging has allowed the widespread distribution of drugs; without it there would be no pharmaceutical industry, and the quality and quantity of life we have come to expect would not exist. Try to imagine a product that you use that comes to you without the use of packaging, perhaps a home-grown tomato from your garden! Packaging serves many functions (1,2) such as

- Protection and containment—packaging is expected to maintain the quality and quantity of a drug product until expiration from the filling line to the patient. Packaging must not interact with nor alter the efficacy of the drug, and leachables levels must not present a toxicity risk to patients. In addition, packaging must meet all pharmacopoeial requirements where marketed.
- Transportation and storage—drug products are transported and stored in packaging containers until used. Stability studies provide information on the finite "shelf life" of a drug product in a specific container.
- Identification—printed components provide the product name, strength, expiry date, dose, precautions and contraindications, and other information to healthcare professionals and patients.
- Compliance—necessary dose quantities can be conveniently packaged to facilitate the delivery of specific quantities of drug per day.
- Delivery—some packaging components, such as the prefilled syringe, provide the additional function of being delivery or administration devices.

In essence, good packaging must be "suitable for use," that is, it must provide the necessary protection, compatibility, safety, and performance. E&L studies measure two of the four suitability requirements: compatibility and safety.

Despite the necessity and positive functions of packaging, there is one significant disadvantage that must be evaluated for every pharmaceutical product. That is, *packaging materials interact with drug products*. In fact an article in Chemical & Engineering News (3) stated that "It is not a question of whether packaging components will leach into a product, it's a question of how much." A statement in the U.S. current Good Manufacturing Practices (cGMPs) acknowledges and reflects this concern (4).

> § 211.94 (a) Drug product containers and closures shall not be reactive, additive, or absorptive so as to alter the safety, identity, strength, quality, or purity of the drug beyond the official or established requirements.

Drug product-packaging interactions are commonly classified into four types:

- Adsorption—some part of a drug product is sorbed or concentrated onto the contact surface of a packing component.
- Absorption—following adsorption, the substance may penetrate the surface of the packaging material and migrate into the material.
- Permeation—further migration may lead to migration of a substance through the packaging to the noncontact surface and beyond.
- Leaching—substances may migrate from a packaging component into the drug product.

All four or any combination of the interactions may and probably do take place in any given packaging situation; however, as long as the interactions do not "alter the safety, identity, strength, quality, or purity of the drug beyond the official or established requirements" during the shelf life of drug product, the packaging is deemed acceptable for use. Although all four interactions are important, only leaching or migration of packaging substances into drug products or related materials will be discussed in this chapter.

Thus far, the two terms, extractables and leachables, have been mentioned but not defined. They are often used interchangeably but they have distinctly different meanings.

- Extractables are chemical substances that are removed from a material by the exertion of an artificial or exaggerated force. That force may be a strong solvent, a high temperature, a long extraction time, or a combination of the three. An extractables test is a *packaging test* performed on a packaging component or material.
- Leachables are chemical components that migrate from a contact material into drug products during storage at "normal" conditions. A leachables test is a *drug test* performed on a drug product or related material to identify and quantify substances that have migrated into it from a packaging component or other related component.

Extractables tests may be performed on a component or material without specific knowledge of what drug may be ultimately in contact with it. Lists of extractables may be generated by manufacturers or suppliers of packaging materials and components using solvent systems that have come to be standardized as common industry practice. A list of extractables represents a list of *potential leachables* that may be targeted for identification and quantification is a leachables study. Leachables are usually a subset of extractables—and not all extractables are leachables. Exceptions will be discussed later in this chapter.

E&L studies have different goals.

- Controlled extractables studies—to identify as many chemical compounds as possible that have the potential to become leachables.
- Leachables stability studies—to identify, quantify, and qualify as many compounds as possible that migrated from packaging materials into a drug product. Qualification is the process of acquiring and evaluating data that establishes the biological safety of an individual impurity or a given impurity profile at the level(s) specified. It may not be necessary to identify, quantify, and qualify a leachable if the amount present in the drug does not present a safety concern. This will be discussed later in the chapter when the work of the Product Quality Research Institute (PQRI) is discussed (5).

Extractables studies should always precede leachables studies since the target compounds for the leachables studies are identified in the extractables studies.

Why have leachables become an issue in the world of pharmaceuticals? The obvious reason is because substances that have migrated into a drug product may "alter the safety, identity, strength, quality, or purity of the drug" (4). Some commonly listed reasons are that leachables

- May interfere with drug product assays (e.g. have the same retention time as the drug substance in a high performance liquid chromatography (HPLC) assay)
- May interfere with medical diagnostic tests
- May cause appearance change in drug product (e.g., color change)
- May increase the impurity level of drug product to an unacceptable level
- May react with one or more of drug product components [e.g., may cause precipitate, pH change, or degradation of the active ingredient; Zn/epinephrine reaction in dental anesthetics (6)]
- May increase toxicity of drug product (e.g., may require identification, risk assessment, qualification, and quantification)

Taking into account all the reasons listed above, the goal of any leachables study is to reduce risk to patients who receive the drug products. A recent example of product that caused adverse affects due to the lack of leachables data was EPREX®, a product of Janssen-Cilag (7). The history of events surrounding EPREX is listed below.

- EPREX in vials contained human serum albumin (HSA), 1994.
- EPREX with HSA was offered in a prefilled syringe containing an uncoated rubber plunger, 1994.
- HSA-free EPREX was introduced in 1998; HSA was replaced by Polysorbate 80 as the protein stabilizer.
- Polysorbate 80 increased the extraction of adjuvant-like leachables from the rubber plunger causing an unwanted side effect called PRCA (pure red-cell aplasia) in EPREX patients.
- Recall occurred and a coated plunger, which reduced the amount of leachables, was substituted for the uncoated closure.

The authors of the paper who describe the EPREX problem recommend that "... an active program of monitoring products for the presence of extraneous molecules is prudent. The necessity of using multiple techniques for the detection, identification, and quantitation of the leachables in this study indicates that no single method is sufficient and suggests that multiple, orthogonal techniques be routinely employed." In this particular case, even if a thorough E&L study was performed, the connection of the rubber leachable with PRCA probably could not have been predicted.

An older study of a dental anesthetic presents a simpler situation (6). In 1981, Astra Pharmaceutical Products, now Astra Zeneca, produced and marketed a dental anesthetic in a syringe cartridge. The cartridge had two rubber components: a thin rubber seal on the needle end, which is pierced by a double-ended needle in the syringe body to allow expulsion of the drug, and a chlorobutyl rubber plunger that seals the opposite end of the cartridge and is pushed forward by a plunger rod to deliver the anesthetic to the patient. The drug product contained a vasoconstrictor, epinephrine, to localize the anesthetic in the area of the dental procedure. Reports from dentists that the anesthetic was less effective than usual in several cases led Astra to the search for the cause. Dentists reported that the anesthetic effect was no longer localized, leading to an investigation of the level of epinephrine. It was known that oxygen reacts with epinephrine and that oxygen could be permeating through or around the thin rubber seal or around the thicker plunger. Neither was occurring; the oxygen level in the cartridges was not out of acceptable limits. Further investigation led to a correlation of high soluble aluminum with low levels of epinephrine. The source of the aluminum was calcined clay, which was a reinforcing agent in the rubber plunger. A change in the calcining temperature of the clay caused an increase in the solubility of the aluminum from 5000 to 25,000 ppm in the dental cartridges. It was determined in subsequent studies (8,9) that aluminum catalyzes the degradation of epinephrine.

In the 1980s, there were no leachables studies as we would do them today, and extractables testing on rubber components was confined to compendial tests such as the USP and the European Pharmacopoeia (Ph. Eur.). In this particular case, there were no material or process changes made by Astra or their rubber supplier, but changes were made in the process of manufacturing the clay by the rubber manufacturer's clay supplier. Extractables tests and leachables tests would have identified aluminum as a significant leachable and alerted Astra to perhaps monitor its concentration in incoming rubber. And certainly quality agreements, which included change controls, between Astra and the rubber supplier, and the rubber supplier and its clay supplier, would have given all parties the opportunity to evaluate the impact of the proposed change in calcining temperature.

The key lessons learned from the Astra studies are

- E&L studies, done prior to marketing, are necessary to identify possible harmful leachables in a drug product.
- E&L studies are not a "one time and over" study; periodic monitoring of packaging components and drug products is necessary.
- Quality agreements, emphasizing change control, are necessary to reduce the chance of process or material changes.
- Information-sharing relationships between user and supplier are key to maintaining quality, not only knowing what is needed from each other but why it is essential.

Leachables from both food and drug packaging was the lead article in an issue of the widely distributed Chemical & Engineering News (3). In it the author highlighted two cases of interaction of protein-based drugs with components of prefilled syringes. In one case, a syringe manufacturer was using an epoxy adhesive to attach the metal needle to the syringe barrel. Unfortunately, a solvent from the partially dried epoxy leached into the liquid drug product, oxidized the protein, and caused it to aggregate. In the second case, a tungsten wire used to make the hole in the tip of the syringe to hold the needle left a tungsten oxide residue that later migrated into the drug product and also caused protein aggregation.

Leachables in a drug product have many sources. Since leachables have several sources, E&L studies must not focus only on the primary drug package. Both drug contact and noncontact materials may be sources. The chief concern of any health authority is safety—how do leachables affect the drug and what direct affect do leachables have on the body (toxic effects)? Figure 1 illustrates the many routes though which leachables may enter the body. These routes are primarily through the drug, but leachables, if that is what we wish to call them, may also enter the body by direct contact with packaging materials such as with a drug patch, catheter, or implanted drug delivery device. In Figure 1, this is represented by the arrow from the material to the body. The other routes are

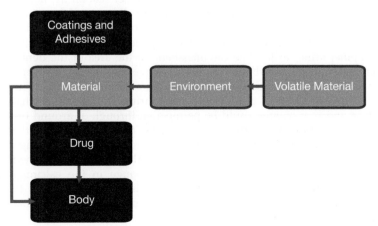

Figure 1 Routes of leachables.

- From material (packaging) into a drug, then into the body
- From coatings and adhesives into packaging materials, then into a drug, then into the body. This occurs most often in flexible and rigid plastic containers with labels where adhesives and ink components can migrate through the plastic container walls into the drug. This does not occur with glass containers.
- From the environment—volatile compounds in the environment can be adsorbed and absorbed by packaging materials and then move into a drug and ultimately the body.
- Volatile materials—volatile materials that are in close proximity to packaging materials may get into the environment and eventually into the body using the route described above. For example, wooden pallets are often fumigated with methyl bromide to rid them of insects. Plastic and rubber packaging materials shipped and stored on wooden pallets could absorb the fumigant if temperature is high and the space is confined.

Consider the case of a glass prefilled syringe with a staked needle (10). The drug is in contact with the following components/materials:

- Glass barrel—glass, although quite nonreactive, is not inert. Metal ions from the glass and pH shifts are the chief concerns. Tungsten, a residue from the manufacturing process, is also a concern. Tungsten wires are used to make the orifice in the tip of the glass barrel where the stainless steel needle will be placed.
- Rubber plunger—rubber is a source of both organic and inorganic extractables. Extractables from both glass and rubber will be discussed in detail later in the chapter.
- Rubber needle shield—the tip of the needle is imbedded in rubber to seal the syringe. Although the area of contact is very small, the needle shield must be considered in an E&L study.
- Stainless steel needle—iron and nickel are chief metal ions extracted.
- Adhesive—an adhesive is used to bind the steel needle to the glass barrel. Contact area is small, but organics may be extracted into the drug product.
- Silicone oil—although not thought of as a packaging component, silicone oil is placed on the inside surface of the glass to reduce the break loose and gliding forces of the rubber plunger. Silicone is often a concern with biotechnology-derived products.

In designing an E&L study there are many choices to be made and each has associated risks. Some of these choices are

- Level of study for a drug product. The U.S. Food and Drug Administration's (FDA) guidance on packaging (11) contains graduated levels of E&L studies for different routes of administration.
- Packaging materials/components to study. It is obvious that a packaging material that will be in direct contact with the drug (primary packaging materials) must always be considered. But, how about labels, or over raps that are not in contact with the drug?
- Extracting solvent systems. How many? Which ones?
- Extraction conditions. How long? What temperature? What quantity/surface area per volume of solvent?
- Minimum quantity to identify, quantify, and qualify.
- Analytical methods to use and validate.
- Participating laboratories. What skill sets? In-house or contract?
- Frequency of leachables measurements over shelf life.

Even after considerable study of the E&L requirements and best practices, those considering an E&L study have the same common questions. These questions are not easily answered and there is not only one correct study protocol. The correctness of a protocol depends on the drug (e.g., route of administration, frequency of use, patient population), the packaging materials (type and degree of potential interaction with the drug), division of FDA or regulatory authority that regulates the drug product, drug company making the submission, and many other variables. The common questions are as follows:

- What do the regulations say? There are several guidances, guidelines, ICH (www.ich. org), and CFR (www.gpoaccess.gov/cfr/retrieve.html) documents to consider.
- How far do we go? Scientists ponder how much effort is expected to identify, or quantify, or qualify a chemical compound that is found in an E&L study. Many extractable compounds may not have published reference spectra or chromatographic retention time data or toxicity data, nor is authentic reference material available, and the quantity extracted may be at the limit of detection. Regulations (11) often contain terms such as "appropriate solvent," "significantly exceed," and "may be advisable," which make definitive interpretation difficult.
- How is a practical science-based E&L study performed?

These and other questions will be answered in this chapter giving the reader enough information to design a "correct" protocol and to avoid inappropriate, non-science-based methods and procedures.

COMMON TERMS DEFINED

The language of E&L studies is still evolving and many common terms have specific meanings when used in the context of E&L. Even the terms "extractable" and "leachable", though very distinct, are often used interchangeably in both written and oral presentations. Common terms are defined here, rather than in an appendix, so that the reader is familiar with the language of E&L before moving into more detailed sections of this chapter. The most important two terms, extractables and leachables, have been defined previously so the reader is referred back to section "Introduction: Origin and Importance of Extractables and Leachables."

Packaging

- Container closure system: Refers to the sum of packaging components that together contain and protect the dosage form. This includes primary packaging components and secondary packaging components.
- Packaging component: Any single part of a container closure system.
- Packaging materials: May refer to packaging components or to materials of construction.
- Primary packaging component: A packaging component that is or may be in direct contact with the dosage form. Examples of primary components are

☐ Ampoules	☐ Bottles [glass, high-density polyethylene (HDPE)]
☐ Flexible bags	☐ Closure liners
☐ Tube with/without liner	☐ Desiccant container
☐ Pouch	☐ Filler (cotton, rayon)
☐ Cap inner seal	☐ Rubber vial stopper
☐ Blister packaging	☐ Glass, plastic vial
☐ Bulk containers	☐ Prefilled syringe

Primary packaging components *are the major source of leachables* because they are in intimate contact with drug products.

- Secondary packaging component: A packaging component that is not and will not be in direct contact with the dosage form. Some examples are as follows:

☐ Overwraps	☐ Cartons
☐ Inks	☐ Labels
☐ Plastic plunger rods	☐ IV tubing clamp
☐ Aluminum and aluminum/plastic crimp seals for vials	

Secondary components *may be a source of leachables*. Adhesive from labels on plastic containers must be considered as potential leachable but not when on glass containers. Volatile substances from secondary components may also migrate into primary components and become leachables.

- Associated component: Component intended to deliver the dosage form to the patient but not stored in contact with the dosage form. Examples are

  ☐ Dosing cups   ☐ Calibrated spoons
  ☐ Sterile empty syringe ☐ Medicine dropper

Since these components are only intended for short-term contact (minutes) with drug products, *leachables are unlikely*; however, if it is possible and likely that drug products could be stored in these components for a significant time (hours or days), such as in a sterile empty syringe or even in a dosing cup, then these components should be evaluated for leachables.

Drug

- Drug: A therapeutic agent or any substance other than food used in the prevention, diagnosis, alleviation, treatment, or cure of diseases.
- Drug product: The dosage form in the final immediate packaging intended for marketing. Leachables studies are performed on the drug product for a time equal to the shelf life.
- Drug substance or active pharmaceutical ingredient (API): An active ingredient that is intended to furnish pharmacological activity or other effect in the prevention, diagnosis, alleviation, treatment, or cure of diseases. The unformulated drug substance that may be subsequently combined or formulated with excipients to produce the drug product.
- Drug product vehicle: The entity (or mixture of entities) that delivers the drug to the site of application. For a liquid dosage form, the drug product vehicle is every part (or component) of the liquid preparation except the drug substance or API. In certain circumstances, such as when the API would interfere with the analysis of extractable substances in the drug product, leachables testing may be performed with the drug product vehicle and not the drug product. To justify this, the analyst would need to demonstrate that the API does not alter the extraction properties of the drug product.
- Solvent: An organic or inorganic liquid used as a vehicle for the preparation of solutions or suspensions in the synthesis or manufacture of a new drug substance. Also, an organic or inorganic liquid used in extraction studies that will extract chemical components that are potential leachables but will not dissolve the material or component being studied.
- Simulated solvent: Solvents commonly used to mimic the extraction properties of foods and beverages to be used for extractables testing prescribed in the food additive regulations. The food simulating solvents are generally water, heptane, and 8% and 50% alcohol. Extraction conditions are based on conditions of use and type of food. In the leachables testing of drugs, the "drug product vehicle," may be used as a "simulating solvent" when the API interferes with analytical testing.
- Degradation product: An impurity resulting from a chemical change in the drug substance brought about during manufacture and/or storage of the new drug product by the effect of, for example, light, temperature, pH, water, or by reaction with an excipient and/or the immediate container closure system (12).
- Identified degradation product: A degradation product for which a structural characterization has been achieved (12).
- Unidentified degradation product: A degradation product for which a structural characterization has not been achieved and that is defined solely by qualitative analytical properties (e.g., chromatographic retention time) (12).
- Specified degradation product: A degradation product that is individually listed and limited with a specific acceptance criterion in the new drug product specification. A specified degradation product can be either identified or unidentified (12).
- Impurity: Any component of the new drug product that is not the drug substance or an excipient in the drug product (12).

- Impurity profile: A description of the identified and unidentified impurities present in a drug product (12).
- Identified impurity: An impurity for which a structural characterization has been achieved (13).
- Potential impurity: An impurity that theoretically can arise during manufacture or storage. It may or may not actually appear in the new drug substance (13).
- Specified impurity: An impurity that is individually listed and limited with a specific acceptance criterion in the new drug substance specification. A specified impurity can be either identified or unidentified (13).
- Unidentified impurity: An impurity for which a structural characterization has not been achieved and that is defined solely by qualitative analytical properties (e.g., chromatographic retention time) (13).
- Unspecified impurity: An impurity that is limited by a general acceptance criterion, but not individually listed with its own specific acceptance criterion, in the new drug substance specification (13).

Extractions and Leachables Studies

- Extraction profile: Analysis (usually by chromatographic means) of extracts from a container-closure system, usually qualitative. A profile is usually presented as a chromatogram or as a table showing the identity, relative peak height, and retention time or as a table.
- Quantitative extraction profile: An extraction profile in which the amount of each substance is determined.
- Qualification: The process of acquiring and evaluating data that establishes the biological safety of an individual impurity or a given impurity profile at the level(s) specified.
- Thresholds [from ICH Q3B(R2): Impurities in New Drug Products] (12):
  ☐ Reporting threshold: A limit above which an impurity needs to be reported.
  ☐ Identification threshold: A limit above which an impurity needs to be identified.
  ☐ Qualification threshold: A limit above which an impurity needs to be qualified.

- PQRI: Product Quality Research Institute—a nonprofit consortium of organizations working together to generate and share timely, relevant, and impactful information that advances drug product quality and development (14). The PQRI has completed and published one E&L study of inhalation products (OINDP) (5) and is currently completing another study of parenteral and ophthalmic products (PODP). (See definitions of OINDP and PODP below). An important objective of these studies is to define toxicological and analytical limits for E&L studies.
- OINDP: PQRI study of Orally Inhaled and Nasal Drug Products such as
  ☐ Metered dose inhalers   ☐ Dry powder inhalers
  ☐ Inhalation solutions   ☐ Inhalation suspensions
  ☐ Spray products   ☐ Nasal sprays

- PODP: PQRI study of Parenteral and Ophthalmic Drug Products such as injectable SVPs (small volume parenterals), injectable LVPs (large volume parenterals), and all ophthalmic products—injectable and noninjectable.
- SCT: PQRI term for Safety Concern Threshold, which is the threshold below which a leachable would have a dose so low as to preset negligible safety concerns from carcinogenic and noncarcinogenic toxic effects. For OINDP products, this threshold was concluded to be 0.15 µg/day (n). For PODP products the threshold may differ.
- QT: PQRI term for Qualification Threshold, which is the threshold below which a given leachable is not considered for safety qualification (toxicological assessment) unless the leachable presents structure-activity relationships (SAR) concerns (5). For OINDP products this threshold was concluded to be 5.0 µg/day. For PODP products the QT threshold may differ also.

REGULATORY REQUIREMENTS

Drug products and medical devices are regulated worldwide for the purpose of protecting all consumers; although guiding principles are intended to achieve the same end result, the legislation is not international. Drug products, medical devices, and their raw materials are obtained globally, in part, or finished product, from both established and emerging economies. Regulatory bodies in each country will have their own set of expectations for the safety, quality, effectiveness, and performance of drugs and medical devices. For instance, in Europe, licensing can be granted at a national and/or European Union level and a number of different regulatory agencies may be involved. Through the ICH process, considerable harmonization has been achieved among the three regions (Japan, Europe, and the United States) in the technical requirements for the registration of pharmaceuticals for human use. These products are licensed through a market application and approval process and the FDA, European Union Medicines and Healthcare Products Regulatory Agency (MHRA), and Japanese Ministry of Health, Labor and Welfare (MHLW) are among the regulatory bodies driving the standards for governance of drugs and medical devices. The regulations for pharmaceuticals, biologics, and medical devices are not totally harmonized but the expectations and process have much in common. In this chapter, the U.S. legislation, FDA guidance documents, and recognized standards will be discussed in relation to qualification of container closure systems for drug products with consideration given to the medical device regulation.

In 1906, the original U.S. Food and Drugs Act was passed by Congress to prohibit interstate commerce of misbranded and adulterated foods, drinks, and drugs. This was later revised in 1938 to the Federal Food, Drug, and Cosmetic (FDC) Act to contain new provisions to extend the control to cosmetics and therapeutic devices. New drugs were also required to be shown safe before marketing. Amendments were made in 1976 and 1990 for medical devices so that all devices are to be divided into classes with varying amounts of control required and indication of safety. The sections of the United States Code (USC) Sections 501, 502, and 505 are associated with container closure systems for drug products; the following transcriptions are noted:

- "a drug is deemed adulterated if its container is composed in whole or part of a poisonous or deleterious substance that may render the contents injurious to health ..."
- "an application shall include a full description of the methods used in the manufacturing, process, and packaging of such a drug. This includes facilities and controls used in the packaging and drug product."

The rulings for drugs were codified in 1978 under the CFR Title 21 parts 210 and 211, more commonly known as current Good Manufacturing Practices; devices were regulated under 21 CFR 820 Quality System Regulation, sections specific to these products are described as follows (15,16):

21 CFR 211.160 General Requirements
   Laboratory Controls shall include the establishment of scientifically sound and applicable written specifications, standards, sampling plans, and test procedures including resampling, retesting, and data interpretation procedures designed to ensure that components, drug product containers, closures, in-process materials, labeling, and drug products conform to appropriate standards of identity, strength, quality, and purity.
21CFR 211.94 Drug Product Containers and Closures
   Device containers should not be reactive, additive, or absorptive as to alter the safety, identify, strength, quality, or purity of the drug beyond the official or established requirements of drug product.
      Standards or specifications, methods of testing, and, where indicated, methods of cleaning, sterilizing, and processing to remove pyrogenic properties shall be written and followed for drug product container and closures.
21 CFR 820
   cGMP requirements are set forth in this quality system regulation. The requirements in this part govern the methods used in, and the facilities and controls used

for, the design, manufacture, packaging, labeling, storage, installation, and servicing of all finished devices intended for human use. The requirements in this part are intended to ensure that finished devices will be safe and effective and otherwise in compliance with the FDC Act.

Applications for drug products and devices are submitted to one of the three FDA centers: Center for Biologics Evaluation and Research (CBER), Center for Devices and Radiological Health (CDRH), and Center for Drug Evaluation and Research (CDER). Each center governs specific to the nature of the product and its intended use. This means the information required for a drug or biological application is similar but may not be necessary for that of a device, and the information for a device may not pertain to that of a drug product. Over time as drug products and administration forms have evolved, combination products have entered the market. A combination product is defined under the FDC Act as

1. A product comprising two or more regulated components, i.e., drug/device, biologic/device, drug/biologic, or drug/device/biologic, that are physically, chemically, or otherwise combined or mixed and produced as a single entity.
2. Two or more separate products packaged together in a single package or as a unit and comprised of drug and device products, device and biological products, or biological and drug products.
3. A drug, device, or biological product packaged separately that according to its investigational plan or proposed labeling is intended for use only with an approved individually specified drug, device, or biological product where both are required to achieve the intended use, indication, or effect and where upon approval of the proposed product the labeling of the approved product would need to be changed, e.g., to reflect a change in intended use, dosage form, strength, route of administration, or significant change in dose.
4. Any investigational drug, device, or biological product packaged separately that according to its proposed labeling is for use only with another individually specified investigational drug, device, or biological product where both are required to achieve the intended use, indication, or effect.

Combination products raise a variety of regulatory and review challenges since the products share many of the same basic features, they are also each somewhat unique.

Drugs, devices, and biological products each have their own types of marketing applications, GMP regulations, and adverse event reporting requirements. When drugs and devices, drugs and biologics, or devices and biologics are combined to create a new product, questions are sometimes raised about how the combination product as a whole will be regulated as there is no special type of marketing application for combination products. Under Section 503 of the FDC Act, a combination product is assigned to a center with primary/lead jurisdiction based on a determination of the primary mode of action (PMOA) of the combination product. A combination product is assigned through the FDA Office of Combination Products (OCP) (17).

The type and amount of container closure information required in a given application can vary and the interpretation of the legislation can be dependent on different factors. Requirements mandated by the FDA are found in applicable monographs (those with numbers under 1000) of the USP/National Formulary (18). Beyond the container closure information specified in the monographs, the FDA recommends additional information to be provided on the basis of guidance documents. The FDA's Guidance for Industry does not suggest a comprehensive list of tests, specific test methods, or acceptance criteria. Batch-to-batch consistency of packaging components and acceptance criteria should be based on good scientific principles for each specific system and product (11). The guidance documents concerning cGMP and the container closure guidance will be discussed here. A list of other related guidance documents will be included in "References".

In 2006, the modernization of the cGMPs was initiated to bridge the 1978 regulation with current understanding of quality systems, harmonize with other widely used quality systems, and establish a framework for a more systematic risk-based approach to manufacturing of pharmaceuticals. The FDA issued a report in 2004 titled, "21st Century" Initiative on the Regulation of Pharmaceutical Manufacturing, which described plans for forthcoming guidance on the new quality system. The Guidance for Industry *Quality Systems Approach to Pharmaceutical CGMP Regulations* was published in 2006 by CDER, CBER, Center for Veterinary Medicines (CVM), and Office of Regulatory Affairs (ORA), providing the framework to instill the philosophy that "Quality should be built into the product, and testing alone cannot be relied on to ensure product quality" (19). The concept was subsequently more fully detailed in the following Guidance for Industry documents developed by CDER/CBER within the expert working group of ICH (20–22):

– 2009 Q8(R2) Pharmaceutical Development
– 2006 Q9 Quality Risk Management
– 2009 Q10 Pharmaceutical Quality System

The concept of design space and building quality through process development and improvements are presented in Q8. The choice and rationale for the container closure system should be consistent with the Common Technical Document (CTD) format. Information on both leachables and extractables should be included in Module 3 (Quality) Manufacturing Process Development section under Container Closure System (3.2.P.2.4.). When warranted, E&L related impurities, the correlations and specifications should be included, if leachables are confirmed through shelf life. Q8 cites "The degree of regulatory flexibility is predicated on the level of relevant scientific knowledge provided." Relevant scientific knowledge is grounded in the principles of risk management, which is described in Q9 and illustrated as shown in Figure 2.

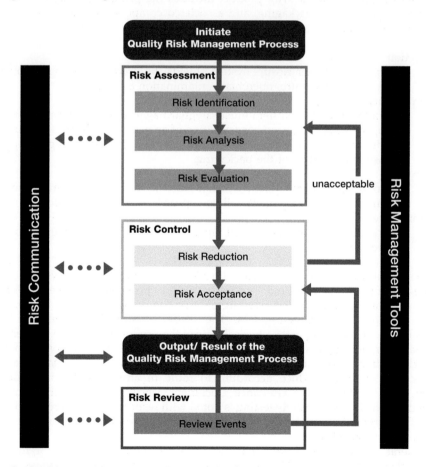

Figure 2   Quality risk management process.

Examples of risk tools such as Failure Mode Effects Analysis (FMEA), Fault Tree Analysis (FTA), Preliminary Hazard Assessments (PHA), and Hazard Analysis and Critical Control Points (HACCP) are referenced. Quantifying the probability of any particular extractable migrating into the drug product and the severity of the impact is the goal of a leachable study.

Three main objectives of Q10 are to (*i*) achieve product realization, (*ii*) establish and maintain a state of control, and (*iii*) facilitate continual improvement.

The ICH Q10 model outlines the pharmaceutical quality system on the basis of the International Organization for Standardization (ISO) quality concepts. Quality systems to support the technical activities for pharmaceutical development, technology transfer, commercial manufacturing, and product discontinuation are explained. ICH Q10 augments the cGMPs by providing details on specific quality elements such as process performance monitoring systems, corrective action/preventative action, change management, and management responsibilities. Implementation of Q10 throughout the product lifecycle should facilitate innovation and continual improvement and strengthen the link between pharmaceutical development and manufacturing activities. Leachables and extractables, although not specifically mentioned, is key in achieving the Q10 objectives. Container closure systems must be manufactured under GMP conditions and satisfy the same quality elements.

The Guidance for Industry: Container Closure Systems for Packaging Drugs and Biologics was published by CDER/CBER in 1999 and provides recommendations for the information to be provided in an application for any drug product. Since the publication of this guidance there have been additional recommendations on inhalation products as well as those of the PQRI. These recommendations will be cited in the "References" section but are too specific to include in this section (5,23,24).

The market package for a drug product includes the primary packaging components, secondary package, external packaging, and associated components. The FDA recommends the packaging to be suitable based on assessments in four main categories:

Protection
> Ensure the container closure system shields the product from light, solvent loss, reactive gases, moisture, microbial contamination and filth.

Compatibility
> The container closure system must safeguard against loss of potency, degradation of drug substance, reduced concentration of an excipient, changes in drug product pH, discoloration of either the dosage form or the packaging component or increase in brittleness of the packaging.

Safety
> The container closure system will not leach harmful or undesirable amounts of substances to which a patient will be exposed when being treated.

Performance
> The container closure system will function and deliver in the manner for which it was designed.

Each suitability category is associated with a level of testing in which level 1 indicates the greatest degree of evaluation required. The safety category provides guidance for E&L studies to determine what chemical species may migrate into the dosage form and the toxicological evaluation of those migrated substances. The concern for package component–product interaction is ranked according to the physical state of the product (liquid vs. solid) and type of liquid (organic, organic-aqueous, and aqueous). For example, inhalation aerosol products that contain highly extracting organic solvents are ranked HIGH while solid oral tablets are ranked LOW. The type of drug product, according to route of administration and concern for interaction, is evaluated in the Guidance Document Matrix as shown in Figure 3. The recommended level (1S–5S) of safety testing is noted with respect to the different types of drug products. The highest level of testing is 1S and the testing recommendations are gradually reduced as 5S is approached.

In general, recommended testing may include any combination of the USP monographs <661> for plastic containers, <381> for elastomeric closures, <660> for glass containers,

Degree of Concern Associated with Route of Administration	Likelihood of Packaging Component-Dosage Form Interaction		
	High	Medium	Low
Highest	• Inhalation Aerosols and Solutions (1S) • Injections and Injectables Suspensions (2S)	• Sterile Powders and Powders for Injection (2S) • Inhalation Powders (5S)	
High	• Ophthalmic Solutions and Suspensions (2S) • Transdermal Ointments and Patches (3S) • Nasal Aerosols and Sprays (1S)		
Low	• Topical Solutions and Suspensions (3S) • Topical and Lingual Aerosols (3S) • Oral Solutions and Suspensions (3S)	• Topical Powders; Oral Powders (4S)	• Oral Tablets • Oral Capsules (4S)

Figure 3 Guidance Document Matrix.

and <87> <88> for biologic reactivity tests, plus generation of qualitative extraction profiles, quantitative extraction profiles, and reference to 21 CFR Indirect Food Additive Regulation (25).

The ICH recommends identification and acceptance criteria for leachables in ICH Q3B (R), Impurities in New Drug Products, which applies only to the reaction products of the drug substance with the immediate container/closure system in amounts ≥0.1% (12). Also, ICH Q6A, Specifications: Test Procedures and Acceptance Criteria for New Drug Substances and New Drug Products, has provisions for extractables specifications when data demonstrates the need that acceptance criteria for extractables from container/closure components are appropriate (26).

The ISO published ISO 10993 Biological Evaluation of Medical Devices, a multiple part standard in which chemical characterization is required to evaluate potential leachable chemicals and their bioavailability (27). This standard is intended for medical devices and has limited utility for container closure systems, although the chemical characterization and specification parts have common elements. The applicant has the overall responsibility to ensure the suitability of the container closure/device system throughout the shelf life of the product. A program to evaluate for leachables during stability studies will indicate which controls will be needed to show that the product is consistent with respect to container closure/device system interaction. In all cases, the guidance documents are not prescriptive and there are other approaches that can be taken to indicate container closure/ device system safety. The information provided in an application should have a science base rational that is data supported following cGMP.

E&L FROM RUBBER COMPONENTS
Rubber has been in commercial use as material for packaging components, especially parenteral packaging, since the early part of the 20th century. It possesses unique physical properties that are important to the functions of the total parenteral packaging system. Even before that, soon after the discovery of the vulcanization process by Charles Goodyear in 1839, the use of rubber in medical applications such as bandages, gloves, tubing, hot water bottles, and syringes was described (28,29).

In this section, rubber as a material for packaging components will be reviewed. The discussion will be divided into three parts:

- Composition of rubber components
- Sources of extractables/leachables
- Reduction of leachables

Composition of Rubber Components

Rubber is a unique material. It can be molded into an almost limitless variety of shapes and forms; it is flexible and conformable; and, when penetrated, it is resealable (28). Rubber is commonly formed into packaging components such as syringe plungers, needle shields, dropper bulbs, and vial stoppers. Syringe plungers conform tightly to the cylindrical syringe barrel and effectively seal the drug product in the syringe without leakage. Vial stoppers are forced and held against the top finish of glass vials by aluminum seals, effectively sealing any irregularities in the glass-rubber interface. When penetrated by needle cannula, rubber reseals the opening once the needle is withdrawn. No other materials, such as glass, thermoform plastic, and metal possess these abilities.

Rubber can be divided into natural and synthetic. Natural rubber was largely used pre-1940 for pharmaceutical packaging components (30). In 1940, butyl rubber was invented. Butyl has better gas barrier properties and more thermal oxidation resistance than natural rubber and therefore was quickly adopted for use in rubber stoppers. In the 1960s, first chlorobutyl and then bromobutyl rubber were introduced, and today most stoppers are made from these polymers. It is not only the barrier and oxidation resistance properties of the halogenated butyls (chloro- and bromobutyls) that have made them the predominate choice for stoppers over both natural and butyl rubber; the additional advantage is that halobutyls can be cured using low levels of "clean" curing agents, including sulfur- and zinc-free ones. This is a great advantage when E&L are a concern. So halobutyls have both physical (moisture and oxygen barrier, resistance to oxidation) and chemical (lower levels of less toxic curing agents) advantages over natural rubber. Today, it is estimated that 80% to 90% of all injection and infusion stoppers are based on halobutyl rubber. A new polymer, brominated isobutylene paramethylstyrene terpolymer (BIMSM), was recently introduced, which is said to yield very low levels of extractables that heretofore have only been possible with polymer-coated stoppers (30). More on this subject later in the chapter.

Rubber is composed of several materials, each of which is necessary for a particular physical or chemical property. Typical key materials are listed below.

Pharmaceutical Rubber Formulation Materials	
Material	Function
• Elastomer	• Base elastomer or polymer
• Curing agent	• Forms cross-links between chains, also known as a vulcanizing agent
• Accelerator	• Affects the type and rate of cross-linking
• Activator	• Modifies the efficiency of curing agents
• Antioxidant	• Reduces oxidation of polymer
• Plasticizer	• Acts as processing aid
• Filler	• Modifies physical properties such as hardness
• Pigment	• Coloring agent

A rubber formulation may contain more than one elastomer; blends of natural and chlorobutyl are common. More than one pigment may be used; mixtures of carbon black and titanium dioxide are used to produce gray-colored rubber components. Similarly, multiple types of other materials may be used also; therefore, a rubber formulation may contain many ingredients, each of which may contribute to leachables in drug products. A typical chlorobutyl pharmaceutical formulation, materials and percentage, is listed below.

Typical Chlorobutyl Rubber	Formulation Used for Pharmaceutical Packaging
Material	Percentage by weight
Chlorobutyl rubber (elastomer)	53.1
Calcined clay (filler)	39.8
Paraffin oil (plasticizer)	4.2
Titanium oxide (pigment)	1.1
Carbon black (pigment)	0.13
Thiuram (curing agent)	0.14
Zinc oxide (activator)	1.0
Butylated hydroxytoluene, BHT (anti-oxidant)	0.53

Sources of Extractables and Leachables

Each of the ingredients in a rubber formulation can be a source of leachables. Examples are as follows:

- Elastomer—monomers; oligomers (short chains of monomer units both cyclic and noncyclic); halogenated oligomers from chloro- and bromobutyl rubber; polymer additives and by-products such as BHT, antioxidants, calcium stearate, and epoxidized soybean oil (ESBO). Some of the monomers and oligomers are volatile and can migrate from rubber stoppers into dry lyophilized drug products during storage (30).
- Fillers—metal ions such as Ca, Al, Mg, Mn, Si.
- Plasticizers—volatile oligomers.
- Pigments—metals ions such as Ti, Fe, and Ca; and polynuclear aromatics from carbon black.
- Curing/vulcanizing agents and accelerators—original chemical compounds and reaction products.
- Activators—metal ions (e.g., Zn) and stearates.
- Antioxidants—original chemical compounds (e.g., amines and phenols) and degradation products.

Many factors affect both the number of chemical compounds and their amounts extracted from rubber into a drug product. These factors are as follows:

- Type of rubber formulation—bromobutyl rubbers are generally "cleaner" (have lower extractables) than other rubber types in the order:
  bromobutyls > chlorobutyls > butyls > natural.
- Number and type of chemical compounds in the rubber formulation—modern rubber formulations may have only 6 to 8 ingredients, while older natural rubber formulations have 10 to 15 ingredients.
- Drug product vehicle—aqueous (pH, ionic strength); organic. Vehicles with higher percentages of organic solvents have more leachables. For example, metered dose inhalers (MDI) utilize chloroflorocarbons (CFC) and hydrofloroalkanes (HFA) for both the vehicle and propellant in OINDP. These organic solvents can extract larger amounts of and greater numbers of organic compounds from rubber packaging components than aqueous solutions commonly utilized for injectable or ophthalmic drugs.
- Rubber surface area to drug product vehicle volume ratio—the larger the surface area of the rubber component exposed to the drug product, the greater the opportunity of extraction. Using the smallest rubber components possible or using film-coated components [e.g., West's Flurotec® (31) or Helvoet's Omniflex® (32) stoppers] are viable strategies to minimize drug-rubber contact and therefore leachables.

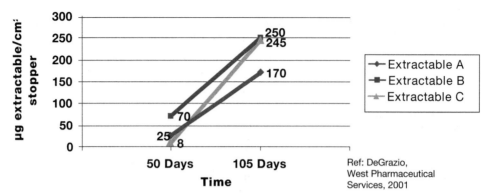

Point to Consider: Extractables may increase in drug product over shelf life.

Figure 4 Extractables over time in aqueous solution.

- Temperature—temperature during terminal sterilization, transport, and storage. The rate of migration of chemical compounds to the surface of a packaging component and the solubility of the compounds in the drug product vehicle increase with temperature. Products that are terminally sterilized by heat are especially vulnerable to higher leachables since the drug product is in direct contact with packaging components at 121°C or higher for 30 or more minutes. Refrigerated (2–8°C) and frozen (−25 to −10°C) products mitigate the rate of migration of packaging extractables. Transport and storage at temperatures higher than that recommended on the label not only affects the stability of the product but also may increase leachables (33).
- Time—the longer the shelf life of a product, the greater the opportunity for increased leachables. This is illustrated in the data presented by DeGrazio in Figure 4.

The analysis of E&L from rubber is particularly challenging for the following reasons:

- The composition of rubber is usually proprietary; therefore, getting information about potential extractables from suppliers is unnecessarily difficult. Drug manufacturers generally must perform extractables studies prior to leachables studies to qualify rubber components. It is recommended that the first step in any E&L study is contact with the supplier to get as much rubber composition information as possible. Having information about what is and what is not in a rubber formulation *before* extractables studies are initiated will save time and money.
- There are many raw materials in a rubber formulation. Refer to section "Composition of Rubber Components" discussed earlier.
- The raw materials are not pure. They contain many impurities that may or may not be known to the rubber manufacturer. Also, these impurities are not commonly listed in the ingredient list found in a Drug Master File. Typical impurities found in the vulcanizing agent, N-t-butyl-2-benzothiazyl-sulfenamide (TBBS), are listed below.
  - $C_{(7-9)}$ alkyl benzyl phthalate
  - Benzothiazyl disulfide
  - t-Butylamine
  - 2-Mercaptobenzothiazole
  - Benzothiazole
  - Mineral oil (*anti-dust agent*)

Inorganic materials, such as fillers, are also complex materials. The composition of a typical kaolin clay is shown below.

- Al_2O_3 44.48%
- SiO_2 52.41%
- Water   5,000 ppm, maximum
- TiO_2 17,900 ppm, typical

- Na_2O	2,800 ppm
- CaO	300 ppm
- Fe_2O_3	5,800 ppm
- Co	200 ppm
- K_2O	1,500 ppm

- Impurities may be present in very small quantities.
- Many compounds, including impurities, are changed chemically during and after the manufacturing process. During the vulcanization or cross-linking process reactive curing agents, accelerators, and antioxidants are chemically changed; antioxidants continue to change postmanufacturing as they react with oxygen to protect the polymer.
- The reaction products of vulcanization are often not known, are present in very small quantities, and pure standards are not available.

These factors make identification, quantification, and qualification very challenging.

The type of rubber or elastomer that was discussed thus far is *thermoset* rubber. Thermosets are polymers that have been chemically cross-linked to form the final structure of the material. Cross-linking and forming of the rubber into a functionally shaped packaging component take place in a mold in the presence of heat and pressure. Once formed, thermosets do not melt and cannot be easily reformed due to the permanence of the chemical cross-links (29). The necessity of a curing system (curing agent, accelerator, activator) in thermosets increases the number of chemical compounds in the rubber and thus the opportunity for them and their reaction products to become leachables.

Thermoplastic rubber, called TPE or thermoplastic elastomer, is another type of rubber. These materials have functional properties similar to thermoset rubber but can be melted and reformed into a different shape if desired like common plastics such as polyethylene. Thermoplastic rubbers are not chemically permanently cross-linked like thermosets. The cross-link in thermoset polymers is a covalent bond created during the curing process. However, the cross-link in thermoplastic elastomer polymers is a weaker dipole or hydrogen bond that can be broken when sufficient heat is applied and reformed when cooled. Because of the absences of a chemical curing system, thermoplastics have simpler chemistries than thermosets and potentially lower levels of leachables. Unfortunately, thermoplastics have found limited use to date as pharmaceutical packaging components because of their tendency to deform during terminal sterilization.

Reduction of Leachables

Leachables are an inevitable companion of packaging, but one can take steps to reduce them to levels that are safe for the purpose intended.

- Choose the "most compatible" rubber formulation. Perform accelerated extractables screening studies with the drug product. There are several approaches to this prescreening.
  - Information due diligence—compare available rubber formulation information with known drug product chemistry. Discuss any likely incompatibilities with the rubber supplier. Also, review the information with toxicologists for any likely concerns.
  - Drug product spiking—prepare a concentrated extract of the rubber formulation by extracting the rubber with the drug product vehicle or a solvent system that mimics the drug product vehicle. Perform the extract at high temperature (e.g., reflux or autoclave) using rubber with a large surface area (e.g., cutting the rubber component into small pieces). Mix a portion of the extract with drug product and observe/analyze product for interactions.
  - Accelerated stability testing—store drug product in contact with rubber component at the highest temperature that the product will tolerate for two to four weeks. Observe/analyze product for interactions.
- Use the smallest possible packaging component to minimize drug-rubber contact. A typical 20-mm vial stopper has more than twice the drug product contact surface area than a 13-mm stopper (20-mm S-127 stopper = 3.65 cm^2 vs. 13-mm V-35 stopper = 1.65 cm^2) (34).

- Limit contact time between drug product and rubber component by limiting the shelf life.
- Pre-extract rubber components before use to reduce the amount of substances available for migration into drug products. This method was very common for natural rubber vial stoppers and syringes when older sulfur-containing cure systems were used. Rubber components were typically autoclaved in the drug product vehicle before use. The introduction of synthetic halobutyl stoppers with cleaner curing systems has become much less common. However, for rubber components used in contact organic solvents, such as valves and o-rings in inhalation drug containers, pre-extraction of components is still commonly used (3).
- Use the best possible contact conditions. Freeze drying, refrigerating, and freezing the drug product will reduce the rate of extraction of impurities from rubber. Terminal sterilization, in which the drug product and rubber packaging component are in contact at high temperatures for a short time ($\sim 121\,^{\circ}$C for 30 minutes), can produce large amounts of extractables compared with normal storage conditions ($\sim$RT for 3 years). When undesirable drug product-packaging interactions are anticipated, avoidance of terminal sterilization in favor of aseptic processing is recommended.
- Use a coated stopper. Stoppers coated with "a more inert than rubber" coating can reduce leachables. Refer to the EPREX example discussed earlier.

     Coated stoppers are commercially available in two types: partially coated and totally coated. In the partially coated type, a thin film of inert polymer is laminated onto the surface of the rubber closure in the molding process. The West Flurotec (31) stopper, which is laminated with a copolymer of ethylene and tetrafluroethylene (ETFE), is an example of a partially coated stopper. The ETFE laminate can be applied to either the bottom plug or product contact area of the stopper or to both the top and bottom of the stopper. The function of the laminate on the top of the stopper is not to reduce extractables but rather to provide a nonstick surface on lyophilization stoppers so that they do not stick to the lyophilizer shelves during stoppering. Illustrations of these stoppers are shown in Figures 5 and 6. The black area represents the rubber and the gray represents the coating.

     The Helvoet Omniflex3G® (32) product is an example of a totally coated stopper. In the Helvoet process, stoppers are coated on all surfaces with a

Figure 5   Plug coated stopper.

Figure 6   Plug and top coated stopper.

Figure 7   Totally coated stopper.

fluoropolymer material after the stopper is molded. An illustration of this stopper is shown in Figure 7.

The West Flurotec film is available on syringe plungers also.
In managing leachables from rubber components it is important to note the following:

- Extractables can migrate into both liquid and solid products (powder, lyophilized).
- Rubber components are composed of several raw materials, each of which has the potential to migrate into drug products and become a leachable.
- Raw materials used in pharmaceutical and medicinal rubber components are not pure and may be composed of several chemical compounds and contain impurities.
- Chemical compounds are changed during and after the manufacturing process. Therefore, leachables found may differ in identity from those used in the rubber formulation recipe.
- The number and quantity of leachables found will depend on the composition of the elastomeric formulation, rubber processing and sterilization cycles, time, and the unique characteristics of the drug product.
- Identity and quantification of extractables from rubber is a complex process requiring expertise in both chemical analysis and rubber chemistry.

E&L FROM GLASS COMPONENTS

Although glass, in most applications, is less reactive with drug products than rubber or plastics, it is not inert. In dealing with E&L there are two major differences between glass and rubber. The first is that glass compositions are more uniform from supplier to supplier. The percentage of each raw material may differ slightly but the materials themselves are quite uniform. The second difference is that glass compositions are usually not proprietary and are readily shared by the supplier with the drug packager. These differences make the identity and quantification of glass E&L much easier than in either rubber or plastics.

The history of glass as a packaging material is a long one; glass containers existed in Egypt around 1500 BC (1). Glass has been used as a pharmaceutical packaging component for several hundred years particularly during the 20th century during the tremendous growth of pharmaceutical industry. Even though many predicted doom for the glass industry when modern plastics became available some 50 years ago, the use of glass for pharmaceutical containers has endured. There are several unique advantages of glass that account for this endurance.

- Excellent chemical resistance, but NOT INERT
- Impermeable to gases
- Easily cleaned, sterilized, and depyrogenated
- Transparent
- Rigid, strong, and dimensionally stable

Glass packaging components are made by either forming them in a rigid mold from molten glass (molded glass) or by forming them from heated extruded glass tubing (tubing glass). Size determines the optimum manufacturing method. Small components are made easily from available small diameter tubing while large components must be made by molding because large diameter glass tubing is not available. The transition point is about 100-mL capacity. Common glass components and their method of manufacture are listed below.

- Ampoules (tubing glass)
  - ○ Inert, low cost, high particulate from opening, not user-friendly
- Vials (tubing or molded)
  - ○ Ease of filling, multiple use, user-friendly
- Bottles (>100 mL, molded)
  - ○ Used for LVPs but being replaced by plastic
  - ○ Smaller bottles used for solid dosage forms

In this section, glass as a material for packaging components will be reviewed. The discussion will be divided into three parts:

- Composition of glass components
- Sources of extractables/leachables
- Reduction of leachables

Composition of Glass Components

Commercial glass is an inorganic product of fusion that has cooled to a rigid state without crystallization (35,36). It is essentially a rigid liquid. Glass may be thought of as a thermoplastic—it is softened by heat, capable of being formed into a wide spectrum of shapes, and can be reheated and remolded into new shapes without degradation of the material properties. The essential difference between glass and common thermoplastic materials, such as polyethylene and polypropylene, is that glass is an inorganic material while the cited plastics are organic materials.

Glass is composed of the following materials:

- Matrix material—SiO_2 or sand
- Fluxing agents that lower the melting point
  - ○ Na_2CO_3 soda ash that converts to Na_2O
  - ○ K_2CO_3 potash that converts to K_2O
- Stabilizers
  - ○ $CaCO_3$ (lime) converted to CaO for hardness and chemical resistance
  - ○ Al_2O_3 (aluminum oxide) for chemical resistance
  - ○ B_2O_3 (boron trioxide) lowers melting point
  - ○ Cerium oxide increases resistance to discoloration by γ-radiation
- Coloring agents
  - ○ Fe_2O_3 (iron oxide) and TiO_2 (titanium oxide) for amber glass
  - ○ Cobalt and copper oxides for blue glass
  - ○ Iron, manganese, and chromium oxides for green glass

Pharmaceutical glasses fall into two types that differ in their essential compositions. These are as follows:

- Soda-lime glass, which is composed of the following:
  - ○ 71% to 75% SiO_2
  - ○ 12% to 15% Na_2O (soda ash)
  - ○ 10% to 15% CaO (lime)

The name is derived from the two compounds that predominate—*soda* ash and *lime*.

- Borosilicate glass, which contains the following:
  - ○ 70% to 80% SiO_2 (silicate or sand)
  - ○ 7% to 13% B_2O_3 (boron trioxide)

- ○ 4% to 6% Na_2O and K_2O
- ○ 2% to 4% Al_2O_3

The name again comes from the two predominate components—*boron* trioxide and *silica* dioxide.

A survey of the compositions of commercial borosilicate glasses (37) demonstrates the subtle differences in properties (Table 1). Flint glass is colorless while amber is yellow-brown in color; amber glass is used for drug products that are sensitive to light.

Glass can be divided into types in many ways—by composition (soda lime/borosilicate), color (flint/amber), forming method (molded/tubing), coefficient of expansion (33/50/90)—but in the world of E&L, the most meaningful method is by chemical resistance. The USP classifies glass (38) into three types according to chemical resistance requirements as show in Table 2.

The glass with the highest resistance is borosilicate glass or type I. It is differentiated from soda-lime glass (type III) by a Powdered Glass (PG) test in which the alkalinity of the glass or its capacity to exchange sodium ions (Na^+) in the glass for hydrogen ions (H^+) in solution is measured by a titration with dilute sulfuric acid. The lower the amount of acid consumed at the endpoint, the higher the resistance of the glass. The USP specification for type I glass is 1.0 mL or less; for type II glass the limit is 8.5 mL or less. Since ion exchange (extraction of sodium and other ions in the glass by replacement with hydrogen ions from the solution) is the principal reaction between glass and aqueous solutions, this test is a measure of the extractability of the glass.

There is a third type of USP glass—type II treated soda-lime glass. Type II glass is created by chemically pre-extracting or "treating" containers made from type III glass. The previously discussed tests for type I and III glass are performed on powdered glass made by crushing glass containers (vials, ampoules, bottles) with a steel mortar and pestle prior to testing. But the Water Attack (WA) test for type II glass is performed on intact containers because only the inside surface is treated to lower the extractability. The limits for type II glass containers are 0.7 mL per 100 mL of test solution for those with a capacity of 100 mL or less, and 0.2 mL per 100 mL for those over 100 mL capacity.

Table 1 Compositions of Commercial Borosilicate Glasses

	Glass "A", flint	Glass "B", flint	Glass "C", amber
Composition %			
Silicon dioxide (SiO_2)	81.0	75.0	70.0
Boron trioxide (B_2O_3)	13.0	10.5	7.0
Aluminum oxide (Al_2O_3)	2.0	5.0	6.0
Sodium oxide (Na_2O)	4.0	7.0	7.0
Potassium oxide (K_2O)	–	–	1.0
Calcium oxide (CaO)	–	1.5	<1.0
Titanium dioxide (TiO_2)	–	–	5.0
Physical properties			
Thermal expansion ($\times 10^{-7}$ in./in./°C)	33	49	54
Annealing point °C	560	565	560
Chemical resistance			
USP Powdered Glass Test (max. limit = 1 mL)	0.26	0.30	0.35

Table 2 USP <660> Types of Glass

Type	Composition	Test	Size, mL	Max. mL of 0.02 N Acid
I	Highly resistant, borosilicate	PG	All	1.0
II	Treated soda lime	WA	100 or less	0.7
			Over 100	0.2
III	Soda lime	PG	All	8.5

USP 31 <660>
Abbreviations: PG, Powdered Glass; WA, Water Attack.

Type I glass is the predominate choice for injectable drug products. Type II is not used as much as previously because of the variability in treatment effectiveness and the environmental impact of the treatment processes. Type III glass is not used for parenterals. This subject will be discussed in the section "Reduction of Extractables."

Sources of Extractables/Leachables

Several factors affect the number and amount of leachables from glass containers into drug products.

- Drug composition
  - Type and concentration of ions
  - pH
- Method of glass container fabrication
  - Molded, tubing, fabrication temperature, glass treatments
- Methods of container processing
  - Terminal sterilization, aseptic processing, depyrogenation
- Container size and shape relative to the drug product volume
- Storage conditions of filled container
  - Time, temperature, orientation of container
- Glass composition
  - Borosilicate, soda lime

Glass–drug product interactions may be divided into several types.

- Ion exchange: It is the predominate method of interaction. Na^+, K^+, Ba^{2+}, and Ca^{2+} are the major extractables from glass via ion exchange.
- Glass dissolution: Phosphates, oxalates, citrates, and tartrates can accelerate the dissolution of glass. Silicates and Al^{3+} are released by the dissolution.
- Pitting: EDTA can form complexes with many divalent and trivalent ions and accelerate dissolution of glass resulting in pitting of the glass surface.
- Adsorption: Proteins such as insulin and albumin are known to adsorb on glass surfaces.

In a comprehensive study (39), Borchert et al. investigated the extractables from borosilicate glass with an accelerated procedure using unbuffered aqueous solutions at pH = 4, 6.5, 8, and 10.4 and buffered solutions at pH = 8 and 10. Accelerated extraction was performed at 121°C for one hour. The authors concluded the following:

- Low levels of extractables were leached from glass with solutions of pH 4 to 8. Significantly higher levels were observed when the glass was exposed to alkaline media (pH > 9).
- Silicon was the major extractable.
- Sodium was another major extractable. In acidic solutions, Na is extracted from the glass by ion exchange (H^+ into glass, Na^+ out of glass); in basic solutions, Na^+ is released during dissolution of the glass.
- Minor extractables observed were K, Al, Ba, and Ca—Al by dissolution and K, Ba, and Ca via ion exchange.
- Other elements extracted were Mg, Fe, and Zn but at the level of detection, that is, <0.1 ppm.
- A positive shift in the pH of the unbuffered extracts was observed when the initial pH of medium was <7. Conversely, a negative shift was observed with medium of pH > 8. The pH of acidic solutions is raised by ion exchange—H^+ ions leave the aqueous medium and become part of the glass. The pH of basic medium is lowered by the consumption of OH^- ion during glass dissolution.
- Treated glass had less extractables than untreated glass.

Borchert's observations are important for anyone using glass containers for aqueous drug products. It is important to note that

- Unbuffered solutions with initial pH values <7 or >8 will shift in pH. Drug products with narrow acceptable pH ranges over shelf life may require a buffer. The reactions of glass in acidic and basic medium are illustrated below.

  Acidic/Neutral (ion exchange)

  $$Na^+(glass) + H_3O^+(sol) \rightarrow Na^+(sol) + H_3O^+(glass)$$

  Basic (dissolution)

  $$2OH^-(sol) + (SiO_2)_x \rightarrow SiO_3^{2-} + H_2O$$

- Glass, although less extractable than rubber and some plastics, is not inert. pH shifts and metal ion extracts are probable.
- With glass containers, it is not typically necessary to precede a leachables study with an extractables study since the metals extracted from glass are very consistent from glass to glass and are well known. Since the purpose of an extractables study is to identify potential leachables, this is not necessary. Also methods of metal ion identification and measurement, such as inductively coupled plasma/mass spectrometry (ICP/MS) and inductively coupled plasma/atomic emission spectroscopy (ICP/AES), are able to analyze many elements in one test.

  With rubber, extracts are not consistent from one rubber to another, so an extractables test is a necessary step prior to a leachables study.

Reduction of Leachables from Glass

There are three primary sources of glass that will reduce leachables—treated glass, glass manufactured using special methods, and coated glass. Combinations of these methods such as chemical treatment of containers made from special glasses are common.

Glass Treatments

Glass treatments are physical or chemical processes used to modify the physical or chemical durability of glass.

- Physical: The most common methods of improving the physical durability of glass are fire polishing and annealing. Fire polishing is a method of smoothing the surface or edges of glass by exposing it to a flame or heat. By melting the surface of the material, surface tension smoothes the surface. This process removes cracks or scratches that make glass containers more susceptible to breakage (1). Annealing is the process of reheating then cooling glass containers at a controlled rate to relieve stresses that are imparted during the forming process. The annealing process takes place in an oven called a lehr (35). The annealing temperature of borosilicate glass is 580°C and 560°C for soda-lime glass (1).
- Chemical: Chemical treatments reduce extractable substances, mainly sodium, from the surface of glass containers. Several types of chemical treatments can be used (see following text), but ammonium sulfate is the most widely used for containers for drug products.

Sulfur treatments

$$SO_2, SO_3, (NH_4)_2SO_4 \rightarrow Na_2(SO_4)$$

Chlorides salts and chlorine gas

$$NaCl, (NH_4)Cl, Cl_2 \rightarrow NaCl$$

Fluorine compounds

$$C_4F_8, CClF_3, C_2F_4 \rightarrow NaF$$

With sulfur treatments, sodium sulfate [$Na_2(SO_4)$] is formed on the surface of the glass container; it is then easily washed away during the container washing process. With chlorine/chloride treatment, sodium chloride (NaCl) is the resulting salt deposited. When fluorides are used, sodium fluoride (NaF) results. All three treatments remove sodium from the surface and convert it to a soluble salt that is easily removed. The chemistry of the ammonium sulfate treatment is shown below. In this treatment, a small amount of an aqueous solution of ammonium sulfate is added to each container (e.g., vial, ampoule). The containers are then conveyed into an oven at approximately 550°C where the ammonium sulfate is converted to ammonium bisulfate and ammonia gas. The ammonia bisulfate reacts with the sodium in the surface of the glass, and in an ion exchange reaction, the sodium in the glass is replaced with hydrogen ions. Sodium sulfate and ammonia gas are by-products of the treatment.

$$(NH_4)_2SO_4 \rightarrow (NH_4)HSO_4 + NH_3$$
$$2Na^+(glass) + (NH_4)HSO_4 \rightarrow Na_2SO_4 + NH_3 + 2H^+(glass)$$

Treatment of glass with ammonium sulfate is most commonly used to convert type III glass to type II ("treated glass") or to improve the chemical properties of type I glass.

$$Na^+(glass) \rightarrow H^+(glass)$$

Type III $\rightarrow$ Type II

or

Type I $\rightarrow$ Type I (improved chemical resistance)

Treatment decreases both the amount of metal ions extracted from glass container surfaces and the pH shift of solutions in contact with glass. Data from a presentation by Aldrich (40) demonstrated the improvement of glass from treatments.

Treated Vs. Untreated Glass Type I glass, 10 mL vials, extraction @ 121°C for 1 hr		
Component	Untreated	Treated
SiO$_2$	20.7 ppm	0.6 ppm
Ba	0.7	<0.1
Al	1.3	<0.1
Na	3.1	0.3

Borchert (41) and Aldrich (40) also showed that treatments only affect the surface of glass containers. In the untreated containers, the percentage of sodium in the matrix of the glass is 15% but higher near the surface (down to 6000 Å). This increase in surface sodium results from migration of sodium to the surface during the formation process, especially with tubing glass. Areas that are heated frequently, such as the bottom and top (shoulder, neck, and finish) of the vial, have higher percentages of surface sodium than the sides of the vial. Treatment reduces the sodium to very low levels near the surface but does not affect the depth below approximately 8000 Å. Data from Aldrich is shown in the following table.

Percentage Na Vs. Depth		
Depth, *A*	Untreated (%)	Treated (%)
1300	25–41	<1
4150	16–19	
6000	17–18	↓
>8000	15	15

Since the bottom and top areas of tubing glass containers contain more surface sodium than the walls, it is important to minimize contact of drug with these high sodium areas. This is done by both using the appropriately sized container for the volume to be packaged (small

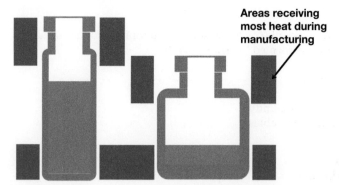

Figure 8 Effects of high sodium zones on the interior surface and filled volume.

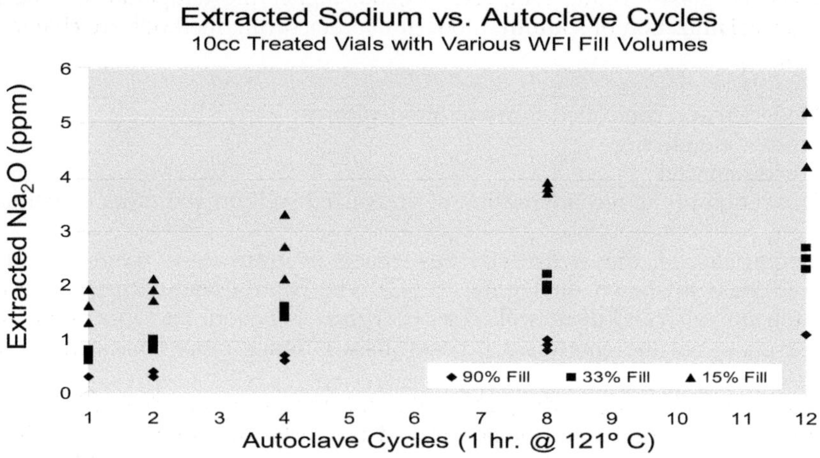

Extracted Sodium vs. Autoclave Cycles
10cc Treated Vials with Various WFI Fill Volumes

Figure 9 Extracted sodium versus fill volume and time.

volumes in small containers) and optimum-shaped containers (height and diameter). As illustrated in Figure 8, tall narrow vials minimize drug contact with the bottom and maximize contact with the low-sodium vial sides (42). However, vials that are too tall and narrow are not stable so the objective is to balance height and diameter.

The effect of fill volume and exposure time on the amount of extracted sodium was investigated by Swift (37). Figure 9 demonstrates the following:

- Extractable sodium increases with time of solution-container contact. In this case, time is measured in autoclave cycles.
- As the fill volume decreases from 90% to 33% to 15% of capacity, the amount of sodium extracted increases. At 15% of fill capacity, a proportionally larger amount of the solution is in contact with the bottom compared with the sides.

Sulfur treatment of glass, despite its disadvantages, is still a widely used method of producing type II glass and of improving the chemical resistance (reducing extractables) of type I glass. Some disadvantages are as follows:

- Can add cost
- Damages production equipment
- Adds stress to washing system
- Contaminates the washing system
- Causes an environmental issue because of disposal of salts
- Treatment can be inconsistent. Excessive treatment causes pitting of glass surfaces.

- Introduces risk of sulfate residues
- No standards or tests for the amount/concentration of ammonium sulfate used
- Restricts user's ability to test "as formed" surface quality

The chief advantages are as follows:

- Significant reduction in surface alkalinity
- Reduction in other metal ion extractables

Special Glass Manufacturing Methods
A method that produces glass with lower extractables is preferred over the posttreatment glass produced with higher extractables. The amount of sodium on the surface of the glass is dependent on the glass-forming temperatures; the higher the temperatures, the more the migration and volatilization of sodium. Modern manufacturing methods are characterized by the following:

- Moderate and controlled forming temperatures
- Process monitoring
- Process control
- Prevention of the recondensation of volatilized sodium and other constituents

A comparison of the hydrolytic resistance of untreated, treated, and control-manufactured glass is shown on Figure 10 (42). The control-manufactured glass is much better than the untreated and uncontrolled process glass and about the same as the treated and uncontrolled glass, but the controlled process glass is much more consistent in hydrolytic resistance.

Special glasses are being developed to meet specific needs such as low aluminum type I glass for injectable nutritional and blood-derived products. An example is SGD's Asolvex® type 1 glass, which SGD claims reduces the aluminum content by a third (43).

Even with properly treated and/or control manufactured glass, the proper-sized container will significantly reduce the amount of extractable substances.

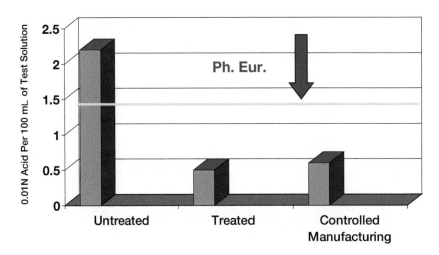

Using the Ph. Eur. Test for Surface Hydrolytic Resistance
Comparison is of 3cc vials
1.3mL of acid titrant is the Ph. Eur. limit for 3cc vials

Figure 10   Comparison of the hydrolytic resistance of glasses.

Coatings for Glass

As with FluroTec-coated rubber (*v*) to reduce extractables, glass coatings have been developed to minimize glass-solution interactions. Schott has developed a coated vial marketed as Type I plus® (44). Schott Type I plus containers are made of pharmaceutical type I glass with a chemically bonded, ultrathin ($\sim$150 nm) layer of pure SiO_2 on their inner surface. Since the SiO_2 layer contains no sodium or other metal ions such as K, Al, Ba, Ca, Mg, Fe, or Zn, pH shifts and extractables are significantly reduced as shown in the data from Schott.

Comparison of Extractables from Type I and Type I plus® Glass			
Metal ion	Type I	Schott Type I plus	Reduction factor
Na	3.5 ppm	<0.01 ppm	>350
Ca	1.1	<0.05	>22
B	3.5	<0.01	>35
Si	5.0	<0.3	>15
Al	2.3	<0.05	>45

Extraction conditions—autoclaving for 6 hr @ 121°C with WFI.

Glass containers remain a viable option to the pharmaceutical packager, but it is important to note that

- Glass is NOT inert
- Extractables and leachables are
  - Si, Na—major
  - K, B, Ca, Al—minor
  - Mg, Fe, Zn—trace
- All type I glasses are NOT equivalent
- Methods are available to reduce extractables and leachables.

E&L FROM PLASTIC COMPONENTS

Plastics as pharmaceutical packaging materials are the "new kids on the block." It was not until the early 1950s, with the full commercialization of polyethylene, that plastic emerged as a packaging material (1). The development of polymer technology has made plastics the material of choice over glass bottles for LVPs. Plastics are also becoming an alternative for SVPs. Plastic containers have the following advantages over glass:

- Less fragile
- Lighter in weight
- More easily fabricated into complex shapes
- Less expensive in some cases

There are also some disadvantages compared with glass:

- Clarity can be an issue with many plastics
- More permeable to gases, water vapor, and secondary packaging chemicals from inks and label adhesives
- Less stable during handling because of lighter weight
- More complex extractable substances
- Not as easily sterilized and depyrogenated
- Can cost more than glass

Today there are more than 50 different plastic materials used in various pharmaceutical and medical applications. As with rubber and glass, extractable materials from plastics are a concern for the pharmaceutical packager.

In this section, plastic as a material for packaging components will be reviewed. The discussion will be divided into three parts:

- Composition of plastic components
- Sources of extractables/leachables
- Factors that affect migration: thermodynamics and kinetics

Composition of Plastic Components

As with rubber and glass, modern plastics are formulated with several types of chemical substances and additives, each of which imparts specific chemical and/or physical properties to the plastic component. The most important of these substances are the following (1,45,46):

- Polymers—impart basic desired properties to component. Classes of plastics used in pharmaceutical applications will be discussed in a later section.
- Fillers—reduce degradation, reduce cost, affect moisture absorption and shrinkage. For example, carbon black, calcium carbonate, talc, clays, silica, and magnesium carbonate.
- Lubricants—ease the movement of the melted polymer against itself during processing and may enhance end-use lubricity. For example, zinc stearate, PE waxes, fatty acids, amides, and polydimethyl siloxane.
- Antioxidants—reduce the degradation of polymers exposed to heat, light, ozone, oxygen, radiation, or mechanical stress. For example, hindered phenols and cresols, secondary amines, phosphites, thioesters, BHT, and butylated hydroxyanisole (BHA).
- Heat stabilizers—protect polymers from the effects of heat, pressure, and shear during the polymerization process or secondary processes. For example, octyl thio-tin complexes, calcium-zinc salts, epoxidized materials, and estertins.
- Plasticizers—impart flexibility, resilience, or softness. Plasticizers are used mainly in polyvinylchloride and may be used in polyvinyl alcohol/acetate copolymers, polyvinyl acetate and polyvinyl alcohol formulations, polymethyl methacrylate, and nylon. For example, dimethyl phthalate (DMP), diethyl phthalate (DEP), dibutyl phthalate (DBP), and di(2-ethylhexyl) phthalate (DEHP or DOP).
- Pigments/colorants—used to color, tint, or hide the color of the base polymer; may affect the physical, chemical, or mechanical properties of the plastic. For example, carbon black and inorganic oxides such as TiO_2, ZnO, Fe_2O_3; fillers can also function as pigments; organic dyes and organometallic complexes such as Phthalocyanine Green G (Pigment Green 7), which is a complex between copper (II) and chlorinated phthalocyanine and are not good choices for pharmaceutical applications where extractables are an important factor.
- Other additives, some of which are not commonly used in pharmaceutical applications, are antistatic agents, catalysts, bactericides, impact modifiers, release agents, brighteners, flame retardants, ultraviolet (UV) absorbers, inhibitors, nucleating agents, and blowing agents.

Polymers

A wide variety of plastics are used by the pharmaceutical industry but fewer are used in packaging and process materials where extractables are a concern. The reader is directed to other texts for broader information on the manufacturing and properties of polymers used in the plastics industry (1,2,47). The following is a selected list of plastics divided by application.

- Vials: Polypropylene, polyethylene terephthalate, polyethylene glycol terephthalate, high-density polyethylene, polycarbonate, cyclic olefin polymer
- Bottles: Polycarbonate, high-density polyethylene, polypropylene, polyethylene terephthalate, polyethylene glycol terephthalate, polycarbonate, cyclic olefin polymer
- Syringe barrels: Polypropylene, high-density polyethylene, polycarbonate, cyclic olefin polymer

- Form-fill-seal vials: Low-density polyethylene
- SVP and LVP flexible bags: Ethylene vinyl alcohol copolymer, ionomers [Surlyn® (48)], polyamides (nylons), low-density polyethylene, linear low-density polyethylene, polyvinylchloride, polyethylene terephthalate
- Single-use process bags (1–100 L): Low-density polyethylene, ethylene vinyl alcohol copolymer, ethylene vinyl acetate, polyamides (nylons), very low density polyethylene, ultralow density polyethylene
- Process tubing: Polytetrafluoroethylene and fluorinated ethylene propylene–lined rubber, silicone, polyvinylchloride [Tygon® (49)], ethylene-vinyl acetate
- Filters: Cellulose acetate, polyethersulfone, cellulose nitrate, polyvinylidine chloride, polyamide, polycarbonate, polytetrafluoroethylene, polypropylene
- Inhalation containers and devices: Polypropylene, polyvinylchloride
- Thermoforms: Polypropylene, polyethylene glycol terephthalate, polymonochlorotrifluoroethylene [Aclar® (50)], polyethylene terephthalate, polyvinylchloride, polystyrene
- Films: Polytetrafluoroethylene, fluorinated ethylene propylene, ethylene tetrafluoroethylene, poly(para-xylene) (Parylene)
- Adhesives: Polyvinyl acetate, ethylene-vinyl acetate

Sources of Extractables/Leachables

Because of the complexity of formulations, the leachability of plastics is virtually impossible to predict a priori. Some common plastics and extractables found are listed in the following table (40,51).

Acetal polyoxymethylene (POM) (Delrin®) (52)	Phenolic antioxidant, nitrogen stabilizer, residual formaldehyde, formic acid, trioxane, calcium, magnesium, silicon, zinc, sodium, phosphorus, oligomers
Polypropylene (PP)	Calcium stearate, butylated hydroxytoluene (BHT), dilaurylthiodipropionate, titanium dioxide, aluminum, titanium, chlorine, potassium, sodium, C_6–C_{16} oligomers
Polyethylene terephthalate (PET)	Cyclic dimers and trimers, terephthalic acid, diethylene and ethylene glycol, acetaldehyde, fatty acids, aluminum, calcium, cobalt, tin, zinc
Polyethylene (PE)	Phenolic antioxidant, phosphite antioxidant, amide lubricant, aluminum, chloride, titanium, zinc, calcium, sulfur, phosphorus, t-butyl alcohol, oligomers, glycerol monostearate
Polyvinylchloride (PVC)	Phthalate esters, sulfur stabilizer, fatty acid, benzene, chlorine, silicon, magnesium, zinc, calcium, aluminum, residual vinyl chloride, HCl, cartenoids, diphenyl polyenes, oligomers, calcium stearate
Polystyrene (PS)	Residual monomer, ethyl benzene, phenolic antioxidant, glycol esters, oligomers

Factors That Affect Migration: Thermodynamics and Kinetics

An article by Jenke (53) on the leaching of substances from plastics identifies both thermodynamic and kinetic factors that impact migration. Thermodynamics defines the absolute maximum equilibrium interaction between a material and solution (amount extracted) and kinetics defines the rate at which that equilibrium is achieved (speed of extraction).

Thermodynamics

Estimates of how much substance (e.g., an antioxidant such as Irganox 1076) will migrate from a plastic (e.g., polyethylene) into an aqueous solution can be calculated from the partition coefficient, E_b (i.e., the ratio of the equilibrium concentrations of the substance in the plastic and the concentration in the aqueous solution). However, these partitions coefficients between plastic and water are rarely available in the literature. They can be calculated, however, from other more readily obtained partition coefficients, $P_{o/w}$, specifically between octanol and water. $P_{o/w}$ values can be obtained from the literature or by experimentation. Once relationship between E_b and $P_{o/w}$ is established for a particular plastic, E_b can be calculated from $P_{o/w}$.

Many factors affect the partitioning process. If the extractable substance is ionizable, its dissociation constant (pK) and pH of the aqueous solution play a role. Concentrations of other ionic and nonionic substances in the aqueous phase (drug product) also affect partitioning. These substances may affect the apparent "polarity" of the drug product; this may affect the level of leachables. The effect of added Polysorbate 80 (Tween 80) on the extraction of antioxidant from a rubber plunger into the Eprex drug product was discussed earlier and is a good example of the effect of polarity on partitioning (7). The utilization of octanol/water partition coefficients to model solution-plastic interactions has been described in detail for several plastic materials by Jenke et al. (54).

Kinetics
Kinetic factors determine the level of migration if equilibrium has not been reached. In long-term contact situations, such as a drug filled into an LVP container, terminally sterilized, and stored before use, equilibrium may be reached. In the case of short contact situations, such as a drug being filled into a sterile empty syringe for immediate injection, it is the kinetics of migration which determines the amount of migration that has taken place.

E&L from plastics is a growing area of research for several reasons:

1. Compared with glass
   - The number of different substances that can migrate is very large since the compositions of plastics are more diverse than the compositions of glass.
   - The migrating substances may be both organic and inorganic.
2. Compared with rubber
   - The drug product contact surface area of plastics (containers and closures) is usually much larger than rubber (closures), increasing the potential for leachables.
3. Growth of plastic applications
   - Flexible plastic applications such as SVP, LVP, and large volume disposable applications (2–1000 L single-use 2D and 3D bags).
   - Blister packaging for tablets and capsules.
   - Vial materials with low extractables and high clarity [e.g., Cyclic olefin polymers (COP/COC) such as Ticona's (Celanese Corp, Florence, KY, USA) Topas and Nippon Zeon's (Zeon Chemicals L.P., Louisville, KY, USA) Zeonex resins].

E&L FROM INKS, ADHESIVES, AND COATINGS
Label components such as inks, adhesives, and coatings are not primary packaging materials since they are not in direct contact with drug products or substances. However, substances from the materials may migrate through the walls of plastic containers and appear as leachables. Examples of unexpected migrations of materials abound. One such report described the migration of a UV protective coating from a label through a hard plastic bottle into the drug product (55); another described the migration of two photoinitiators, 1-benzoylcyclohexanol and 2-hydroxy-2-methylpropiophenone, from ink on labels through HDPE bottles into a solid dosage form product (56).

Inks
Inks may be composed of the following types of substances:

- Oligomers—polyesters, epoxy compounds
- Photoinitiators—α-hydroxyketones, α-aminoketones, phenylglyoxylates
- Monomers—acrylates
- Stabilizers—4-methoxyphenol
- Pigments—carbon black, talc, organic pigments

The composition of inks for food products is regulated by the FDA in several parts of 21 CFR (57). 21 CFR 178.3297 concerns colorants for polymers. Other components of inks may be the subject of separate regulations such as 21 CFR 177.1520 for olefins in inks.

Adhesives

Even more so than inks on labels, label adhesives are in direct contact with the drug product container. And like inks, adhesive components can migrate through plastics. Pressure-sensitive label adhesives are typically formulated from acrylics or modified acrylics, wetting agents, and biocides. There are also adhesives based on rubber/resin blends. Each of these materials may be composed of several chemical substances that make the formulation quite complex, and, as a rule, it is very difficult to obtain a comprehensive list of adhesive ingredients from label suppliers making the extractables/leachables process difficult.

Adhesives for food applications is regulated by the FDA under 21 CFR 175.105 (Adhesives) (58). Section 175.105(a)(2) states that manufacturers of finished food packaging must ensure that the adhesive is either separated from the food by a functional barrier or is subject to limits of GMP practices. Of course, there are no CFR limits for drugs.

Identification of Extractables from Labels

Because of the difficult in obtaining credible and comprehensive information about the extractable substances from labels suppliers, pharmaceutical companies must perform a controlled extraction study to identify extractable substances. An example of extractables protocol for labels is described below.

Label Extractables Protocol

Before beginning any study, the label supplier should be requested to supply as much information on the label composition as is possible. Discussion with the suppliers/manufacturers of the ink, adhesive, and paper portions of the label are also desirable. Review of suppliers/manufacturers literature (paper and web based) will prepare the user with the proper questions for suppliers. Any upfront knowledge of the label composition will provide dividends in time and money saved during the extraction study.

A controlled extraction study of a packaging label involves exposing a sample to an appropriate solvent system at elevated temperatures to accelerate the extraction process followed by chemical analysis. The label, which is not intended in ordinary applications to be ever exposed to solvents, will not only be subject to extraction but likely will partially dissolve in some solvents. At least two solvents are recommended—water or an aqueous system that mimics the drug product (e.g., buffer at drug pH) and 2-propanol. The extraction is performed by soaking the label in a solvent for a fixed time at a controlled temperature or by refluxing the solvent over the label. Extraction profiling of the label is divided into three parts: nonvolatiles, volatiles, and semivolatiles.

Nonvolatile Profile

Label solvent extracts are analyzed for residual substances. An UV spectrum of the extract will provide identity information on the possible unsaturated extractables. A portion of the solvent extract is evaporated to dryness and prepared for analysis. Nonvolatile residue (NVR) is calculated on the basis of label weight. This provides generalized information on possible extractables when comparing labels. Infrared (IR) spectroscopy or Fourier Transform IR (FTIR) spectroscopy of the residue provides general identification of the significant functional groups of the extracts. Qualitative trace elements in the residue are then analyzed by scanning electron microscopy and energy dispersive X ray (EDX). Residual extracts are then analyzed by ICP/MS.

Semivolatile Profile

After extraction, the extract is separated by chromatographic methods and identified by retention information or by mass spectrometry (MS). Typical methods used may be (51):

- Gas chromatography/mass spectrometry (GC/MS)
- Liquid chromatography/mass spectrometry (LC/MS)
- Liquid chromatography/diode array detection (LC/DAD)
- Gas chromatography/flame ionization detection (GC/FID)

- Liquid chromatography/ultraviolet detection (LC/UV)
- Inductively coupled plasma/mass spectroscopy (ICP/MS)
- Inductively coupled plasma/optical emission spectroscopy (ICP/OES)

Volatiles Profile
Samples of the labels are profiled without solvent extraction by head space GC/MS. Labels are cut into small pieces and placed into headspace vials along with a small amount of WFI. The headspace vials are heated for an appropriate amount of time and temperature to generate any volatile substances and then analyzed by chromatography with an MS detector.

Identification of extractable and volatile substances is made by comparison to standard libraries and comparison to reference standards when available. Results should be shared with the label supplier to validate, where possible, the origin of the substance is identified. Protocols for leachables studies will be discussed later in the chapter.

PACKAGING STANDARDS AND COMPENDIA TESTS
In this section, the compendia tests on rubber, glass, and plastic from the USP, Ph. Eur., and, to a lesser extent, the Japanese Pharmacopoeia (JP) will be reviewed.

Compendial test procedures are specific for certain types of materials, yet the data does not typically provide compound specific information. While this appears to be an enigma, compendial methods denote that a standardized approach can be used to provide general information to identify certain characteristics of the materials and the physicochemical nature of the materials. There are specifications associated with many of the monographs and those specified materials must be compliant to indicate suitability for use with a pharmaceutical product. Standardized tests indicating biological responses to the component materials as well as certain functional tests are also included in the compendia. Data from compendia tests are limited and considered as first-line information to be acquired when qualifying a container closure system. Complete stability, compatibility, and safety assessment of the container closure system are necessary to ensure it is appropriate for the intended use. An example of container closure systems materials types covered in the USP and Ph. Eur. monographs are shown in Table 3 (59).

Materials that are in contact with pharmaceutical container closure systems must comply with the appropriate pharmacopoeia monographs as required by regulatory agencies around the world. Unfortunately, there are no global specifications for container closure systems and requirements are enforced according to the terms of each country. Even though the materials of construction may have general characteristics that are standard, these attributes are not necessarily consistent on a global level. A single container closure system may be composed of several different materials types, and these materials types can be intended for different uses. The monographs in the Ph. Eur. are based on aspects associated with specific materials together with the applications so that rubber closures, ophthalmic containers, single-use syringes, lubricants, containers for parenteral, intravenous infusions, and parenteral nutritionals as well as systems used with blood products are to be tested according to the specified monograph. The extraction procedures, analysis, and specifications vary according to each

Table 3 USP/Ph. Eur. Material Types

Material	USP	Ph. Eur.
Elastomers	X	X
Glass	X	X
Polypropylene	X	X
Polyethylene	X	X
Polyvinyl chloride	–	X
Polyethylene-vinyl acetate	–	X
Polyethylene terephthalate	X	X
Silicone oil/elastomers	–	X

monograph within Ph. Eur. and even more so compared with USP monographs. The specifications in the USP monographs have been standardized on the basis of the materials used in containers in general and elastomeric closures for injection. The USP monographs have a broader application, but as mentioned previously, compendia testing is first-line assessment and the data provided is not definitive enough to qualify a container closure system for its intended use. The scope of compendia test standards are related to the chemical and biological characteristics for individual material formulations with functional tests associated with performance aspects of a particular system. There are efforts to harmonize compendia but the impact of changing test criteria and specifications for marketed container closure systems are far reaching and consensus is challenging so the process for harmonization is slow moving.

Rubber

The ISO publishes a set of procedures for elastomers, which encompass the majority of the USP, Ph. Eur., and JP test methods. ISO standards are voluntary and provide baseline information but again limited with respect to a complete assessment for any particular material and use. A listing of ISO standards for elastomers is below (60):

ISO 8871 Elastomeric parts for parenterals and for devices for pharmaceutical use

Part 1: Extractables in aqueous autoclavates
Part 2: Tests for identification and evaluation
Part 3: Determination of particles
Part 4: Biological requirements and test methods
Part 5: Functional requirements and testing

The multitude of standardized testing for the chemical and biological attributes of elastomeric closures currently required by the USP, Ph. Eur., and JP (61) is summarized in Tables 4 to 6. The ISO tests that correlate to the compendia tests are also noted. Although the tests between the three pharmacopoeias and ISO may have some commonalities, the extraction

Table 4 Compendia Extraction Tests

Test	USP	Ph. Eur.	JP	ISO
Appearance	X	X	X	X
Absorbance	X	X	X	X
Acidity/alkalinity	X	X	X	X
Ammonium	X	X	–	X
Reducing substances	X	X	X	X
Extractable zinc	X	X	X	X
Residue on evaporation	–	X	X	X
Volume sulfides	X	X	–	X
Heavy metals	X	X	–	X
Turbidity/color	–	–	–	X
Conductivity	–	–	–	X
Foam	–	–	X	–

Table 5 Compendia Nonextraction Tests

Test	USP	Ph. Eur.	JP	ISO
Total Cd and Pb	–	–	X	–
IR pyrolyzate	–	X	–	X
Resistance to steam	–	–	–	X
Ash	–	X	–	X
Density	–	–	–	X
Hardness	–	–	–	X
Elasticity	–	X	–	X

Table 6 Biological Tests

Test	USP	Ph. Eur.	JP	ISO
In vitro cell culture	X	–	–	X
In vivo systemic inj.	X	–	X	X
In vivo intracutaneous	X	–	–	X
Pyrogen	–	–	X	X
Hemolysis	–	–	X	–
Bioburden	–	–	–	X

methods and procedures can be unique, thereby making it a challenge to show compliance with all three regulatory bodies. So the question lingers, what data is meaningful, useful, and scientifically sound for the initial testing phase?

The compendia summary tables indicate ISO has the most comprehensive set of tests, but these methods are unique and are not mandated as those of the pharmacopoeias. The ISO tests are parallel to those of the pharmacopoeias such that mainly initial information on the chemistry and biological response is provided and a full study is still needed to show suitability of the container closure for its intended purpose, and it is conceivable that some of these tests may not be relevant and/or redundant. The specificity of standardized tests presents a challenge when qualifying materials on a global basis because it is difficult to standardize acceptance criteria when test procedures are not consistent. The sampling, extraction conditions (solvents and exposure), and analysis conditions (techniques and conditions) vary, and results are specific to those conditions. The sampling in some cases is intended to evaluate the container closure material's formulation and others, the actual container closure item or configuration (component or system). The USP and Ph. Eur. chemical and biological tests for elastomeric closures can be done on each formulation and the functional tests on each product (item-formulation combination). The JP chemical and biological tests for elastomeric closures must be done on each product due to the weight/weight ratio used in preparation of extracts. The JP method favors large items; large items are more likely to meet JP specifications than smaller items.

Requirements for functionality and cleanliness must also be considered when assessing container closure suitability, and standards have been developed for the functional evaluation of elastomeric closures that include specifications for penetrability, fragmentation, self-sealing capacity, and container closure integrity. Standards for visible (>25 μm) and subvisible (>2 μm and <25 μm) have also been developed.

The pharmacopoeia monographs provide a wide-range of test procedures and intended for materials used within a certain context. As a result of some of the generalities, often these tests can be applied to materials that may not fall into the intent of the specification. There are also new or combinations of materials being used in container closures system that may not have existed when specifications were set. As old materials maybe discontinued or new materials enter the market, updates of the compendia may not keep pace, and standardization will become more challenging. The USP monograph <381> Elastomeric Closures was only recently updated to include provisions for the required physicochemical, functionality, and biological testing relative to the types of coating on closures as well as responsibilities of suppliers and end users. This section is also more similar to Ph. Eur. and states test limits for type I (aqueous preparations) and type II (typically nonaqueous preparations).

Classification schemes have been established, which integrate the significance of the data and guide in the initial selection of materials. Categories for plastic, elastomers, and glass materials have been developed to differentiate suitability for a particular application.

Glass

There are three classifications for glass; the USP and Ph. Eur. classify glass into type I, II, and III. Type I is highly resistant borosilicate glass used for parenteral preparations of all pH values. Type II is treated soda-lime glass and is also used for parenteral application of all pH values where stability studies have demonstrated suitability. Type III is a soda-lime glass much less durable and only allowed for parenteral use if stability studies were found to be acceptable. In addition to the type I, II, and III designations, ISO has distinguished glass in the same manner as Class HC 1, 2, and 3. The classification for glass is based on the characteristic of the solubility of glass in water when autoclaved. The USP test for solubility is referred to as Chemical Resistance and Water Attack; in Ph. Eur., Hydrolytic Resistance; and in JP, Soluble Alkali. The procedures between pharmacopoeias are different, but the solubility for all is measured on the basis of titration of water, after exposure to glass, with a weak acid to detect the amount of alkali (base) present. There are other pertinent tests and specifications for glass in each of the pharmacopoeias, but these are not factored in to the classification scheme.

Plastics

Pharmaceutical products and container closure systems continue to evolve to meet the needs of patients and caregivers. Packaging is no longer limited to the protection and storage of a drug product, with the rising demand for innovative delivery and administration devices, the boundaries for regulation between container closure systems and devices have blurred. The FDA, European Union and Health Canada have designation for classes of medical devices, but the USP classes are intended to qualify the materials. The USP addresses materials requirements for container closure systems and devices as well as ISO; the Ph. Eur. and JP do not deal directly with medical devices.

Regulatory controls for device materials are grounded in biological reactivity tests, and the degree of testing is linked to the material classifications. Plastics are assigned the USP class designation of I to VI on the basis of results of the biological reactivity data.

Injection tests are used to assign the class designations I, II, III, V to plastics; implantation tests must be used to assign USP Classes IV and VI. The numerical class increases relative to the duration (risk) of contact between the body and device. In the category of implantable devices, exclusive use of Class VI is mandated.

The USP chapters over 1000 are not mandated but recommended, and USP<1031> has established a set of recommendations for "Biocompatibility of Materials Used in Drug Containers, Medical Devices, and Implants" that describes tests and classes required for medical devices and implants based on the following:

- Similarity and uniqueness of product relative to a previously marketed (predicate) product
- Extent contact between product and patient, etc.
- Duration of contact
- Material composition of product

Plastics must meet the requirements of the USP <87> Biological Reactivity Tests in vitro test (cell culture) to be suitable for a drug container. No further testing is necessary for containers to establish biocompatibility. The USP <1031> tests are designed to detect the nonspecific, biologically reactive, physical, or chemical characteristics of medical products or the materials used in their construction. ISO has available a more comprehensive set of test procedures, ISO 10993 for "Biological Evaluation of Medical Devices."

The requirements for bacterial endotoxins must be met for medical devices listed in USP Chapter <161>, Transfusion and Infusion Assemblies and Similar Medical Devices. The requirements apply to sterile and nonpyrogenic assemblies and devices in contact directly or indirectly with the cardiovascular and lymphatic systems and cerebrospinal fluid such as but not limited to solution administration sets, extension sets, transfer sets, blood administration sets, intravenous catheter, implants, dialyzers and dialysis tubing and accessories, heart valves, vascular grafts, intramuscular drug delivery catheters, and transfusion and infusion assemblies.

In summary, portions of devices made of plastics or other polymers meet the requirement of Biological tests—Plastics and Other Polymers under USP Containers <661>. Portions of devices made of elastomeric materials should meet the requirements of Elastomeric

Closures for Injection <381>. If a class designation is needed for plastics the requirements under <88> Biological Reactivity Tests, in vivo, apply. Compliance with compendial standards is necessary for regulatory approval but not sufficient to indicate the suitability of a container closure or device. Additional compatibility, functionality, and performance tests are necessary to prove suitability with a specific drug product and application.

PROCESS WORK FLOWS FOR MANAGING PACKAGING EXTRACTABLES AND LEACHABLES

There is a logical progression to the framework for identification and control of leachables involving a set of steps to guide the degree of qualification relative to the phase of development. The qualification of leachables is associated with not only the discovery and amount of leachables but also the toxicological impact to the patient. There are various routes that can be taken to qualify a suitable container closure system and the drug product type, route of administration, duration of exposure, and patient population are among the variables to be considered. The qualification process is detailed, spanning a long period of time, and a team of analytical chemists, toxicologists, quality/regulatory professionals, engineers, and procurement specialists would facilitate development of a process map. The first step and most vital step of the process is to identify the primary container closure components and drug product contact materials and other critical components to be evaluated for extractables. These materials would then be assessed for potential extractables, starting by obtaining supplier information and results of compendia tests for materials that have an applicable monograph. The compendia test will not provide adequate information to correlate to patient safety, so the next step would entail developing a study design to obtain a chemical profile of the potential extractables for all the critical materials.

The protocol for an extractable study should employ multiple solvents having varying propensity to extract constituents from the container closure materials using aggressive conditions. The conditions for extractions should be adequate to provide a chemical profile but not so extreme as to create anomalous results. The extracts should then be analyzed using multiple techniques to detect organic and inorganic constituents of the container closure system. After a chemical profile is obtained, the data can be compared with the supplier information and evaluated for any compounds of concern. In the initial stage of the evaluations, much of the data will be qualitative or at best semiquantitative having only tentative identifications. It is not always evident if there is a toxicological concern until positive identification is made and the compounds are measured. Once extractable compounds are identified and measured, an assessment for toxicological impact can be attained and alternative materials may be considered if necessary. The measurement of extractables in container closure systems should be made using well-characterized methods to have a level of confidence to guide in the decision-making process. Measurements should be direct, provide high assurance of reproducibility, have purpose, and be sufficient and timely to provide a meaningful evaluation of quality (62). The methods should be fully validated if they are to be used to control the container closure materials.

Selection of the extractable compounds to be evaluated in a leachable study will need to have careful consideration; an extractable will not necessarily become a leachable. It is also conceivable that an extractable may form an interaction product or degradation product once in contact with the drug product. Migration of extractables or interaction products may occur under certain conditions relative to the drug product that may take place over a period of time. It is necessary that the leachable methods have the appropriate sensitivity and specificity as well as be free of interferences from the drug product. These methods must be validated under the guidelines provided by the regulatory agencies. The drug products can then be set-up on stability and evaluated at the specified time points. Assessments of the toxicological impact should be made throughout the studies and a correlation between E&L should be made to enable control of leachable compounds. A summary of these steps is illustrated in Figure 11. The process flow diagram describes a 14-step process, each step being a building block for the next (51). Another approach for a process map is shown in Figure 12 (51).

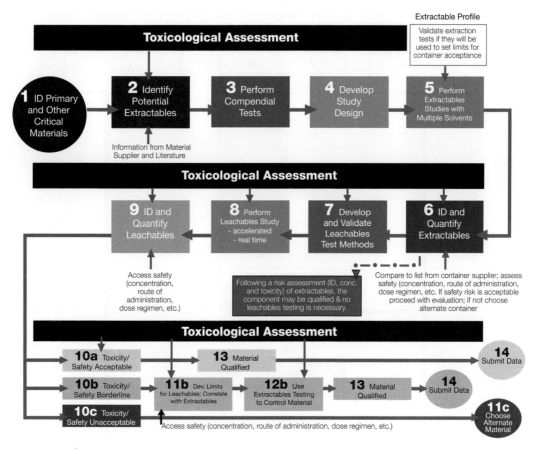

Figure 11 Extractables and leachables management.

MANAGING E&L FROM SINGLE USE AND PROCESS COMPONENTS

There is a potential for pharmaceutical products to be contaminated from contact materials during any phase of production or storage. While it is true that the guidance documents use the term container closure systems, evaluation of these systems are not limited to only the primary containers; secondary and ancillary materials can also contribute to leachables as well as any materials that may be in intermediate contact with the pharmaceutical product during manufacture. According to GMPs, 21CFR 211.65, packaging and the *equipment shall be constructed so that surfaces that contact components, in-process materials, or drug products shall not be reactive, additive, or absorptive so as to alter the safety, identity, strength, quality, or purity of the drug product beyond the official or other established requirements.*

The process materials and equipment used in manufacturing biopharmaceutical products fall into this category and can introduce leachables, albeit the intermediate or upstream nature of the processing materials. Certain bioprocess conditions may serve to filter or concentrate a given extractable compound introducing a leachable into the drug product with potential to cause harm to a patient. This presents the challenge to the manufacturer of biopharmaceutical products such that the critical materials to be evaluated for extractables must be understood early in the pharmaceutical development process so that this can be incorporated into leachable studies. Extractables, therefore, potential leachables from components of manufacturing systems and process materials may include filters, capsules, tubing, pumps, films connectors, and fittings for bioprocess containers (BPCs), single-use bags,

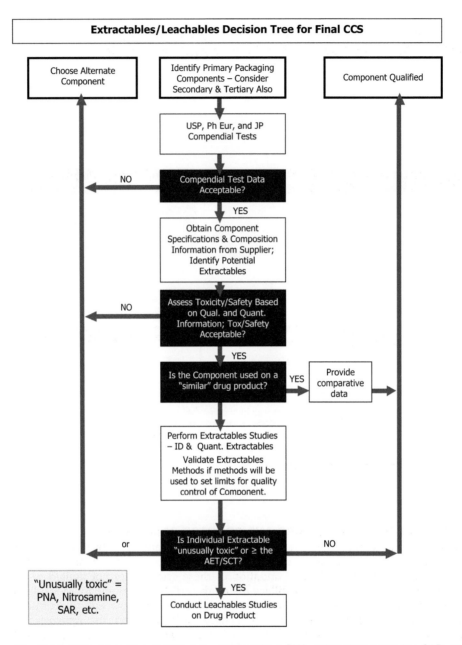

Figure 12 Extractables leachables decision tree. *Abbreviations*: PNA, polynuclear aromatics; SAR, structure-activity relationships; SCT, Safety Concern Threshold; AET, analytical evaluation threshold; CCS, container closure system.

and other product-contact materials (PCMs). Typical conditions of drug product exposure to a final container closure system compared with that of a disposable or process component is shown in Table 7.

The likelihood of interaction of single use or process components with a biopharmaceutical product will depend on four major conditions: (*i*) the direct contact to the actual drug product, (*ii*) indirect contact of solutions or materials that are precursors to the drug product, (*iii*) immediate contact of the drug product or precursor that is not processed further after contact, and (*iv*) remote contact of the drug product or precursor that is processed further after contact (63). The degree of safety qualification for the components used in the processing and

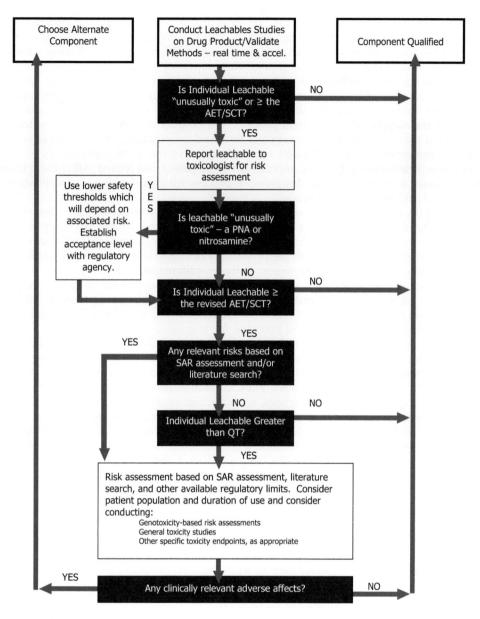

Figure 12 Continued

Table 7 Drug Product Exposure to Process Components

Final container closure	Disposable and process components
Contact time 18 mo. to 4 yr	Contact time seconds to years
Only contacts drug product	May contact drug product and precursors
Surface area to volume rations are in a relatively narrow range	Surface area (SA) to volume (V) rations are in a relatively wide range. (As size increases SA/V decreases)
Temp. are limited to freezing to autoclave, approx. −20 to +121°C; storage −20 to +25°C	Temp. may be from −80 to +121°C
Narrow range of materials—glass, plastic, rubber	Wide range of materials—metals, glass, more types of plastics and rubbers (cellulose, neoprene, silicones, nylons)

manufacture of biopharmaceuticals can be best managed by first considering which components have the highest risk for interaction with the pharmaceutical product. Contamination of a biopharmaceutical product can pose a safety risk to a patient causing a toxic or allergic effect; contamination can also change the properties of the drug product having a negative impact to the product and putting the patient at risk by not receiving intended therapy. The risk of constituents migrating from the contact material into the drug product or the precursors must take into consideration the compatibility of the component materials, proximity of the component to the final product, the product or precursor composition, the surface area of the contact material, the contact time and temperature. Other issues of concern are possible pretreatments and intended use such as exposure of the components to steam sterilization or γ radiation and if the components will be rinsed and reused. All of these factors combined create a dilemma when designing a process for qualification and validation of surface contact and in-process materials.

To select the appropriate materials for evaluation, the risk assessment tools, described in ICH Q9, can be incorporated to judge the critical disposable and bioprocess components of interest. Risk scores can be assigned based on conditions such as those listed in Table 8. A risk score can be developed on the basis of the probability of extractables occurring relative to the severity of harm caused from contamination of the biopharmaceutical.

Risk scores are generally based on predictive models developed for particular materials in a particular system and would need to be developed for each specific application. Although there will always be a degree of uncertainty in the risk values, an informed decision can be made by considering each material and potential for migration.

Once the surface contact materials from processing equipment and components are evaluated for the potential to contaminate the biopharmaceutical product, a decision can be made as to whether an alternative material should be considered or to proceed with an extraction study. According to the 1999 FDA Guidance for Industry: Container Closure Systems for Packaging Drugs and Biologics, an extraction study would employ at least three extraction solutions followed by the analysis of extracts using multiple analytical techniques. Table 9 shows solutions that may be considered for an extraction study of a bioprocess container. The use of multiple solvents should include clean solvents to provide a chemical profile as well as solution simulating the drug product. Conditions of exposure should be exaggerated to indicate worst case and provide data for the chemical profile. On the basis of the extractables data constituents would be evaluated for potential impact to patient safety as

Table 8 Considerations for Risk Evaluation

Proximity to API	Area of exposure
Type of solvents(s)	Known extractables
Length of contact	Known toxicity
Temperature of exposure	Resistance to extraction

Abbreviation: API, active pharmaceutical ingredient.

Table 9 Solutions for Extracting Bioprocess Containers

High pH solutions
  3M NaOH, 6M urea, 2M Na bicarbonate, 4M NaCl, 1M Tris Base
Mid pH solutions
  WFI, 1M Na phosphate buffer @ pH = 7, 1M succinate, 10M ethylene glycol
Low pH solutions
  2M Na carbonate, 4M guanidine, 2M HCl
Solvents
  50% EtOH/WFI, 10% DMSO
Solutions specific to company drug products
  4M acetic acid, NaCl/Na phosphate/DMSO 0.5M/0.2M/3%, 10 mM histidine/0.50M arginine

well as to the drug product, at this point another decision can be made to seek alternative material or to proceed to a leachable assessment (51).

The target analytes for leachable studies, derived from the extraction of the surface contact materials from processing equipment and components, can be combined with those from the primary container closure system to understand the required sensitivity, those that are in common or may interfere with the drug product or other target analytes before developing the leachable study plan. Several methods may need to be developed and validated to encompass all of the constituents of interest. Leachable measurements must be accurate and precise having potential to indicate the presence of interaction products associated with the drug product formulation. Like the final container closure system, disposables and process components also require extractables assessment and leachables control. Manufactures of processing materials and equipment may provide baseline information on extractables but the materials suitability for one process may not be valid for a different biopharmaceutical product. The ultimate proof of suitability relies on user-specific studies. The degree of scrutiny for single use and processing components depends on several factors, and a risk assessment is commonly used to identify and prioritize studies to qualify and control the materials and components.

EXTRACTABLE STUDIES: MATERIAL AND TESTING CONSIDERATIONS
Extraction studies are conducted to achieve a greater understanding of the materials that are in contact and critical to a pharmaceutical product in an effort to protect the product and patient from adverse effects. There are different study approaches for conducting extraction studies depending on the intended outcome of the study. The choice of extraction conditions and analysis techniques can be relative to the goal of obtaining qualitative profiles, quantitative profiles, predictive modeling, and/or actual conditions of exposure. A comprehensive extractable study can entail a combination of all the above objectives. The FDA guidance on container closure systems classifies the degrees of concern for likelihood of interaction between the container closure system and drug product from high to low depending on the route of administration. Inhalation and injectable dosage forms are among the highest level of concern. The PQRI Leachables and Extractables Working Group has published guidelines for inhalation products titled *Safety Thresholds and Best Practices for Extractables and Leachables in Orally Inhaled and Nasal Drug Product*. This recommendation document, available at www.pqri.org, details a systematic comprehensive approach for investigation and control of leachables by employing controlled extraction studies followed by correlation to leachables and control. These recommendations incorporated threshold values to answer the question of "how low to go," but these thresholds currently apply only to inhalation products. Investigations for extractables in container closure systems would be more rigorous as concern for interaction increases and future recommendations for parenterals and ophthalmics are planned (64).

The solvents and exposure conditions used for extractions studies considers not only the nature and use of the drug product, but the physical and chemical nature of the material under investigation. Typical materials used in the container closure systems for different dosage forms are shown in Table 10.

The critical components for evaluation is the first and most important step since qualifying container closure systems is a long process, and taking steps backward to include an overlooked critical component will cause a disappointing delay.

There are certain factors that are relevant to selecting critical components for evaluations and include patient contact, product contact, device performance, type of secondary packaging, and if there are ancillary components. With respect to other components of concerns, the intermediate package, bulk containers, and process materials may also be germane. The FDA recommends that a stronger extracting solvent than the drug product would be used to obtain a qualitative extraction profile.

The composition of the critical components along with information on the drug product matrix will drive the type of solvents to employ and most suitable exposure conditions. Information from the supplier and downstream suppliers will aid in developing an extraction protocol. An understanding of the base material, additives and processing aids, polymerization and fabrication processes, as well as a type of cleaning, pretreatment, storage and shipping

Table 10 Example Container Closure Components

Dosage form	Components	Example material
Inhalation	MDI/DPI components, canisters, valves, gaskets, blister packs, bottles, actuators, mouthpiece, pumps, closures, liners, label/inks	Polyolefins, SBR and EDPM rubber, thermoplastic elastomers, polyacetal, polyesters, polyamides, acrylics, epoxies, paper/paperboard, metals, glass
Injectable	SVP < 100 mL/LVP > 100 mL cartridge, syringe, vial, ampoules, flexible bag, closures/plungers, injection ports, needles, adhesives, inks, overwraps	Polyolefins, butyl rubber, EPDM rubber, polyvinyl chloride, polyurethanes, polycarbonate, acrylics, polyamides, polystyrene, thermoplastic elastomers, silicones polyesters, epoxides, cellophane, fluoropolymers, styrenics, paper/paperboard, metals, glass
Ophthalmic	Bottles, droppers, screw caps, liners, tips, tubes/liners, labels/ink	Polyolefins, acrylics, vinyls, epoxies, polyamides, thermoplastic elastomers, polyesters, cellophane, glass, paper/paperboard, metals
Transdermal	Adhesives, membranes, barrier films, reservoir, coatings, blister packs, preformed trays, overwraps, substrates, topical aerosol components	Polyolefins, acrylics, vinyls, polyamides, polyesters, styrenics, rubber material, thermoplastics, metal
Associated components	Nebulizers, dosing spoons, dropper, dosing cups	Polyolefins, glass, rubber, thermoplastic polyesters

Abbreviations: MDI, metered dose inhalers; DPI, dry powder inhalers; SVP, small volume parenterals; LVP, large volume parenterals.

will all factor into the decision for not only extraction but also the appropriate analytical techniques.

A container closure will be suitable if it protects the drug product, functions properly, and found to be compatible with the dosage in addition to ensuring harmful chemicals will not leach into the product. The compatibility of a container closure has many variables; the complexity of the process for assessing risk is represented in the Ishikawa diagram Figure 13 (65).

Results from extractable studies should be representative of appropriate sampling. All lots are not created equal and the variability must be realized to set specifications and acceptance criteria. The sample to surface ratio should be adequate so that required sensitivity can be achieved and is consistent with the sample preparation techniques. A container closure system may be multicomponent, multilayered, coated, or have surface treatments or pretreatments that would also have a bearing on the sampling plan.

The conditions of extractions may serve different purposes and there are three general types: (*i*) accelerated extraction that is intended to reduce experimental time to reflect actual use; (*ii*) exaggerated or aggressive extractions, which are conditions that are intended to maximize the amount of extractables; and (*iii*) simulated extractions, which are conditions intended to mimic actual use such as those used in the CFR for indirect food additives (66).

As a rule of thumb, suitable extraction solvents would have the following properties (67):

- Range of boiling points
- One of similar extracting properties to drug product vehicle
- High purity and relatively nonreactive
- Easily and safely handled and readily available

The probability that constituents will migrate from the container closure materials into the drug product are related to diffusion of the entity from the polymer and solubility in the drug product. The purpose for the extraction may be intended to provide a qualitative profile or materials control methods, in this case exaggerated or aggressive extraction conditions would be optimal. If the intent is to provide a predictive model, accelerated or simulated extraction

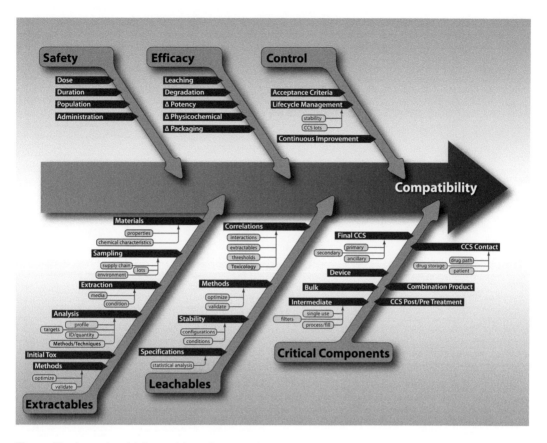

Figure 13 Assessing risk for container closure system.

techniques would be more practical. Examples of data representing exaggerated extractions under reflux conditions, accelerated extractions at 50 and 70°C, and simulated extractions for sterilization at 50°C are shown in Figure 14.

Aggressive and exaggerated extraction techniques would include reflux, Soxhlet, autoclave, microwave, accelerated extractors, and sonication. The solvents would be harsh such as hexane, ethanol, isopropanol, or chlorinated solvents. Whereas simulated extraction may only have water or diluted alcohol. The nonvolatile residue of the extracts, infrared spectroscopy, total organic carbon, or other broad-based information can be acquired to aid in developing conditions for more sensitive and selective methodology. Identification of the extractables is usually acquired by mass spectrometry; gas and liquid chromatography are typically employed for trace organic compounds. Inductively coupled plasma spectroscopy and ion chromatography are commonly used techniques for detection of inorganic species. Multiple analytical techniques should be employed to allow the most comprehensive profile. Complementary techniques and authentic standards can confirm species identifications. Once identifications are confirmed, methods can be optimized and validated for measuring the species of interest.

As it can be seen from the GC/MS profile chromatograms, run under identical conditions, a simulated extraction would not provide any information on a chemical profile of a container closure component; the aggressive and exaggerated conditions provide a wealth of information, but there is danger in creating anomalies using harsh conditions.

Not all of the data generated during the exploratory phase of an extraction will be useful to correlate to leachables, but it is prudent to have too much data rather than not enough. Interaction, hydrolysis, and degradation products may also occur and would not be evident in

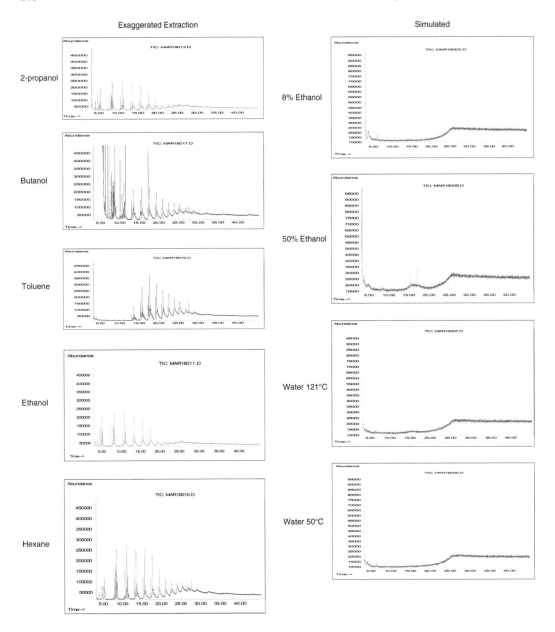

Figure 14   GC/MS qualitative profiles.

the extractables data. For this reason, the leachable methods would need to be adequate to detect unspecified species at the stability time points. In any event, the information required in drug product application is regulatory policy and the expectation is that it can change on a case-by-case basis.

LEACHABLE STUDIES: DRUG PRODUCT AND TESTING CONSIDERATIONS

Potential leachables are indicated from the component parts evaluated in the extraction studies. A comprehensive analysis of appropriately prepared extracts should detect, when present, residual starting materials from the polymerization process, primary or secondary additives, extractable contaminates from known or unknown sources, processing aids, and additive impurities, oxidation or breakdown products. Leachables can also arise from a reaction of an extractable with drug product or secondary components that may have been overlooked when selecting components for evaluation.

The data and analysis conditions from the extraction studies can be used to develop leachable methods and include development of optimal conditions for the analytical techniques to measure target potential leachables at required sensitivity. Sample preparation trials can be carried out on the drug product control to optimize the leachable methods. The method should be evaluated to indicate suitability for validation by performing spiking, recovery, repeatability, and linearity studies. Assessment of the proposed method can then be accomplished by analysis of the initial drug product samples in contact with packaging materials and accelerated and shelf life stability samples stored in the final package. The leachable methods should be validated according to regulatory guidelines before routine and stability testing are performed.

Several lots of drug product, stored at different orientations, should be evaluated to realize variability and provide adequate information that can be used to: (*i*) determine maximum leachable levels and establish acceptance criteria if necessary, (*ii*) perform a risk assessment of leachable species on the basis of actual stability time points, and (*iii*) provide the ability to correlate leachable data to extractables to determine packaging specifications if appropriate. An extraction study should indicate greater concentrations of extractables compared with leachables. This means that methods to measure extractables should be valid and reliable. A correlation can be established if the leachables detected can be quantitatively linked, directly or indirectly to an extractable. The maximum leachable levels can be predicted based on achieving asymptotic levels of extractables. It is conceivable that routine analysis and control of the packaging components could ensure acceptable levels of leachables over the shelf life of the product. In the end, the container closure system suitable for one drug product may or may not be suitable for another drug product.

REFERENCES

1. Dean DA, Evans ER, Hall IH. Pharmaceutical Packaging Technology. New York: Taylor & Francis 2000.
2. Soroka W. Fundamentals of Packaging Technology. 2nd ed. Herndon, VA: Institute of Packaging Professionals, 1999.
3. Everts S. Chemicals leach from packaging. Chem Eng News 2009; 87:11–15.
4. US Code of Regulations, 21 CFR 211.94(a).
5. Safety Thresholds and Best Practices for Extractables and Leachables in Orally Inhaled and Nasal Drug Products, Report of the PQRI Leachables and Extractables Working Group, September 8, 2006. Available at: http://www.pqri.org/pdfs/LE_Recommendations_to_FDA_09-29-06.pdf.
6. Milano EA, Waraszkiewicz SM, DiRubio R. Extraction of soluble aluminum from chlorobutyl rubber closures. J Parenter Sci Technol 1982; 36:116.
7. Pang J, Blanc T, Brown J, et al. Recognition and identification of UV-absorbing leachables in EPREX® pre-filled syringes: an unexpected occurrence at a formulation-compound interface. J Pharm Sci Technol 2007; 61:423–432.
8. Milano EA, Waraszkiewicz SM, DiRubio R. Aluminum catalysis of epinephrine degradation in lidocaine HCl with epinephrine solutions. J Parenter Sci Technol 1982; 36:232.
9. Milano EA, Williams DA. The formation of aluminum-epinephrine complex and its effect on the addition of bisulfite to epinephrine. J Parenter Sci Technol 1983; 37:185.
10. Fries A. Drug delivery of sensitive biopharmaceuticals with prefilled syringes. Drug Deliv Technol 2009; 9:22–27.
11. Guidance for Industry: Container Closure Systems for Packaging Human Drugs and Biologics— Chemistry, Manufacturing, and Controls Documentation. FDA, 1999. Available at: http://www.fda. gov/downloads/Drugs/GuidanceComplianceRegulatoryInformation/Guidances/UCM070551.pdf.
12. Guidance for Industry: Q3B (R2) Impurities in New Drug Products. International Conference on Harmonization (ICH), July 2006. Available at: http://www.fda.gov/RegulatoryInformation/ Guidances/ucm128032.htm.
13. Guidance for Industry: Q3A (Rev 2) Impurities in New Drug Substances, FDA, June 2008. Available at: http://www.fda.gov/downloads/RegulatoryInformation/Guidances/ucm127984.pdf.
14. Product Quality Research Institute (PQRI), 2009. Available at: http://www.pqri.org.
15. US Code of Federal Regulation Title 21, parts 210 and 211.

16. US Code of Federal Regulation Title 21, part 820.
17. Guidance for Industry: Current Good Manufacturing Practice for Combination Products (Draft Guidance), US Department of Health and Human Services Food and Drug Administration, OCP, September 2004.
18. United States Pharmacopeia-National Formulary (USP-NF). Rockville, MD: United States Pharmacopeial Convention, Inc.
19. Guidance for Industry: Quality Systems Approach to Pharmaceutical Current Good Manufacturing Practice Regulation; Draft Guidance for Industry US Department of Health and Human Services Food and Drug Administration, CDER/CBER/CVM/ORA, September, 2004.
20. Guidance for Industry: Q8 (R2) Pharmaceutical Development Revision 2, US Department of Health and Human Services Food and Drug Administration, CDER/CBER, ICH November, 2009.
21. Guidance for Industry: Q9 Risk Management, US Department of Health and Human Services Food and Drug Administration CDER/CBER, ICH November, 2009.
22. Guidance for Industry: Q10 Pharmaceutical Quality System, US Department of Health and Human Services Food and Drug Administration, CDER/CBER, ICH April, 2009.
23. Guidance for Industry: Draft Metered Dose Inhaler (MDI) and Dry Powder Inhaler, Chemistry Manufacturing and Controls Documentation, US Department of Health and Human Services Food and Drug Administration, CDER, July 2002.
24. Guidance for Industry: Nasal Spray Inhalation Solution, Suspension, and Spray Drug Products: Chemistry Manufacturing and Control Documentation, Guidance for Industry; US Department of Health and Human Services Food and Drug Administration, CDER, July 2002.
25. US Code of Federal Regulations, title 21, Indirect Food Additive Regulation.
26. European Medicines Agency ICH Q6A Specifications: Test Procedures and Acceptance Criteria for New Drug Substances and New Drug Products: Chemical Substances, May 2000 CPMP/ICH/367/96.
27. ISO 10993 Biological Evaluation of Medical Devices, Parts 1, 9, 13, 14, 15, 17, 18. Available at: http://www.iso.org/iso/catalogue.
28. Smith EJ, Nash RJ. Elastomeric closures for parenterals. In: Avis KE, Lieberman HA, Lachman L, eds. Pharmaceutical Dosage Forms: Parenteral Medications. Vol. 1, 2nd ed. New York: Dekker, Inc, 1992.
29. Gurley T. The leachable challenge in polymers used for pharmaceutical applications. Rubber World 2008:22–26.
30. Wong WK. Impact of elastomer extractables in pharmaceutical stoppers and seals—material supplier perspective. Rubber World 2009; 20–29.
31. Flurotec® is a registered trade mark of West Pharmaceutical Services, Lionville, PA, USA. Available at: www.westpharma.com.
32. Omniflex® is a registered trade mark of Helvoet Pharma, Alken, Belgium. Available at: www.helvoetpharma.com.
33. Technical Report No. 39 (Revised 2007): Guidance for Temperature-Controlled Medicinal Products: Maintaining the Quality of Temperature-Sensitive Medicinal Products Through the Transportation Environment. Bethesda, MD: Parenteral Drug Association, 2007.
34. West Pharmaceutical Services, Lionville, PA 19341. Available at: http://www.westpharma.com.
35. Bacon FR. Glass containers for parenterals. In: Avis KE, Lieberman HA, Lachman L, eds. Pharmaceutical Dosage Forms: Parenteral Medications. Vol. 1, 2nd ed. New York: Dekker, Inc, 1992.
36. ASTM Standard C162-05. Standard Terminology of Glass and Glass Products. West Conshohocken, PA: ASTM International, 2005. Available at: http://www.astm.org.
37. Swift RW. Extractables from Glass Containers. Presented at: the PDA Forum on Extractables, Rockville, MD, November 12–13, 2001.
38. Containers-Glass, Chapter 660, United States Pharmacopeia-National Formulary (USP-NF). Rockville, MD: United States Pharmacopeial Convention, Inc.
39. Borchert SJ, Ryan MM, Davison RL, et al. Accelerated extractable studies of borosilicate glass containers. J Parenter Sci Technol 1983; 43:67–79.
40. Aldrich DS. The Upjohn Company, 1994.
41. Borchert SJ, Maxwell RJ. ESCA depth profiling studies of borosilicate glass containers. J Parenter Sci Technol 1990; 44:153–182.
42. Asselta R. Neutraplex—Tubular Glass vials with Improved Hydrolytic Resistance for Pharmaceuticals. Comar Inc. Presentation, February 2001.
43. SGD Pharma, New York, NY, USA. Available at: www.sgd-pharma.com.
44. Schott NA, Inc., Elmsford, NY. Type I plus® Brochure, 2009. Available at: www.us.schott.com.
45. Paskiet DM. Extraction profile methodology for pharmaceutical and medical packaging. In: Pharmaceutical and Medical Packaging 2000. Vol. 10. Copenhagen: Hexagon Holding ApS, 2000.
46. Kiang PH. Parenteral Packaging: Rubber, Glass, Plastic, and Metal Seals. Bethesda, MD: PDA Training and Research Institute, 2001.

47. Solomon DD, Jurgens RW, Wong KL. Plastic containers for parenterals. In: Avis KE, Lieberman HA, Lachman L, eds. Pharmaceutical Dosage Forms: Parenteral Medications. Vol. 1, 2nd ed. New York: Dekker, Inc, 1992.
48. Surlyn® is a registered trademark of DuPont, Wilmington, DE, USA. Available at: www2.dupont.com/Surlyn/en_US/.
49. Tygon® is a registered trademark of Saint-Gobain, Akron, OH, USA. Available at www.tygon.com/.
50. Aclar® is a registered trademark of Honeywell, Morristown, NJ, USA. Available at: www51.honeywell.com/sm/aclar/.
51. Paskiet DM, Smith EJ. Assessing Packaging and Processing Extractables/Leachables. Bethesda, MD: PDA Training and Research Institute, October 2009.
52. Delrin® is a registered trademark of DuPont, Wilmington, DE, USA. www2.dupont.com/Plastics/en_US/Products/Delrin/Delrin.html.
53. Jenke D. A Discussion of the Physiochemical Factors that Regulate the Leaching of Organic Substances from Plastic Contact Materials into Aqueous Pharmaceutical Solutions. American Pharmaceutical Review 2009; Web Article, July/August 2009.
54. Jenke D, Couch T, Gillum A, et al. Modeling of the solution interaction properties of plastic materials used in pharmaceutical product container systems. PDA J Pharm Sci Technol 2009; 63:294–306.
55. Packaging predicaments. Packaging Technol 2007:16. Available at: http://www.pharmtech.com.
56. Fang X, Cherico N, Barbacci D, et al. Leachable study on solid dosage form. Am Pharm Rev 2006; 9:58–63.
57. Baughan JS, Barnett AB. The Regulation of Printing Inks in the United States. Available at: http://packaginglaw.com. Accessed February 2002.
58. Baughan JS, Boot K. FDA Regulation of Adhesives: Not Necessarily a Sticky Situation. Available at: http://packaginglaw.com. Accessed April 2002.
59. European Pharmacopoeia. 6th ed. European Directorate for the Quality of Medicines, 2006. Available at: http://www.edqm.eu/store.
60. ISO 8871 Elastomeric parts for parenteral and for devices for pharmaceutical use-Part 1 and Part 2.
61. The Japanese Pharmacopoeia. 15th ed. (JP15). Available at: http://jpdb.nihs.go.jp/jp15e/.
62. Manufacturing, Measurements & Quality Systems: Seeking Order in Complexity Richard L. Friedman Director, Division of Manufacturing & Product Quality Center for Drug Evaluation & Research, Presentation at PQRI Conference on Advancing Drug Product Quality and Development, December, 2009.
63. Jenke D. PharmaEd Container Closure Systems Conference, Philadelphia, September 2008.
64. Proposal for Reporting and Qualification Thresholds for Leachables in Parenteral and Ophthalmic Drug Products. PODP Leachables and Extractables Working Group, March, 2007, Product Quality Research Institute (PQRI).
65. Paskiet DM. Accelerating Assessment, Pharmaceutical and Medical Packaging, May 2009.
66. Hojnicki J. A Systematic Approach to the Extraction Process. Presented at: PDA Leachables and Extractables Conference, Extractables Puzzle, May 24, 2005.
67. Norwood D, Granger A, Paskiet D. Extractables and Leachables in Drugs and Packaging. Encyclopedia of Pharmaceutical Technology. 3rd ed. London: Informa Healthcare, 2007.

11 | Process analytical technology and rapid microbiological methods

Geert Verdonk and Tony Cundell

INTRODUCTION

In September 2004, the Food and Drug Administration (FDA) Guidance for Industry PAT—A Framework for Innovative Pharmaceutical Development, Manufacturing, and Quality Assurance was issued to encourage pharmaceutical manufacturers to develop and implement effective and efficient innovative approaches in providing quality pharmaceuticals to the public (1). The linkage of rapid microbiological methods (RMMs) to Process Analytical Technology (PAT) is largely based on real-time release, which is the ability to evaluate and ensure the acceptable quality of in-process and/or final product on the basis of the collection and analysis of in-process data. As stated in the FDA guide, the PAT component of real-time release typically includes a valid combination of assessed material attributes and process controls. Material attributes such as bioburden, endotoxin content, and sterility could be assessed using direct and/or indirect process analytical methods. The combined process measurements and other test data gathered during the manufacturing process could serve as the basis for real-time release of the final product and would demonstrate that each batch conforms to established regulatory quality attributes. The FDA considers real-time release to be comparable with alternative analytical procedures to the compendial microbiological tests for final product release. It is notable that the guidance document stated that real-time release as defined in this guidance builds on parametric release of terminally heat sterilized drug products, a practice in the U.S. large-volume parenteral industry since 1985. In real-time release, material attributes such as formulation, bioburden, container size, and load pattern, as well as process parameters such as sterilization parameters, are measured and controlled.

In this chapter, the authors will attempt to define the role of RMM in PAT and discuss the application of RMM to aseptic filling, biopharmaceutical upstream and downstream processing, environmental monitoring and control in clean rooms; the selection, development, validation, and implementation of RMM for PAT applications; industry, regulatory, and compendial guidelines for RMM; regulatory approval of RMM and the future of RMM in parenteral medication manufacturing.

TRADITIONAL MICROBIAL TEST METHODS

Unexpectedly to some, the standard-setting organization for drug products marketed in the United States is not the U.S. Federal Food and Drug Administration but the U.S. Pharmacopoeial Convention Inc., an independent standards organization, empowered by the U.S. Federal Food, Drug, and Cosmetic Act as the official drug standard-setting organization in the United States for drug products. The U.S. Pharmacopoeial Convention publishes and maintains the U.S. Pharmacopoeia (USP), National Formulary (NF), and USP Reference Standards and sets the official tests and quality standards for both drug products and pharmaceutical ingredients. In Europe and Japan, the compendia, that is, European Pharmacopoeia (Ph. Eur.) and Japanese Pharmacopoeia (JP) are government-controlled organizations that play a similar role to the USP.

Traditional USP microbial testing methods, as referee tests, rely on the growth of microorganisms in culture media for detection, enumeration, and selective isolation. These traditional methods continue to be used because of their long history of use, simplicity, effectiveness, low cost, and suitability for use in all microbiological testing laboratories. However, serious questions can be raised if the continued use of these traditional methods is the right strategy to improve quality and efficiency in the pharmaceutical industry. Those traditional methods were originally designed for the detection of human pathogens and not for the microbiological quality control of pharmaceutical processes and products. The drivers of the microbial testing should be the critical microbiological quality attributes associated with a specific drug product and the risk assessment of the potential for microbial contamination of that drug product and resulting patient infection. The next few paragraphs will discuss the industry experience with compendial microbial testing.

Bioburden Testing

Nonsterile drug substances, pharmaceutical excipients, and drug products are evaluated for bioburden using microbial limit or microbiological examination tests. On November 8, 2005, at the Pharmacopoeial Discussion Group meeting in Chicago, Illinois, representatives of the three major compendia, that is, JP, Ph. Eur., and USP signed off on the harmonized microbiological examination tests. The USP published the General Test Chapters <61> *Microbiological Examination of Non-sterile Products: Microbial Enumerations Tests* and <62> *Microbiological Examination of Non-sterile Products: Tests for Specified Microorganisms*, and the General Informational Chapter <1111> *Microbiological Quality of Non-sterile Pharmaceutical Products* in the Second Supplement to USP 29/NF 24 in June 2006 with an official date of August 1, 2007. On November 14, 2006, the USP announced on their website a postponement of the official implementation date to May 1, 2009, to allow companies more time for method qualification, change control, and regulatory. The companion Ph. Eur. chapters are 2.6.12 *Microbiological Examination of Non-sterile Products (Total Viable Aerobic Count)*, Ph. Eur. 2.6.13 *Microbiological Examination of Non-sterile Products (Test for Specified Micro-organisms)*, and Ph. Eur. 5.1.4 *Microbiological Quality of Pharmaceutical Preparations*. These referee tests are clearly unsuitable for PAT applications due to their extended incubation times, relative insensitivity, and low precision, and even have limitation as release test methods as they may not detect all objectionable microorganisms that could be present in a nonsterile drug product.

Sterility Testing

Sterility testing was traditionally been conducted by inoculating a microbiological broth with an aliquot of the test material and scoring growth by the detection of turbidity. The compendial sterility tests have been harmonized in terms of media, growth-promotion requirements, suitability tests, incubation conditions, number of containers and amounts of material tested, and observation and interpretation of the results. Limited local requirements from the different pharmacopoeias were included in the compendial tests and these will be removed in May 2009. The membrane filtration test is the preferred test over the direct inoculation test as it has the capacity to test the entire contents of a product container and inhibitory substances may be rinsed from the membrane. The details of the tests may be found in USP Chapter <71> *Sterility Tests* and Ph. Eur. 2.6.1 *Sterility*. The incubation period for the test is at least 14 days, making it clearly unsuitable for a PAT application.

Bacterial Endotoxin Testing

Bacterial endotoxins are pyrogenic materials, for example, lipopolysaccharide, present in the cell wall of gram-negative bacteria. Bacterial endotoxins, if present in injectable products, can lead to dose-related adverse reactions in patients receiving injections ranging from chills to fever to death. A threshold pyrogenic dose is 5 EU per kg of body weight for a parenteral administration. In terms of weight and not potency, this is about 1 ng per kg of body weight for *Escherichia coli* and 50 to 70 ng/kg for *Pseudomonas aeruginosa* in both rabbits and humans. For *E. coli*, this represents some 10,000 whole cells per kg. The in vivo rabbit pyrogen test was replaced by the in vitro *Limulus* amebocyte lysate (LAL) endotoxin test in the mid-1970s making the test suitable for both in-process and finished product testing. This test is largely responsible for the elimination of pyrogens from parenteral drug products. As different sources of endotoxin have differing potency, the standard was assigned potency in endotoxin units (EU).

The compendial bacterial endotoxins assays and reference standards have been harmonized in terms of test methods, that is, gel-clot, turbidimetric (end-point and kinetic) and chromogenic (end-point and kinetic) assays, reagents, reference standard, calculation of endotoxin limits for drug products, suitability testing, and assay validation. It should be noted that the gel-clot method is semiquantitative in that it determines the lowest two-fold dilution where clot formation occurs. Despite this limitation, in the event of a dispute as to the endotoxin content of a product, the referee test is the longer established gel-clot method. The details of the tests may be found in USP <85> *Bacterial Endotoxins Assay* and Ph. Eur. 2.6.14 *Bacterial Endotoxins*. Also, the 1987 FDA *Guideline on Validation of the LAL Test as an End-Product Endotoxin Test for Human and Animal Parenteral Drugs, Biological Products, and Medical Devices* contains details on the assay validation requirements. In general, a kinetic turbidimetric or chromogenic method would be preferred to the gel-clot method to fully quantify the bacterial endotoxin and remove the subjectivity of the gel-clot method using dilution one-half, one-quarter, or one-tenth of the maximum valid dilution. As the incubation period is one hour or possibly less, endotoxin screening has a high potential as a PAT application.

Other Testing

Other tests conducted during parenteral manufacturing may not be compendial. A bioburden evaluation of a drug substance, excipient, in-process material, presterile bulk solutions, packaging component, or nonsterile drug product is a noncompendial procedure to evaluate the number and type of microorganisms per unit weight, item, or unit surface area of the material. Typically in product development, a bioburden evaluation is a non–Good Manufacturing Practice (non-GMP) screening test that may not be fully validated or have regulatory status undertaken as part of a risk assessment during formulation and manufacturing process development. For example, as part of sterilization process development, the numbers, cellular morphology, cell size, staining reactions, and spore-forming capabilities of the predominant microbial population associated with the material would be determined to establish the appropriate sterilization parameters for sterile filtration, steam sterilization, or dry heat sterilization. When bioburden testing is used in routine production, it would be considered a GMP test and would be fully validated and included in regulatory filings.

For aseptically filled injectable products, emphasis would be given to the numbers and size of the microorganisms in a presterile bulk solution and the size retention, bulk solution volume, and filtration area subject to sterile filtration. With moist or dry heat sterilization, the numbers of spores and their relative resistance, that is, D-value, to the sterilization process would be considered.

With presterile bulk solutions, the bioburden requirements would be more conditional on the bulk volume, nominal pore size, and the filter size than the nature of the product. The 2004 FDA Aseptic Processing Guidance document overemphasizes toxicogenic materials, especially bacterial endotoxins, derived from the presterile filtration bioburden. The rating of a sterilizing filter is the retention of 10^7 colony-forming units of the challenge organism *Brevundimonas diminuta* per square centimeter of filter surface. It should be noted that the current EU guidelines for presterile bulk solutions are 10 cfu/100 mL, and tandem sterilizing filters are typically employed (2).

WHAT ARE RAPID MICROBIOLOGICAL METHODS?

A RMM is an alternate microbiological test that is completed in shorter time than the classical tests that depend on incubation for microbial growth to detect microorganisms as either colonies on a plate or turbidity in a broth. It may involve reducing the incubation time for plate count by at least half, processing a sample to obtain a result in two to three hours or a direct analytical method. The latter two approaches are typically not growth-based, hence move toward real-time analysis.

As pharmaceutical microbiologists, our primary objectives are to determine which microorganisms, if any, are in our pharmaceutical ingredients, intermediates, plant environment, or drug products; if present, how many microorganisms and what microorganisms they are and their potential impact, to help the quality unit make decisions to proceed with manufacturing and release product to the market. The test methods are classified as detection, screening, enumeration, and identification (3). Examples from the compendial microbial tests are sterility testing (detection/qualitative), absence of specified microorganisms (screening/qualitative), and microbial count (enumeration/quantitative). In addition, there is the noncompendial microbial identification (identification/qualitative).

The classification systems for rapid methods proposed in the 2002 PDA (Parenteral Drug Association) Technical Report No. 33 are based on how the technology works, for example, growth of microorganisms, viability of microorganisms, presence/absence of cellular components or artifacts, nucleic acid methods, traditional methods combined with computer-aided imaging, and combination methods (4,5). Similar, but slightly different, classifications may be found in compendial chapters discussing the validation of alternative microbiological test methods (6,7).

Growth-Based Technologies
These methods are based on measurement of biochemical or physiological parameters other than turbidity or colony formation, used in classical methods that reflect the growth of the microorganisms. Examples include ATP bioluminescence, colorimetric detection of carbon dioxide production and measurement of change in head-space pressure, impedance, advanced imaging, and biochemical assays.

Viability-Based Technologies
These types of technologies do not require growth of microorganisms for detection. Differing methods, including vital staining and fluorogenic substrates, are used to determine if the cell is viable or nonviable, and, if viable cells are detected, they can be enumerated. Examples of this technology include solid-phase cytometry and flow fluorescence cytometry.

Cellular Component or Artifact-Based Technologies
These technologies look for a specific cellular component or artifact within the cell for detection and/or microbial identification. Examples include fatty acid profiles, matrix-assisted desorption ionized—time of flight (MALDI-TOF) mass spectrometry, enzyme-linked immunosorbent assay (ELISA), fluorescent probe detection and bacterial endotoxin LAL test.

Nucleic Acid–Based Technologies
These technologies use nucleic acid methods as the basis of operation for detection, enumeration, and/or identification. Examples include DNA probes, ribotyping polymerase chain reaction (PCR), and ribosomal DNA-based sequencing.

A SURVEY OF RAPID MICROBIOLOGICAL METHODS
In most cases, RMMs may be divided into classes on the basis of their principle of detection. In this survey of RMMs, a ranking is made on the basis of successful implementation in the pharmaceutical industry (Tables 1 and 2). Note: This is provided as useful information to the reader and is not intended to be an endorsement from the authors of this chapter. Other systems may be available or become available that are not included in the table.

Table 1 Some Representative RMM Frequently Implemented in the Pharmaceutical Industry

System	Supplier	Technology	Major application
ChemScan RDI	Chemunex, Princeton, New Jersey, U.S.	Solid-phase LASER fluorescence scanning microscopy	AVC
MicroPRO (RBD 3000)	AATI, Ames, Iowa, U.S.	Fluorescence flow cytometry	AVC, P/A
RapiScreen/AkuScreen	Celsis, Chicago, Illinois, U.S.	ATP bioluminescence	P/A
BacT/ALERT	bioMerieux, Durham, North Carolina, U.S.	CO_2 colorimetric detection	P/A
Pallchek	Pall Corp. East Hills, New York, U.S.	Membrane filtration ATP bioluminescence	P/A
BACTEC 9000	BD, Corkyville, Delaware, U.S.	CO_2 detection fluorescence	P/A
Endosafe PTS	Charles River Laboratories, Wilmington, Massachusetts, U.S.	Handheld chromogenic LAL endotoxin assay	Bacterial endotoxin assay

Abbreviations: AVC, aerobic viable count; P/A, presence/absence; LAL, *Limulus* amebocyte lysate.

Table 2 Some Representative RMM with the Potential to Be Implemented in the Pharmaceutical Industry

System	Supplier	Technology	Major application
Q-PCR: Micro Compass	Lonza, Basel, Switzerland	RT-PCR	AVC
BacTrac 4300	Sy-Lab, Vienna, Austria	Impedance	AVC,P/A
Soleris Pathogen Detection Systems	Biosys (Centrus), Kingsport, Tennessee, U.S.	Optical biochemical	AVC
RABIT	Don Whitley Scientific, Shipley, England	Impedance	AVC
PyroSense	Lonza, Basel, Switzerland	Chromogenic LAL, recombinant Factor C based	On-line endotoxin detection
Biovigilant Air Monitoring System	Biovigilant, Tucson, Arizona, U.S.	Direct cell detection	On-line air monitoring
Growth Direct	Rapid Microbial System, Bedford, Massachusetts, U.S.	Autofluorescence Advanced imaging	AVC
Kikkoman ATP Swabs for Hygiene Testing	Kikkoman, Tokyo, Japan	ATP detection	Surface monitoring

Abbreviations: AVC, aerobic viable count; P/A, presence/absence.

It can be concluded that some of the most successful RMMs are the ChemScan, AkuScreen, and BacT/ALERT systems. More companies use these RMMs for in-process controls than product release. The latter is often product dictated. Rate of success of implementation is determined by the ability to focus and reserve manpower on the qualification and validation work.

WHAT IS A REAL-TIME MICROBIOLOGICAL METHOD?

In general, decision makers (i.e., physicians, production managers, and quality units) claim that the microbiological testing laboratories in the hospitals, food production sector, and pharmaceutical industry are the rate-limiting steps for patient treatment and product release. As microbiologists, we recognized the truth in their criticisms that microbial tests are imprecise with long incubation times. In Table 3, typical incubation times are shown for a range of microbial tests.

Microbiology laboratories count the time in days or even weeks to obtain a result. Furthermore, the results may need to be interpreted, reviewed, and approved before they can be reported. And that is not all! The time to ship the samples to the laboratory must be considered. It is a simple addition calculation: *Time to report = Time to ship the sample to the laboratory + administrative time + analysis time + incubation time + verify time + approval time + time to report the result.* Product release cycle times are protracted and are the sum of all these sequential activities. That means seven (7) items to work on to speed up the overall testing process. With RMM, in most cases, only the analysis time and incubation time is considered.

Table 3 The Incubation Requirement for Microbial Tests Used in Drug Manufacturing

Test	Incubation time
Total aerobic microbial count	3–5 day
Total yeast and mold count	5–7 day
Sterility tests	14–18 day
Absence of specified microorganisms tests	18–72 hr
Limulus amebocyte lysate endotoxin tests	1 hr
Microbial identification, phenotypic	3–5 day
Microbial identification, genotypic	1 day
Preservative efficacy tests	7–28 day
Mycoplasma test	28 day

It is important not to forget the other time-consuming factors in the analytical process when considering RMM implementation.

Other important differences that we recognize is that among in-process RMM testing for production process control, RMM testing for troubleshooting, and RMM testing for product release. All three may have different goals. Some definitions are in order before discussing the goals. The normal way we perform microbial analysis (when the sample is taken to the microbiology laboratory) is called *off-line testing*, if an analysis takes place near the production line but the sample is taken out of the production process, it is called *at-line testing*, and the last one is *in-line testing*, where there is a continuous analysis ongoing in the production process.

Conventional microbial testing, in most cases, is *off-line testing* with a few cases *at-line testing* (depending on the manufacturing infrastructure). If we examine RMMs, they also belong to these two categories with some exceptions that have the potential to be used *in line* (see Table 2 for an overview of different RMMs).

What determines now whether a RMM can be used *off line, at line, or in line*? In most cases, it is the underlying principle of the technique. For this discussion, RMMs can be subdivided into different categories on the basis of their detection principle: (*i*) detection of early growth, (*ii*) viability-based testing, and (*iii*) detection of microbial cell components. RMMs based on the early detection of growth principle are the slowest; the other two will be faster depending on the kind of application. Some examples: Detection of CO_2 production is a growth-depended technique that may be used for sterility testing. This application is unlikely to be an *in-line* application because of the aseptic handling that is inherent to the sterility test. In best case, it could be an *at-line* application. Detection via flow cytometry has a viability-based detection principle. Although it is not on the market, we can imagine that an *in-line* application could be possible to detect and count microorganisms via a laser detection principle. In fact, there are some techniques available that are potential *in-line* detection systems based on viability cell detection (Biovigilant, Tucson, Arizona, U.S.). The last category: Detection of cell component has many applications: detection of DNA, fatty acids, ATP, etc. In most cases, it requires a sample preparation that automatically converts it to an *off-line* application. There are several examples of *at-line* detection of bacterial endotoxin that may be used in parenteral manufacturing. They are the Endosafe PTS (*at line*), Charles River Laboratories, Wilmington, Massachusetts, U.S., and the PyroSense System (*on line*), Lonza, Basel, Switzerland.

Is the conclusion that the only real-time RMM is a system that is based on viability cell detection in an *in-line* PAT application? (time to report = real-time result) In principle, the answer is yes. However, in most cases, it is not possible to use the viability cell detection principle *in line*. What is the best possible option for the production departments and microbiology laboratories that serve them? The most practical option would be *at-line* testing with a viability-based cell detection principle (time to report < 30 minutes). However, because the viability-based cell detection systems have the technical limitations of a lack of sensitivity (limit of detection/quantification) and specificity (differentiating between cells and particulates), we end up with an *at-line* testing option of detection of early growth/cell component to eliminate ambiguity (time to report 24–48 hours).

It must be emphasized that with RMMs, the objective of the testing determines what kind of system is needed. For RMM testing for product release for the market, an *off-line* testing system is the right choice because there is no need for testing at the production floor. RMM testing for troubleshooting, in contrast to product release, can be both *at-line* testing and *off-line* testing. RMM testing for in-process testing would be preferably done *at line*. With the latter, the difficulty and workability of a test method determines the *at-line* or *off-line* application of a test.

THE APPLICATION OF RMM TO ASEPTIC PROCESSING

Aseptic processing may be divided into: (*i*) aseptic bulk processing most often employed with biologics and (*ii*) aseptic filling and lyophilization with both biologics and small molecules. On the basis of a risk assessment, critical control points can be established and, if necessary, monitored to minimize the risk of microbial contamination and loss of environmental control (8). This monitoring would be more effective if conducted in real time to provide the opportunity to take corrective action to reduce the possibility of contamination.

The following microbial tests may be used during in-process monitoring:

- Microbial limits and bacterial endotoxin testing of incoming pharmaceutical ingredients and packaging components
- Microbial counts and bacterial endotoxin testing of water for pharmaceutical use, buffers, and other intermediates
- Presterile filtration bioburden monitoring
- Biological indicator monitoring
- Sterility testing of sterile bulk drug substances
- Microbial monitoring of air, surfaces, and personnel in clean rooms

Bacterial Endotoxin Testing

As pointed out earlier, with endotoxin monitoring, two major innovations are notable. They are handheld bacterial endotoxin monitoring units (Endosafe) that are used by manufacturing personnel to test water for injection points of use immediately prior to delivering ingredient water and at-line monitoring systems (PyroSense) that continuously monitor endotoxin levels in a water-for-injection loop at preset time intervals. These instruments can mitigate risk of using endotoxin-contaminated water.

Water Testing

Microbial counts are used to monitor pharmaceutical water systems for alert and action levels to identify possible out-of-trend conditions that require corrective action. The monitoring can identify potential point-of-use, loop, or entire water system problems. As the European requirements specify the use of membrane filtration with R2A agar incubated at 30 to 35°C for at least five days, excursions are identified long after the ingredient water has been used. RMMs that have been used for monitoring water systems include the Milliflex Rapid System (Millipore Corp, Bedford, Massachusetts, U.S.) based on membrane filtration, ATP bio-luminescence and advanced imaging, the Scan RDI system (AES-Chemunex, Princeton, New Jersey, U.S.) based on membrane filtration, a fluorogenic substrate and solid phase LASER scanning microscopy, and the MicroPro System (AATI, Ames, Iowa, U.S.) based on vital stain flow cytometry. These systems may be used to obtain microbial counts within the order of 18 hours, 3 hours, and 30 minutes, respectively. Of these technologies, only the flow cytometry system meets the definition of real-time, at-line testing suitable for a PAT application, although the method may be too insensitive (level of quantification on the order of 100 bacterial cells per mL) for many applications that depend on enumeration and not just screening for gross contamination.

Bioburden and Sterility Testing

For aseptically filled injectable products, emphasis would be given to the numbers and size of the microorganisms in a presterile bulk solution, the volume of bulk solution to be filtered, and the size retention and filtration area of the sterilizing filter. The rating of a sterilizing filter is the retention of 10^7 colony-forming units of the challenge organism *B. diminuta* per square centimeter of filter surface. As noted earlier, the current EU guidelines for presterile bulk solutions are 10 cfu/100 mL, and tandem sterilizing filters are typically employed in Europe. With tandem sterilizing filters, monitoring the bioburden of the bulk solution challenging the second filter may be eliminated. To demonstrate that the bulk solution meets this requirement, a 100-mL sample would be tested using a membrane filtration method. Given the stringent requirement, RMMs must have a limit of detection and quantification commensurate with the 10 cfu/100 mL limit as well as a rapid turnaround time. This severely limits the options available for bioburden monitoring.

A possible option is to use a RMM as a presence/absence test for water for injection, low-pyrogen purified water, and in-process material to screen out samples that contain no microorganisms where processing would continue and concentrate on additional enumeration of those sample that contain microorganisms.

Sterility testing of sterile bulk drug substances and sterile bulk solution prior to aseptic filling is typically conducted using a 10-mL sample inoculated into broth and incubated for at least 14 days. With sterile drug substances that are being stored for future use, there is no time constraint for sterility testing unless there is a need to reprocess the drug substance, to prevent product loss, if it is found to be not sterile. Sterile bulk sterility tests are legal requirements for biologics marketed in the United States, using the tests according to 21 CFR 610.13.

The MicroCompass™ Detection system (Lonza) based on detection of universal sequences of RNA using a one-step real-time reverse transcriptase PCR assay and MGB™ Eclipse probe technology is a promising new technology. Universal sequences detected are based on ribosomal 16S rRNA (bacteria) and 18S rRNA (yeast and molds). The sensitivity is 50 fg of RNA or as little as 100 cfu. This technology has a detection limit that has sensitivity on the edge of bioburden limit.

Environmental Monitoring

Microbial monitoring of air, surfaces, and personnel in clean rooms is conducted during each manufacturing shift. The results are delayed for five to seven days due to the incubation of the microbiological culture media. As environmental monitoring is by far the largest microbial testing in an aseptic filling facility, the automation of the sampling, incubation, and reading of plates would increase the efficiency and timeliness of the monitoring. A technology that will achieve this goal is the Growth Direct System (Rapid Microbial Systems, Bedford, Massachusetts, U.S.) based on the early detection of microcolonies on plates using advanced imaging.

A technology that will achieve real-time environmental monitoring is the Biovigilant Air Monitoring System that is capable of counting both viable and nonviable particles in a clean room setting. This may be used as a PAT application detecting high-efficiency particular air (HEPA) filter failures, isolator system leaks, human interventions generating airborne microorganisms, and the ingress of microorganisms from supporting areas that would enable immediate corrective action such as line clearance, changes in clean room behavior, and even aborting aseptic filling operations.

THE APPLICATION OF RMM TO BIOPHARMACEUTICAL UPSTREAM AND DOWNSTREAM PROCESSING

In the bioprocessing, microbiological control plays an important role. The definition of the bioprocessing is important. Bioprocessing is the manufacture of therapeutic proteins using mammalian, bacterial, yeast, or other living (plants, insects) cells. This process can be divided into two parts: (*i*) upstream processing, in which the cell culture step takes place, and (*ii*) downstream processing, where the protein is recovered and purified using a range of biochemical purification techniques, especially large-scale column chromatography.

The scale of the bioprocess has increased in the last 10 years. It started with small-scale culture <10 L but increased to larger volumes of the order of 15,000 L. The challenge to prevent contamination of those giant fermentors is huge. Financial risks are high (50 euro/L medium, which means that only the costs of one contaminated fermentor can be of the order of 750K euro).

Looking at the downstream processing, we see the same kind of evolution in scale. It started with small columns and currently large columns, and their associated resins are used that are expensive to maintain and difficult to replace once contaminated.

The golden rule in bioprocess industry is the following:

1. Prevent contamination from input materials and equipment.
2. Detect a contamination as fast as possible.
3. Monitor your process on critical control points.
4. Take corrective action as soon as possible to isolate the incidence and find the root cause analysis.

Sterile media and equipment is achieved using validated sterilization processes and released by the use of a validated rapid microbial method. To be useful, RMMs must generate real-time results within the processing area and not a microbiology laboratory.

If the decision is taken to implement RMM in bioprocessing operations, a series of steps have to be taken to prove the PAT concept. Most important is the first step: The selection of "the most valuable sample," or in risk analysis terminology, the critical control point. These are the key samples that mark a critical step in the process. For example, before the inocula are transferred from a smaller to the next larger fermentor, it is wise to take a sample before the processing reaches the scale of 15,000 L. It goes without saying that all the input materials (media, buffers, cells, compressed air, etc.) are critical samples. If a contamination occurs, RMMs are very useful instruments to troubleshoot the process. The first 24 hours after a contamination occurs is vital. The longer it takes to collect and analyze data, the more difficult it will be actually to find the root cause of the contamination and take corrective action. An important tool for root cause analysis is also a rapid identification technology. The identity of a microbial contaminant can help to find the root cause. Rapid identification, that is, within one day can be very useful. Automatically, a genotyping based technology will be the method of choice, for example, 16s rRNA sequencing, due to its rapidity and accuracy.

THE ROLE OF RMM IN ENVIRONMENTAL MONITORING AND CONTROL

What is the role of environmental microbiological monitoring? In general, monitoring is performed to get insight into the microbiological quality of the manufacturing environment. Depending on the classification of the production environment, critical locations are selected and are sampled by contact plates, settle plates, or active air monitoring. Monitoring can be divided into monitoring of surfaces, air, and personnel. The specifications of the monitored places depend also on the classification of the area and the criticality of the operation. Strict limits are used in a grade A area (ISO 5) (<1 cfu/settle plate), whereas grade B (ISO 6–8) or lower classified areas have less stringent limits. As incubation times are long for monitoring media (3–5 days), the results represent the past history of that sampled area and not the current status. That is widely recognized in the industry; hence, we follow the trend of the microbiological cleanness with respect to sampling times. As soon as an adverse trend is detected in the microbiological quality of a sampling location, corrective action is taken such as additional disinfection, retraining of the personnel, or screening for changes in the environment. Immediate action to an out of limit in monitoring in general is difficult because of the time lag in the actual monitoring action and the time the result is known.

The results that are obtained with the current monitoring techniques give, as expected, a relative value. Monitoring efficiency depends on the type of surface, the contact time, the type of media, and incubation time. This also adds up to the relative value of environmental monitoring, and stresses the importance of performing trend surveys to assure control of the microbiological quality of the environment. What is then the role of RMMs in environmental monitoring? The conventional methods give a good insight into the microbiological quality of the environment; however, they have the disadvantage that manufacturing errors, for example, a wrong disinfection procedure, are detected at a later point or not even detected at all. That could be the benefit of RMM in environmental monitoring. A timely corrective action can be performed and the risk of production in a dirty environment is diminished. RMMs contribute to the validated state of the production process. The link to the actual batch of product that is being produced is difficult to make with environmental monitoring. If production takes place in a microbiologically dirty environment, the chance of getting a contaminated product is higher. If RMM is used, it may be easier to link the actual microbiological measurement to the microbiological quality of the product. Parametric release could be easier using these RMM technologies.

At this moment, there is no definitive RMM for environmental monitoring available which gives results the same day. Direct cell detection by ChemScan/ScanRDI technology was tested by some companies for air monitoring but is not a widespread application because of the low throughput and cost in testing with this technology. ATP measurement could be the method of choice, as instrumentation is available that can process many samples and the technology has been successfully used for hygiene monitoring in the food industry. However, the sensitivity is insufficient to measure low microbial counts on the very clean surfaces that are common in pharmaceutical production. The ultimate RMM for environmental monitoring should give results within 30 minutes and is quantitative and very easy to operate in a clean room environment.

INDUSTRY, REGULATORY, AND COMPENDIAL GUIDELINES FOR RMM

Since May 2000, when the PDA Technical Report No. 33 was published, a number of regulatory and compendial documents have been issued that were strongly influenced by the technical report to address the selection, purchase, implementation, and regulatory submission of alternate microbiological methods including RMMs. They include the following.

PDA Technical Report No. 33

The PDA was the first organization to develop guidance for the evaluation, implementation, and validation of RMMs (4). Guidance information was published as Technical Report No. 33. This document was developed by a committee of individuals from industry, regulatory agencies, compendial groups, and instrument vendors and chaired by one of the authors of this chapter. This guidance provided definitions in microbiological terms for validation criteria similar to the information in USP <1225> for chemistry methods.

USP Informational Chapter <1223> on Validation of Alternative Microbiological Methods

The USP Information Chapter <1223> defined the validation criteria to be used for RMMs, along with definitions of these criteria in terms of microbiology, in contrast to chemistry as found in USP <1225> (6). The proposal also identifies how to determine which criteria are applicable to different technologies, on the basis of the type of testing being performed.

GMPs for the 21st Century

The FDA initiated a program to modernize requirements for pharmaceutical manufacturing and quality. This modernization included encouraging early adoption of new technologies, facilitation of industry application of modern quality management technologies, encouraging implementation of risk-based approaches in critical areas, ensuring that policies for review of a submission, compliance, and facility inspection are based on state-of-the-art technologies, and enhancing the consistency and coordination of FDA regulatory programs. This resulted in an initiative titled "Pharmaceutical cGMPs for the 21st Century—A Risk-Based Approach" in 2004 (9).

FDA Guidance on Aseptic Processing 2004

In 2004, FDA published an updated guidance document on aseptic processing of pharmaceutical products. It includes a provision for the use of alternative microbiological test methods. This guideline was titled "Guidance for Industry Sterile Drug Products Produced by Aseptic Processing—Current Good Manufacturing Practice" (10).

Ph. Eur. Chapter on RMM

In 2006, the Ph. Eur. published 5.1.6. *Alternative Methods for Control of Microbiological Quality* (7). This chapter provided an overview of some RMMs available and potentially applicable to pharmaceutical processes, and how they may be used for microbiological control of products and processes. It also provides guidance on how to choose and validate an appropriate method using the ATP bioluminescence technology as an example.

THE SELECTION, DEVELOPMENT, VALIDATION, AND IMPLEMENTATION OF AN RMM FOR PROCESS ANALYTICAL TECHNOLOGY APPLICATIONS

The implementation of an RMM in the production area is a considerable challenge, although it is becoming easier compared with the situations five years ago. This process can be divided in different steps to be taken, which are important to follow to assure a successful implementation. The goal of implementation of each rapid method can be different (like earlier mentioned). A reduction of cycle time is a common goal. In this case, the testing will be conducted on the end product of the production process. Another goal is risk mitigation for microbial contamination in the production process (preventative) and troubleshooting failures to determine the root cause (reactive). In this case, the RMM is assurance against microbial contamination and will safeguard the production process. The following steps should be taken:

1. Discuss in detail with the manufacturing the details of the production process and select the most valuable sample or critical control points.
2. Select the most suitable detection method (growth based, direct cell detection, or detection of cell components) that is compatible with the nature of the sample, the expected contamination, and the sensitivity to be achieved.
3. Select the instrumentation that fits the best for the sample and the technology.
4. Select an equipment supplier.
5. Perform pilot or proof-of-concept testing to prove that the instrumentation fits the specific application. Perform method suitability testing for a range of test materials.

6. Purchase the instrumentation and perform the equipment validation, that is, Installation Qualification (IQ), Operational Qualification (OQ), and Performance Qualification (PQ) using vendor supplied document whenever possible.
7. Perform method suitability testing at least on three independent batches.
8. Assemble all the GMP documentation (Standard Operating Procedures, calibration programs, regulatory submissions, and change controls).
9. Implement in routine testing.

REGULATORY APPROVAL OF RMM

With the FDA, three avenues are possible for the approval of RMMs. A New Drug Application (NDA) submission for an RMM may be used with a new product and an NDA supplement for existing product, filing a comparability protocol, or using the PAT initiative pathway. The FDA prefers the comparability protocol approach (11) as it accommodates the fact that the FDA approvals are typically drug product specific, and a comparability protocol gives the FDA the opportunity to review your method validation plan prior to executing it for a range of drug products. In general, it is advisable to discuss the application and validation strategy with the regulatory agency in advance.

The most important RMM validation issue is equivalence to the current method. Other standard validation issues include accuracy, sensitivity, precision, and linearity of response. Microbiologist should use supplier-generated validation protocols whenever possible. IQ is best timed with the delivery of the equipment to your laboratory. OQ will demonstrate the functionality of the equipment while PQ will be directly related to your application and products. Remember it is acceptable to include supplier-generated reports and publication from peer-reviewed journals within your validation report so you may avoid repeating the generation of preexisting data. Validation protocols and reports must include the validation rationale, acceptance criteria, and deviations from protocol or acceptance criteria, and the documents must be reviewed and approved by the quality unit.

THE FUTURE OF RMM IN PARENTERAL MEDICATION MANUFACTURING

What is the future of RMMs in parenteral drug manufacturing? The major trends are (*i*) the move away from traditional growth-based methods to RMMs on the basis of vital cell staining, ATP, or nucleic acid concentration, (*ii*) the move from the microbiology laboratory to the production floor as the site of the microbial testing, and (*iii*) the use of RMM to PAT applications by the real-time testing in-process samples.

REFERENCES

1. FDA. Guidance for Industry—PAT A Framework for Innovative Pharmaceutical Development, Manufacture and Quality Assurance. Rockville, MD: Government Printing Office, 2004.
2. CPMP/QWP/486/95 Note for Guidance on Manufacture of the Finished Dosage Form, April 1996.
3. Cundell AM. Microbial testing in support of aseptic processing. Pharm Technol 2004; 56–66.
4. Parenteral Drug Association. Technical Report 33, evaluation, validation and implementation of new microbiological testing methods. J Pharm Sci Technol 2000; 54(3):35.
5. Moldenhauer J. Rapid microbiological methods and the PAT initiative. Biopharm Int 2005; 18(12):31–46.
6. USP <1223> Validation of Alternative Microbiological Methods. US Pharmacopeial Convention, Rockville, MD.
7. European Pharmacopoeia 5.1.6. Alternative Methods for Control of Microbiological Quality. European Directorate for the Quality of Medicines, Strasbourg, France.
8. Cundell AM. Risk-Based Approach to Pharmaceutical Microbiology. In: Miller MJ, ed. Encyclopedia of Rapid Microbiological Methods. Bethesda, MD: Davis Horwood/PDA, 2005.
9. FDA. Pharmaceutical cGMPs for the 21st Century—A Risk-Based Approach, Final Report—Fall 2004. Department of Health and Human Services, U.S. Food and Drug Administration, 2004.
10. FDA. Guidance for Industry Sterile Drug Products Produced by Aseptic Processing—Current Good Manufacturing Practice, Department of Health and Human Services, U.S. Food and Drug Administration, 2004.
11. FDA. Draft Guidance for Industry—Comparability Protocols—Chemistry. Manufacturing and Controls Information, Department of Health and Human Services, U.S. Food and Drug Administration, 2003.

12 | Quality assurance
Michael Gorman

INTRODUCTION

Quality assurance is particularly important in aseptic manufacturing. This type of manufacturing must strictly follow carefully established and validated methods of preparation and procedures. Quality needs to be built into the operations and process and not be placed on just the end-product testing. Quality must be applied to facilities, preparation of materials, and to all aspects of processing.

Key quality systems and key aspects of those quality systems as they apply to aseptic processing are discussed in this chapter. In general, all operating conditions and treatment of materials should be such as to prevent microbial contamination and follow a proven control strategy (1). The output from the control systems of the operating conditions and treatment of materials should then be assessed as part of the lot disposition process (Fig. 1).

To maintain the sterility of the components and the product during aseptic processing, the control strategy needs to include the following: environment, personnel, critical surfaces, container/closure sterilization and transfer procedures, maximum holding period of the product before filling into the final container, and sterilizing processes (1).

In general, two basic areas can be defined—physical assets and process systems. These two basic areas need systems to collect and track information to ensure sterility assurance. The physical assets include facilities, equipment, and utilities (e.g., air handling systems, compressed air, nitrogen, steam generator, and water). The process systems include key points to the quality systems like training, material management, calibrations, validations, processes, batch records, investigations, quality control laboratory, environmental monitoring, cleaning equipment/facilities, and quality information management. The role of quality assurance in product development for an aspect process will be briefly presented.

The output from the control systems for routine monitoring of the physical assets (not in use and during production), coupled with the output from the process systems associated with the production batch, should be included in the quality information management system for assessing status of each lot produced by aseptic processing.

PHYSICAL ASSETS

The physical assets should be designed to support the specific type of production and to reduce the chance for contamination of the product. Many aspects of the physical assets have been discussed in detail throughout other chapters. This section will focus on quality aspects of physical assets related to design preferences and control. Quality plays a key role with physical assets by aligning with operations and provides guidance for design, systems for monitoring, change control, and qualifying. Also, operations and quality need to work closely together to resolve investigations related to the physical assets.

Facilities

For aseptic processing, the facility layout should control the flow of materials and personnel with respect to environment quality needed for the stage of processing. For example, the facility should have a cascade of room classifications from less to more as the process flows toward aseptic requirements. Air locks should be used to separately transfer materials and the flow of personnel into the critical aseptic processing areas to prevent the chance of contamination (Fig. 2).

The construction materials for the production areas should be chosen for durability to allow for frequent cleaning/sanitizing. In clean areas, all exposed surfaces should be smooth, impervious, and unbroken to minimize the shedding or accumulation of particles or microorganisms (2). The wall and room designs should not have areas to collect dust or cause difficulties for cleaning. Examples of adequate design features include seamless and

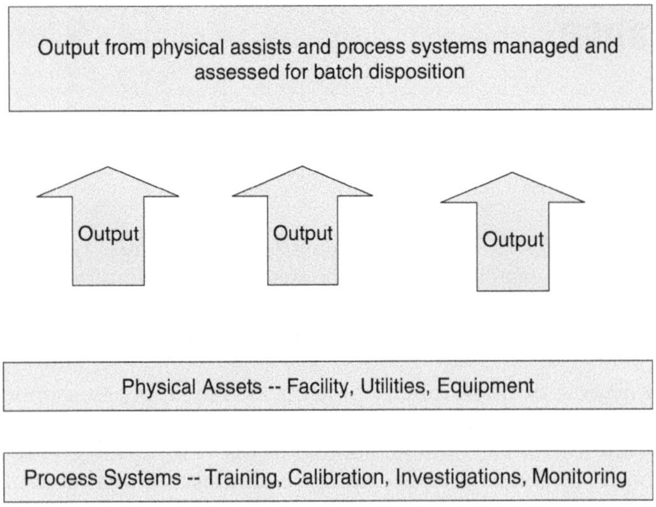

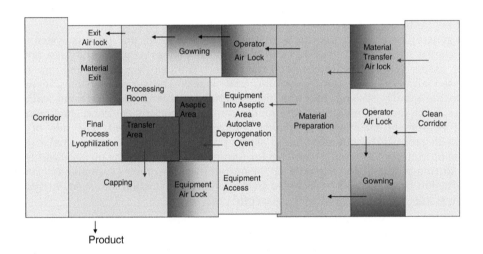

Figure 1 Overview of aseptic control.

Figure 2 Basic facility diagram.

rounded floor to wall junctions as well as readily accessible corners. Ceilings and associated HEPA filter banks should be designed to protect sterile materials from contamination (3). False ceilings should be sealed to prevent contamination from the space above them. Sinks and drains should not be located in areas used for aseptic manufacture.

In quality control microbiology, sterility testing area should have the same or better quality environment as the aseptic processing area. This is done to minimize the potential of false positives during testing.

The facility should be routinely inspected for the need of wall, floor, and ceiling repairs. These inspections should be documented and repairs preformed promptly to keep the facilities in a good state of control to prevent the chance for product contamination. In general, the facility should be inspected before each batch and thoroughly inspected and repaired at defined frequencies (e.g., every 6 months). These inspections and repairs proactively keep the facilities in good working order to prevent contamination of the product.

Clean area control parameters should be supported by microbiological and particle data obtained during qualification studies. Initial clean room qualification should include an assessment of air quality under as-built, static conditions and dynamic conditions. It is

Table 1 Air Classifications

Clean area classification (0.5 μm particles/ft^3)	ISO designation[a]	> 0.5 μm particles/m^3	Microbiological active air action levels[b] (cfu/m^3)	Microbiological settling plates action levels[b,c] (diam. 90 mm; cfu/4 hr)
100	5	3,520	1[d]	1[d]
1,000	6	35,200	7	3
10,000	7	352,000	10	5
100,000	8	3,520,000	100	50

All classifications based on data measured in the vicinity of exposed materials/articles during periods of activity.
[a]ISO 14644-1 designations provide uniform particle concentration values for clean rooms in multiple industries. An ISO 5 particle concentration is equal to Class 100 and approximately equals EU grade A.
[b]Values represent recommended levels of environmental quality. You may find it appropriate to establish alternate microbiological action levels because of the nature of the operation or method of analysis.
[c]The additional use of settling plates is optional.
[d]Samples from Class 100 (ISO 5) environments should normally yield no microbiological contaminants.

important for area qualification and classification to place most emphasis on data generated under dynamic conditions (i.e., with personnel present, equipment in place, and operations ongoing). Table 1 summarizes clean area air classifications and recommended action levels of microbiological quality (4).

The facility should be designed to meet room classifications appropriate for each stage of manufacturing. The facility is key for maintaining appropriate environmental conditions to protect the product from contamination for routine aseptic manufacturing. The facility needs to be properly maintained, monitored, and used for intended purpose.

Equipment

Aseptic processing equipment should be appropriately designed to facilitate ease of sterilization (5). Equipment should be designed to be easily assembled and disassembled, cleaned, sanitized, and/or sterilized. Fixed equipment (e.g., large mixing tanks) should be properly designed with attention to features such as accessibility to sterilizing agent, piping slope, and proper condensate removal. Additionally, the effect of equipment design on the clean room environment should be addressed. For example, horizontal surfaces or ledges that accumulate particles should be avoided. Equipment should not obstruct airflow and, in critical areas, its design should not disturb unidirectional airflow (3).

In the aseptic processing area, smoke studies should be used to verify unidirectional airflow. Videotaping smoke studies provide thorough evidence showing air flow patterns. If changes to equipment or facilities are needed, air flow patterns need to be carefully assessed and recorded.

Equipment shall be constructed so that surfaces that contact components, in-process materials, or drug products shall not be reactive, additive, or absorptive so as to alter the safety, identity, strength, quality, or purity of the drug product. Ideally, product contact surfaces should be disposable or made of materials that are product dedicated (e.g., 316 stainless steel). In multiproduct facility, key product contact surfaces should be dedicated to a product. For instance, filling needles should either be disposable or dedicated to a product. This is done to prevent the chance of cross-contamination.

Adequate cleaning, drying, and storage of equipment will aid in controlling bioburden and prevent contribution of endotoxin load. If adequate procedures are not used, endotoxins can be introduced into the process by the equipment (3). Records should be kept showing cleaning schedules and the performance of the cleaning procedures.

Equipment surfaces that contact sterilized drug product, or its sterilized containers or closures, must be sterile so as not to alter purity of the drug (5). Where reasonable contamination potential exists, surfaces that are in the vicinity of the sterile product should also be clean and free of microorganisms. The validation of cleaning procedures is important to show removal of microorganisms, processing materials, and cleaning agents.

Monitoring devices should be used whenever feasible. Equipment monitoring provides proof that the equipment functioned properly during use. The output from the monitoring devices should be recorded to provide assurance for the proper equipment performance during manufacturing.

Records need to be kept for equipment showing routine and nonroutine maintenance, usage, and calibration of monitoring devices. If equipment does not operate within intended limits, an investigation should be preformed.

Utilities

Utilities for an aseptic processing facility should be designed to prevent contamination. Utilities actually bring processing materials into contact with the product. These materials should be sterilized. For example, the compressed air system may introduce air into a lyophilizer before the product is stoppered. Thus, the air becomes the headspace of the product vial. Quality aspects will be illustrated for the following utilities: air systems (HVAC), compressed air system, nitrogen gas supply, water, and steam generator.

HVAC

The main purpose for the HVAC system is to provide clean air into the processing areas. The HVAC system needs to be designed to deliver particulate- and microbial-free air. Most systems contain prefilters with >95% efficiency filters with terminal or final filters >99.9% efficiency (HEPA). In the aseptic areas, HVAC systems should deliver single-pass air. Therefore, the system should not recirculate air and the air supply should consist of 100% fresh makeup air. This is done to prevent cross-contamination.

The HVAC system should be capable of keeping the processing areas very cool for operator comfort. Typically the environment should be around <65°F and <60% RH. The main reason for this type of temperature and humidity control is to keep the operators, who are generally double gowned, comfortable and free from perspiration to decrease shedding.

Monitoring systems should target continuous monitoring for temperature, humidity and pressure differentials across filters and pressure differentials between rooms. The continuous monitoring should have appropriate ranges. If conditions fall outside of set ranges, an investigation should be triggered with an assessment to the impact on the product.

Compressed Air System

Like the HVAC system, the compressed air system should be designed to provide essentially a source for sterile air. The air system should be monitored at frequencies to show that air is delivered free from contaminates like microorganisms and hydrocarbons. At use-points that come into contact with the product terminal, sterile filters should be used. These filters should be tested for integrity. Records should be maintained for the proper routine performance and lot performance of the air system.

Nitrogen Gas

Nitrogen gas is often used during the production process to control equipment and sometimes used to produce an environment free from oxygen. The nitrogen gas supply should be tested for identity and moisture. Often a plant may use a bulk liquid nitrogen tank coupled with evaporators to supply nitrogen gas. In these systems, each charge of the bulk nitrogen tank should be tested at a minimum for identity. If by accident the wrong liquid was loaded into the bulk tank, this could cause major damage to the nitrogen system and contaminate the production facility. The nitrogen system needs to be routinely monitored for performance. Routine and nonroutine maintenance need to be documented. Additionally, on key locations throughout the nitrogen system, point-of-use sterilizing filters should be used and integrity tested.

Water

Other chapters outline water systems in detail for aseptic processing. From the quality perspective, the water system should be monitored before use to ensure that the appropriate quality is used during processing. Ideally, the water system should be continuously monitored for key parameters like pressure, temperature, conductivity, and total organic carbon. Additionally, the water system should be sampled throughout key points in the system and points of use. Records need to be kept for routine and nonroutine maintenance. To ensure the proper control of the water system, the monitoring data should be analyzed by trending, and reviewed routinely. During the course of monitoring a water system, alert and action limits need to be established. If a limit is exceeded, an appropriate action/investigation should be preformed.

Steam Generator

Steam systems should be supplied with clean water that is free from hydrocarbons, salts, and microorganisms, ideally, water-for-injection quality. The steam quality needs to be routinely tested throughout the distribution system and at key points of use. Like the other utility systems, records should be kept for the maintenance and performance of the steam generator.

QUALITY SYSTEM

Quality assurance needs to remain proactive in aseptic processing by providing guidance to operations for developing systems. A proactive quality system for aseptic processing has a rigorous monitoring, evaluation, and response/corrective action component. Proactive quality needs the right systems in place to react before major problems happen. The monitoring aspects of the quality system should be evaluated for trends and reviewed frequently by quality and operations management. Quality should evaluate and assess the output from the physical assets and quality systems for each batch manufactured. Components of the process system that are discussed in this chapter are the following: training, material management, calibration, validation, process, batch records, investigations, quality control laboratories, environmental monitoring, cleaning equipment/facility, and quality information management.

Training

Each employee has a responsibility to the company to ensure records and training activities are current. All regulations have requirements for training and qualifications of personnel. For example, 21 CFR 211.25(a) states that "Each person engaged in the manufacture, processing, packing, or holding of a drug product shall have education, training, and experience, or any combination thereof, to enable that person to perform the assigned functions." (5) Training shall be in the particular operations that the employee performs and in current good manufacturing practice on an ongoing basis.

   Another point about training that extends to each employee is contained in 21 CFR 211.28(a), stating that "Personnel engaged in the manufacture, processing, packing, or holding of a drug product shall wear clean clothing appropriate for the duties they perform. Protective apparel, such as head, face, hand, and arm coverings, shall be worn as necessary to protect drug products from contamination." (5) Additionally, 21 CFR 211.28(b) states that "Personnel shall practice good sanitation and health habits." (5) These types of regulations are particularly important for aseptic manufacturing to protect the product from contamination from the employees.

   A well-designed, maintained, and operated aseptic process minimizes personnel intervention (e.g., isolator or barrier use). As operator activities increase in an aseptic processing operation, the risk to finished product sterility also increases. To ensure product sterility, it is critical for operators involved in aseptic activities to use aseptic technique at all times.

   Appropriate training should be conducted before an individual is permitted to enter the aseptic manufacturing area. Fundamental training topics should include aseptic technique, clean room behavior, microbiology, hygiene, gowning, patient safety hazards posed by a nonsterile drug product, and the specific written procedures covering aseptic manufacturing area operations.

   After initial training, personnel should participate regularly in an ongoing training program. Supervisory personnel should routinely evaluate each operator's conformance to written procedures during actual operations. Similarly, the quality control unit should provide regular oversight of adherence to established, written procedures and aseptic technique during manufacturing operations. Some of the techniques aimed at maintaining appropriate levels of sterility assurance include the following:

- Contact sterile materials only with sterile instruments
- Move slowly and deliberately
- Keep the entire body out of the path of unidirectional airflow
- Approach a necessary manipulation in a manner that does not compromise sterility of the product
- Maintain proper gown control

Written procedures should adequately address circumstances under which personnel should be retrained, requalified, or reassigned to other areas. Training activities should be clearly documented in records for each employee.

Material Management
Material management needs attention with respect to aseptic processing. The main focus for material management needs to always insure that the integrity of the material delivered to the aseptic process has not been comprised. When materials are received they should be carefully inspected for the condition of the containers for damage and any possible breech of container integrity. The materials should be placed into a state of quarantine until released by quality according to specifications/procedures.

Samples for release testing need to be carefully removed under aseptic conditions to prevent any possible chance of contamination of the material during the sampling procedure. The sampling needs to be performed in an environment of the same or better classification to which the material will be charged into the process. The material needs to be delivered to the production areas in a controlled manner to prevent any possible chance of mix-up or contamination. Records need to show complete accountability, traceability, and handling of the material.

Calibrations
Calibrations should focus on monitoring devices for equipment and facilities. As previously discussed, the calibration devices should be routinely reviewed and the information recorded. Monitoring devices are integral for documenting the performance of the process in relation to sterility assurance.

Monitoring devices need to be calibrated to tolerances that allow for reliable accuracy over the monitoring range of measurement. For example, a thermocouple should not be calibrated with a tolerance of $\pm 2°C$ if the accuracy of the measurement needs to be $\pm 0.1°C$. Also, the calibration should span the range of measurement that the monitoring device will routinely record.

Records need to be kept for monitoring devices. The records need to clearly show calibration results as well as any adjustments and maintenance made to the device. Monitoring devices should be routinely verified before use in manufacturing. For example, a balance should be checked for accuracy by weighing a check weight and recording the results. If a device is found out of tolerance, corrective actions should be taken. Also, an assessment should be documented for what the impact of the out-of-tolerance device had on the facility and processes.

Validations
Other chapters have described key technical aspects about validations. From a quality perspective, validations should be done on facilities, utilities, and equipment. For aseptic manufacturing, validations need to clearly show that the item will routinely perform in a way needed to assure product integrity.

As mentioned previously, the facilities should be proven to provide an environment suitable for the specific type of manufacturing. Typical parameters for validation of facilities are temperature, relative humidity, pressure differentials, and particulate matter (viable and nonviable).

Equipment validation should thoroughly confirm that the performance is appropriate for the process/product. Standard equipment should follow the traditional validation plan of supplier information, installation qualification, operational qualification, and performance qualification. The qualification process should prove that the monitoring and control aspects of the equipment are suitable and in a state of control for the process.

Custom designed equipment should consider following a validation plan that ensures equipment is designed correctly for intended use. An approach for customized equipment is design qualification, factory acceptance testing, installation qualification, operational qualification, and performance qualification. Equipment should be routinely requalified on a routine basis defined by procedures or when significant changes are made.

Sterilizing equipment cycles should be validated to the specific load or cycle to support the process. Additionally, sterilizing cycles need to be routinely revalidated, or if a change occurs to the equipment or the utilities, revalidation should be considered.

Process

Process validation in aseptic manufacturing has two key aspects—can the process reliably manufacture product and maintain sterility. Validation should prove that following the parameters outlined in a control strategy, the process can manufacture product that has the safety, identity, strength, quality, and purity required. The reliability of the manufacturing process traditionally is shown from three validation lots.

To ensure the sterility of products sterilization, aseptic filling and closing operations must be adequately validated (5). The goal of even the most effective sterilization processes can be defeated if the sterilized elements of a product (the drug formulation, the container, and the closure) are brought together under conditions that contaminate any of those elements.

An aseptic processing operation should be validated using a microbiological growth medium in place of the product, media fill. Normally a media fill includes exposing microbiological growth medium to product contact surfaces of equipment, container closure systems, critical environments, and process manipulations to closely simulate the same exposure that the product itself will undergo during the process. The sealed containers filled with the medium are then incubated to detect microbial contamination. Results are then assessed for the potential of a unit of drug product to become contaminated during actual operations (e.g., start-up, sterile ingredient additions, aseptic connections, filling, and closing). Environmental monitoring data from the process simulation can also provide useful information for the processing line evaluation.

A media fill program should incorporate the contamination risk factors that occur on a production line and accurately assesses the state of process control. Media fill studies should closely simulate aseptic manufacturing operations incorporating, as appropriate, worst-case activities and conditions that provide a challenge to aseptic operations. Media fill programs should address applicable issues such as

- Run time
- Representative interventions, routine and nonroutine
- Lyophilization, when applicable
- Aseptic assembly of equipment
- Number of personnel and their activities
- Representative number of aseptic additions or transfers
- Shift changes, breaks, and gown changes (when applicable)
- Type of aseptic equipment disconnections/connections
- Aseptic sample collections
- Line speed and configuration
- Weight checks
- Typical environmental conditions
- Run size
- Container closure systems

A batch record should be followed for media fill studies. Additionally, documentation should be created that notes production conditions, operations, and simulated activities. A video recording can be very useful during media fills. The recording can be used as a record of the event and referred to during training exercises.

In general, a microbiological growth medium, such as soybean casein digest medium, should be used. Use of anaerobic growth media (e.g., fluid thioglycollate medium) should be considered in special circumstances when a nitrogen environment is required for the process. The media selected should be demonstrated to promote growth of gram-positive and gram-negative bacteria, yeast, and mold. The QC laboratory should determine if indicator organisms sufficiently represent production-related isolates. Environmental monitoring and sterility test isolates can be substituted (as appropriate) or added to the growth promotion challenge.

The records from the media fill study should be carefully reviewed in the same way a production batch record would be reviewed. If any aberrant result is observed, an investigation should be initiated.

Batch Records

Batch records are the basic production record. Batch records should provide clear directions to execute the process as well as be the collection point for appropriate information throughout the process. The batch record should have adequate information and verification of collected information to reliably produce the desired product. During aseptic manufacturing, output from environment, facility, equipment, and personnel should be collected. This output should be assessed and compared to proven limits.

During the production run, if any value is collected and is outside of set ranges, this aberrant value should be investigated. The investigation needs to be referenced in the batch record. Any aspect of aseptic manufacturing should be investigated and assessed for impact to product before the lot disposition decision is made.

Following the execution of the batch record, typically the record is peer reviewed by a lead operator. Once the peer review is completed, the manufacturing authority needs to review the record for completeness and accuracy. Any questions or comments should be resolved by the operators. Following the manufacturing review, quality should review the record and verify that all collected data meets the control strategy requirements.

Investigations

Quality assurance needs to approach investigations from a science and risk-management prospective. Investigations tend to be a huge learning opportunity for most operations. The focus of an investigation should be on science and risk to generate an understanding of root cause and formulate a corrective and preventive action. Basically, when an aberrant result/ trend is observed or a nonroutine event occurs, an investigation should take place to understand, learn, and make corrections.

In aseptic manufacturing, an investigation should occur when any aberrant result is obtained or unexpected event takes place, from physical assets and/or from process systems. The initial part of the investigation should assess what lots are impacted by the aberrant result and hold all lots in question until the investigation is fully understood and appropriate corrective actions are taken.

In general, investigations usually take the following steps:

- Discovery of an investigational situation
- Confirmation of the need for an investigation
- Notification of the investigation—hold product and operations
- Clearly record the cause/reason for the investigation
- Information collection
- Formation of hypothesis for why the aberrant result was obtained
- Conformational testing of hypothesis
- Validate hypothesis
- Assess impact to product
- Formulate corrective action
- Test corrective action
- Implement corrective action

The basic concept of investigation process is to follow the scientific model and learn more about the process/facility capabilities and to formulate a decision point on the initial aberrant result or unexpected event.

Quality Control Laboratories

Regulations generally state that the quality unit has the authority to approve or reject all components and materials used in processing and products produced. 21 CFR 211.22(a) states that (5)

> There shall be a quality control unit that shall have the responsibility and authority to approve or reject all components, drug product containers, closures, in-process materials, packaging material, labeling, and drug products, and the authority to review production records to assure that no errors have occurred or, if errors have occurred, that they have been fully investigated. The quality control unit shall be responsible for approving or rejecting drug products manufactured, processed, packed, or held under contract by another company.

The quality control laboratory needs to have the same level of control as the manufacturing operations. The laboratory should be able to perform testing and provide very accurate results. The laboratory systems should be able to collect, store, and handle samples without compromising the integrity of the sample or having any mix-ups. Production operational points should be applied to the operations of the quality control laboratory (Table 2).

As shown in Table 2, many of the operational concepts about manufacturing apply to the quality control laboratory. The focus may be slightly different in that operations focus is on product while the laboratory focus is on the test result. But the concepts are comparable and when working together allow for the production of a quality product.

Environmental Monitoring

In aseptic processing, one of the most important laboratory controls is the environmental monitoring program. This program provides key information on the state of control of the aseptic processing environment during operations as well as routine steady state. Environmental monitoring may be able to identify potential routes of contamination, allowing for implementation of corrections before product contamination occurs.

Evaluating the quality of air and surfaces in the clean room environment should start with a well-defined written program and scientifically sound methods. All environmental monitoring locations should be described in procedures with sufficient detail to allow for reproducible sampling of a given location surveyed. Procedures should also address elements like the following:

Table 2 Comparison of Operations with Quality Control Laboratory

	Operations	Quality control laboratory
Training	Employees need appropriate training and experience to perform assigned responsibilities	Same
Material management	Materials are handled to insure appropriate integrity is maintained and to prevent mix-ups	Samples need to ensure integrity, storage, and traceability in laboratory systems are maintained
Calibration/validation	Production equipment and monitoring devices need to show appropriate level of control	Laboratory equipment need to be treated in a way to insure reliability of results
Process	The operations to produce a sterile drug product	The activities to produce reliable test results
Records	Batch records provide directions and a collection point for all process information	Test records are kept to provide accuracy for testing
Investigations	Focus areas are performing the process and product	Focus areas are method performance and test result
Environment	Production environment needs to be clean, monitored, and kept in a way so as not to contaminate the product	The laboratory environment needs to have appropriate conditions to ensure samples can be handled without causing contamination
Equipment	Clean, maintained, calibrated/qualified	Same

- Frequency of sampling
- When the samples are taken (i.e., during or at the conclusion of operations)
- Duration of sampling
- Sample size (e.g., surface area, air volume)
- Specific sampling equipment and techniques
- Alert and action levels
- Appropriate response to deviations from alert or action levels.

The monitoring program should cover all production shifts and include air, floors, walls, and equipment surfaces, including the critical surfaces that come in contact with the product, container, and closures. Locations that present the most microbiological risk to the product need to be a key part of the program. Data needs to be collected to ensure that the microbiological quality of the critical areas shows whether or not aseptic conditions are maintained during filling and closing activities.

Environmental monitoring data needs to be analyzed looking for trends. From a practical point of view, if the data has all 0 values then a review of sampling and testing needs to occur. If the data shows more positive values in an area, a review of the cleaning procedures needs to occur. In a robust sampling and environmental monitoring program the data will show positives in a more random fashion. But, in the aseptic areas (ISO 5), the data should confirm the required conditions.

The collective output from the environmental monitoring program needs to be carefully evaluated on a routine frequency. Additionally, environmental monitoring data needs to be assessed during the routine manufacturing of one batch.

Cleaning Equipment/Facilities

Cleaning and sterilizing are important activities for aseptic manufacturing. Equipment cleaning procedures should be validated and routinely verified. Critical product contact surfaces need to be sterilized before using in the manufacturing process. Some keep points to consider from a quality perspective of an aseptic cleaning validation program are the following:

- Training of operators
- Sampling methods to account for process materials and microorganisms
- Sanitizing agent contact times need to qualify for effectiveness (e.g., do a small study on coupons of process surfaces spiked with known levels of microorganisms and hold agent for contact time to verify the absence of microorganisms)
- Equipment/material hold times before use
- Transfer and setup of equipment

Ideally, whenever possible in aseptic processing, disposable or single-use critical product contact items should be used. If disposable items cannot be used, then dedicated equipment should be used to protect the product from cross-contamination. If dedicated equipment cannot be used, then the importance of a rigorous cleaning validation and verification plan is extremely critical.

The equipment and facilities should be verified that they have been cleaned and are within the allowable hold times before use. This information should be recorded in the batch records.

Quality Information Management

The principle philosophy of quality information management begins early in product development. The combination of ICH Q8, Q9, and Q10 has provided a road map of key features of information management and how the organization should use that information (6–9). Early in product development, the design space for processing parameters needs to start being developed. As the procedure goes through the development process, refinements are made and knowledge is gained. This information needs to be captured and used to develop the design space and process control strategy.

For aseptic processing, key elements of the control strategy and process knowledge are the following:

- Process hold times
- Product contact surfaces
- Container closure assurance
- Confirmation of material handling
- Sterilization/sanitization procedures of equipment
- Equipment hold times
- Sterilization cycles
- Equipment normal operating parameters
- Confirmation of sterility assurance for the process
- Product interactions with filters and process surfaces

Once the control strategy is set, information should be collected for each batch. The information should be compared with the historical information collected. If any parameter is outside of the normal operating ranges for the process, an investigation should occur to understand why the aberrant result was obtained.

On a set frequency, the information collected according to the control strategy should be reviewed. This information needs to be evaluated for trends over a number of batches. Ranges should be assessed for applicability to quality aspects of the process. Related or repeated events should be assessed and corrective and preventive actions should be done to minimize reoccurrence.

Quality information management systems may include the following:

- Building Management System
- Laboratory Information Management Systems
- Document Information Management Systems
- Equipment Information Management Systems for calibration and validation
- Batch Records
- Deviations and Investigation
- Material Management Systems

A key aspect of the information management systems is change control. The systems need to evolve as process knowledge and systems gain more information. During the change control process, a key point is what impact will the change have to the process, as well as aseptic processing impact. Any impact to aseptic processing needs to be carefully assessed and tested to ensure the appropriate sterility assurance levels are maintained during the process.

The fundamental important point of quality information management is that information is to be collected and this information assessed for the state of control of the entire process. This is the fundamental philosophy behind quality assurance science. Each batch should be assessed against the entire information set collected, information from the physical assets and process systems outputs.

QUALITY ASSURANCE ROLE IN PRODUCT DEVELOPMENT

Quality assurance has an important role in product development. Quality needs to be able to make lot decisions on the basis of the information available about the process. In early development, with only having manufactured the development drug once or twice, a lot may not be known about the process. For aseptic processing, facility and process controls also apply and must be in place so that batch results and process observations represent the specific product/process for which little is known at the outset.

Guidance documents from health authorities have been developed helping refine approach to clinical manufacturing (Fig. 3) (10,11). Quality needs to draw upon all experience and provide input into development team about paths forward when issues occur during manufacturing. As the process is developed, quality can play a key role to the development

ICH Q8	ICH Q9	ICH Q10	Annex 13 (EU GMPs)	FDA Guidance
Pharmaceutical Development	Quality Risk Management	Pharmaceutical Quality System	Manufacture of Investigational Medicinal Products	cGMP for Phase 1 Investigational Drugs
• Parameters for Manufacturing Control (i.e., Design Space) • Enhanced knowledge of product performance	• Facilitates decision making • Proactive means to identify and control quality issues	• Facilitates continual improvement over product life cycle • Maintains a state of control	• Procedures need to be flexible as process is developed. • Protects trial subjects • Maintain batch consistency	• Facilitates the initiation of the human clinical studies • Protects trial subjects

Figure 3 Fit-for-purpose regulatory guidance documents.

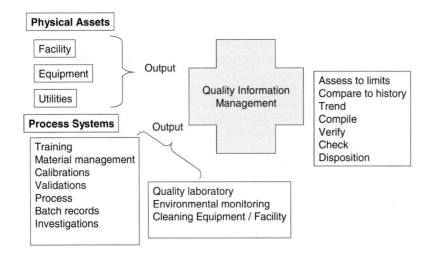

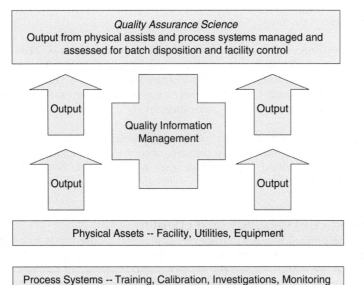

Figure 4 Information flow diagram of aseptic control strategy.

team by helping managing the information collected. This information can be used to help define the design space for the process. Once the process is close to becoming commercial, a control strategy should be prepared. The control strategy should define all monitoring and control parameters for the process.

An important aspect in product development when the product requires to be manufactured by aseptic techniques sterility assurance aspects should be the same across all phases of development. The difficulty of a development drug is the lack of process experience. If the development drug has only been made once or twice, the development team needs to use experience and education to make decisions.

CONCLUSIONS

Every person involved with parental drug manufacturing has the responsibility to assure the quality of the product produced. As stated previously, many of the technical aspects are described in detail in other chapters. The aim of this chapter was to illustrate quality aspects throughout the process. Additionally, within aseptic manufacturing, certain monitoring and information need to be collected on a routine basis to continually assess the state of control of the complete operation. The basis for assessing the state of control is to have rigorous and defined information flow processes (Fig. 4). Once the information is collected, quality assurance should have the ability to assess, evaluate, and make appropriate decisions to ensure the product has the required safety, identity, strength, quality, and purity.

Quality Assurance Science is the process of bringing all of the information together, evaluating the information, making decisions, refining systems, and applying process knowledge. This process begins in the early stages of drug development when not a lot of specific process information is known, but it is important to allow for development to progress, building the process knowledge. However, even in early development, process sterility assurance requirements should be largely the same at all stages of development and routine commercial manufacturing.

REFERENCES

1. Quality Assurance of Pharmaceuticals, A Compendium of Guidelines and Related Materials. Vol. 2. Geneva: World Health Organization, 2007.
2. Eudralex Volume 4, EU Guidelines to Good Manufacturing Practice, Medicinal Products for Human and Veterinary Use, Annex 1, Manufacture of Sterile Medicinal Products, Brussels, November 25, 2008.
3. FDA. Guidance for Industry—Sterile Drug Products Produced by Aseptic Processing—Current Good Manufacturing Practice. U.S. FDA, September 2004.
4. ISO 14644-1. Clean rooms and Associated Controlled Environments, Classification of Air Cleanliness.
5. CFR. Code of US Federal Regulations Part 21.
6. ICH. Guidance for Industry Q8 Pharmaceutical Development, ICH, May 2006.
7. ICH. Guidance for Industry Q8 (R1) Pharmaceutical Development Revision 1, Draft, November 2007.
8. ICH. Guidance for Industry, Q9 Quality Risk Management, ICH, June 2006.
9. ICH. Guidance for Industry, Q10 Pharmaceutical Quality System, Draft, May 2007.
10. Good Manufacturing Practices, Annex 13, Manufacture of investigational medicinal products. Vol. 4. July 2003.
11. FDA. Guidance for Industry, CGMP for Phase 1 Investigational Drugs. U.S. FDA, July 2008.

13 | Application of Quality by Design in CMC[a] development

Roger Nosal, Thomas Garcia, Vince McCurdy, Amit Banerjee, Carol F. Kirchhoff, and Satish K. Singh

Abstract: This chapter will summarize conceptual development of the Quality-by-Design (QbD) approach from the platform principles outlined in the ICH, Q8R, Q9, and Q10 guidelines. The chapter will characterize contemporary definitions of the important elements of QbD and provide examples of the application of QbD in the technical development and management of pharmaceutical products throughout their life cycle in alignment with regulatory expectations. The application of QbD affords the opportunity to capitalize on experience and knowledge using a systematic scientific and risk-based approach to understand the *variability* of quality and material attributes and process parameters with the purpose of improving quality assurance in the safety and efficacy of the product for the patient.

INTRODUCTION

In August 2002, the FDA announced an initiative, *Pharmaceutical Current Good Manufacturing Practices (CGMPs) for the 21st Century* (1,2). The intent of this initiative was to modernize FDA's regulation of the quality of pharmaceutical products by implementing science-based policies and standards. Companies have also been encouraged to use risk-based assessments, in particular when identifying product quality attributes, and adopt integrated quality systems throughout the life cycle of a product. A number of guidance documents have been published related to this initiative (3–8).

The movement toward science-based regulations has not been limited to the United States. The International Conference on Harmonisation of Technical Requirements for Registration of Pharmaceuticals for Human Use (ICH) issued two draft guidelines, further focusing on how to incorporate Quality by Design (QbD) into the preparation of Common Technical Documents (CTD). ICH Q8R addresses Section 3.2.P.2 Pharmaceutical Development (9) and ICH Q9 discusses the use of risk assessment (10). ICH Q10 was subsequently issued to address pharmaceutical quality systems (PQS) (11). Together, the development and adoption of these guidelines stimulated several companies to formally embrace the concepts and apply them to develop their products. While many elements associated with QbD, that is, risk assessments, design of experiments (DoE), operational control strategies, etc., have been employed well before the adoption of the ICH guidelines, application was frequently not systematic, concerted, or prospective, but rather retrospective in response to issues or problems encountered during development or after commercial launch. In addition, provisions in traditional regulatory guidelines did not offer regulatory incentive to pursue or provide additional scientific details describing the breadth of process understanding and product knowledge beyond empirical results from direct manufacturing experience. Consequently, companies were reluctant to pursue a QbD approach or introduce supplemental studies on process capability for fear of unnecessarily increasing regulatory "burden" and potentially delaying regulatory approvals.

In 2005, the FDA launched a pilot program (12) that encouraged pharmaceutical companies to submit science-based New Drug Applications that contained elements of FDA's vision for CGMPs for the 21st century that aligned with the recently issued ICH Q8R and Q9 guidelines. The pilot was largely successful in that it engaged regulators, inspectors, and industry scientists in a meaningful exchange of how to prosecute QbD concepts with real projects and products. In addition, the industry response to the pilot program transformed a largely theoretical opportunity into actual regulatory applications describing the use of concepts,

---
[a]Chemistry, Manufacturing, and Controls

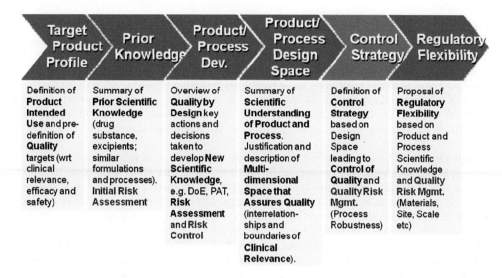

Target Product Profile	Prior Knowledge	Product/ Process Dev.	Product/ Process Design Space	Control Strategy	Regulatory Flexibility
Definition of **Product Intended Use** and pre-definition of **Quality** targets (wrt clinical relevance, efficacy and safety)	Summary of **Prior Scientific Knowledge** (drug substance, excipients; similar formulations and processes). **Initial Risk Assessment**	Overview of **Quality by Design** key actions and decisions taken to develop **New Scientific Knowledge**, e.g. DoE, PAT, **Risk Assessment and Risk Control**	Summary of **Scientific Understanding of Product and Process.** Justification and description of **Multi-dimensional Space that Assures Quality** (interrelation-ships and boundaries of **Clinical Relevance**).	Definition of **Control Strategy** based on Design Space leading to **Control of Quality** and **Quality Risk** Mgmt. (Process Robustness)	Proposal of **Regulatory Flexibility** based on Product and Process Scientific Knowledge and Quality Risk Mgmt. (Materials, Site, Scale etc)

Figure 1 Outline of approach to application of Quality by Design.

which generated a variety of concrete approaches (13–23). The differences in those approaches highlighted the need for clarity and further conceptual refinements of the concepts and their application, and raised several questions:

- How to adequately and appropriately characterize commitments versus data in a regulatory application?
- How is prior knowledge substantiated?
- What level of detail is required to justify risk assessments?
- How should design space be presented and conveyed to demonstrate quality assurance?
- How can modeling be used to justify commercial manufacturing process changes?
- How should control strategy connect drug product quality attributes to process parameters and material attributes?
- Is there an attenuation of regulatory latitude for postapproval optimization and continual improvement?
- What roles do regulatory assessors and inspectors have respectively in assessing QbD relative to the pharmaceutical quality system?

Far from suppressing progress, these, among many other questions, stimulated regulatory authorities and industry to pursue clarification. A fair measure of subsequent progress has improved the consistent application and value of these concepts. The QbD approach, which is outlined in Figure 1 (24), is sequentially consistent with, and intrinsically similar to, a traditional development paradigm. However, risk- and science-based development of product and process design criteria are addressed in a systematic and prospective manner. The objective is to achieve product knowledge and process understanding.

Perhaps, most importantly, the application of a QbD approach and investment in robust pharmaceutical quality systems are expected to reduce *unexpected variability* in manufacturing processes and unanticipated failures in product quality, thereby improving quality assurance of products.

RELEVANCE OF QUALITY BY DESIGN
Value of QbD
During the last several years, the words "Quality by Design" have become synonymous with pharmaceutical process improvement. QbD has seemingly assumed pervasive and even mystical proportions in the media as it has evolved from a conceptual initiative to a "panacea"

for improving pharmaceutical development and manufacturing. However, while many large innovator pharmaceutical companies have embraced QbD principles and conceptual approaches to product development and process understanding, much of the rest of the pharmaceutical industry, including companies that manufacture generic medicines and excipient and raw material suppliers, have remained tentative (25,26). And so, comprehensive adoption of the concepts may still be an aspiration within the industry even where modest incremental efforts to introduce elements of the principles during development or retrospectively to improve manufacturing processes for approved commercial products have increased.

With respect to traditional pharmaceutical development paradigms, contemporary and compelling benefits to legitimately account for the immediate and short-term costs of investing in process understanding are, admittedly, conspicuous in their absence. The resistance to embrace the paradigm shift and adopt the QbD approach has been perceived as operationally cost prohibitive. To date, the return on investment for QbD is largely anecdotal, circumstantial, or academic (27). Accordingly, examples of meaningful return on investment for adopting and implementing QbD principles have been limited and modest, lending credence to the contention that there is no financial incentive to adopt QbD or, alternatively, QbD may be selectively useful only for certain products (28). Certainly, the paradigm shift toward introducing systematic risk assessments that leverage prior knowledge and stimulate studies to understand the dynamic nature of manufacturing processes requires change from a minimalist, empirical, and reactive orientation to a holistic, scientifically designed, and prospective approach.

The intrinsic advantages of investing in process understanding increases confidence and assurance of product quality. Tangible benefits, reductions in manufacturing costs associated with improved efficiencies and expeditious innovations, reduction in manufacturing recalls, failures or extraneous investigations attributed to uncertainty are largely realized over the long term as the life cycle of a product matures. Table 1 provides a list of expected benefits, most of which have been realized by application of QbD principles by companies who participated in the FDA pilot program (15,16,21–23). The proliferation of studies reflecting the application of QbD is a credible testament to its inherent value.

From a business perspective, the logic for adopting a QbD approach is not obvious, especially when current regulatory requirements neither prescribe QbD nor provide immediate incentive for investment. The relatively high probability of product attrition during early clinical development does not justify ancillary investments in establishing design space when adherence to current statutory expectations is satisfactory for achieving regulatory approval. After all, the adoption of QbD is optional (9). However, the principles promulgated in ICH Q8R, Q9, and Q10 are not new. Pharmaceutical companies have applied the elements of QbD during development and postregulatory approval to develop robust manufacturing processes and/or assess process consistency for several years before ICH Q8R was conceived. What is relatively novel is the systematic and mechanistic approach using prior knowledge to assess

Table 1 Anticipated Benefits of Quality by Design

Reduce	Improve
Uncertainty and risk	Process understanding
Recalls, technical anomalies, quality investigations, manufacturing failures	Innovation and process improvement
Manufacturing costs	Quality assurance
Need for repetitious process validation exercises	Regulatory flexibility
Quantity of postapproval regulatory submissions and/or regulatory expectations	Application of technology, e.g., PAT, modeling, scale-down models
Nonscientific regulatory and/or compliance exercises	Regulatory review criteria
	Scientific/technical literacy
	Capitalization experience
	Development efficiency
	Global harmonization

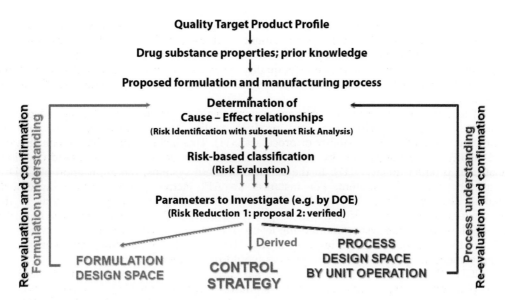

Figure 2 (*See color insert*) Process flow diagram describing the approach to developing process understanding and building quality into formulation and manufacturing process design.

risk, establishing process boundaries to understand and reduce variability and developing holistic control strategies in a prospective focus toward continual improvement.

WHAT DOES QUALITY BY DESIGN MEAN?

The paradigm shift or change in orientation toward adoption and application of QbD principles has been acknowledged by several companies who have incorporated the approach as an integral part of their business. For example, once clinical proof of concept has been demonstrated for a product, development project teams embark on systematic and scientific/ risk-based design of the commercial product formulation and process, where elements of QbD are integral to development as illustrated in Figure 2 (29).

From company to company, the sequence may vary in detail and formality, but the elements of the approach are essentially the same. The therapeutic profile of a medicinal candidate and its preliminary quality criteria provide the definition of the product and its intended use. The properties of the drug substance are confirmed and analyzed and, in conjunction with "prior knowledge," form the basis for understanding material attributes that lead to formulation design. An assessment of the formulation in consideration of prior relevant knowledge and experience may reveal functional relationships between material properties and quality attributes that may warrant experiments to establish important properties and characteristics of the formulated product and their influence on quality. Likewise, design of the manufacturing process with subsequent assessment in conjunction with prior relevant knowledge and experience may reveal functional relationships between process parameters and quality attributes. These functional relationships may lead to experiments that establish design space, whose boundaries can contribute to the understanding and development of control strategy. The process is iterative and can provide a useful understanding of the product and manufacturing process from which subsequent improvements can be planned and executed.

PHARMACEUTICAL QUALITY SYSTEM

Investing in efforts to increase product knowledge and process understanding through continual improvement across the product life cycle is paramount to pharmaceutical innovation. Adopting a QbD approach serves the patients by improving confidence in the assurance and consistency of product quality; reorients regulatory scrutiny on scientifically and technologically relevant commitments and data and balances regulatory expectations for appropriate, continual life cycle improvement; and benefits industry by promoting and progressing scientific and technological innovation. Of course, the application of QbD concepts is predicated on a robust pharmaceutical quality system as described in ICH Q10. In fact, the connectivity between scientific and technological development and routine manufacture is imperative.

The competent quality management system ensures "effective monitoring and control systems for process performance and product quality, thereby providing assurance of continued suitability and capability of processes" (11). This includes knowledge management and quality risk management "by providing the means for science and risk based decisions related to product quality" (11). In the absence of these systems, the application of QbD is effectively rendered impotent. For instance, an API (active pharmaceutical ingredient) manufacturer may have developed an acute understanding of how perturbations in the parameters of the process affect a critical quality attribute (CQA) such as genotoxic impurity levels. However, if the system monitoring levels of those impurities during the process is not robust, the control of those impurities may be suspect. In addition, any subsequent improvement or optimization of the process would be vulnerable to failure.

ICH Q10 reinforces adherence to regional GMP requirements, ISO standards, and the ICH Q7 guideline. In principle, the quality standards described in ICH Q10 are the foundation on which science- and risk-based QbD approaches are "enabled and qualified" (11). In particular, continual improvement relies entirely on robust technology transfer, change management, and knowledge management systems to ensure appropriate continuity. A design space defined by the combination of parameter boundaries that governs roller compaction, lyophilization, hydrogenation reactions, performance of a dry powder inhaler, or a packaging operation is only as good as the systems used to assure consistency of any of those processes. Where the pharmaceutical quality system aligns most directly with the application of QbD is in the establishment of the control strategy. A comprehensive control strategy is generally not confined to those measureable attributes in a specification, but can include the relevant systems governed by PQS. In fact, control strategy is defined in ICH Q10 (11):

> A planned set of controls, derived from current product and process understanding that assures process performance and product quality. The controls can include parameters and attributes related to the drug substance and drug product materials and components, facility and equipment operating conditions, in-process controls, finished product specifications, and the associated methods and frequency of monitoring and control.

The aspects of a PQS that should be addressed in a regulatory submission have not been definitively established. In general, and from experience to date, regulatory authorities expect to understand the criteria used to assess risk, both for parameters and attributes that are important or critical and worthy of study and/or control, as well as justification for noncritical attributes and parameters (13,14). In addition, an understanding of how the manufacturing process and design space are monitored and the approach to change management are useful for regulatory assessors to confirm that systematic and robust processes maintain control.

QUALITY TARGET PRODUCT PROFILE

The importance of understanding the therapeutic expectations for the product cannot be underestimated. The connection between the quality of the product and its impact on safety and efficacy for the patient is paramount. Certainly, physical and chemical properties of the product that influence quality attributes required to ensure safety and efficacy are important. However, other relationships between product quality and patient need should also be considered. For example, patient compliance, which is necessary to ensure efficacy, may be linked to a CQA like the integrity of the product package or ease of patient use. A portion of the intended patient population may have tolerability or allergic responses to specific components typically used in a formulation, that is, lactose intolerance, which may influence formulation design or warrant the use of less desirable substitutes.

Fundamentally, while product quality is inherently assured by how robust a company's pharmaceutical quality systems are, that is, preventing contamination, maintaining manufacturing consistency, training of personnel, etc., science- and risk-based approaches to developing product knowledge and process understanding ultimately assure product quality is aligned with safety and efficacy for the patient. The quality target product profile (QTPP) is therefore a direct reflection of product attributes that warrant attention and may be critical to assuring appropriate product quality. As presented in ICH Q8R (9) "the quality target product profile forms the basis of design for the development of the product" and could include:

- Intended use in clinical setting, route of administration, dosage form, delivery systems
- Dosage strength(s)
- Container closure system
- Therapeutic moiety release or delivery and attributes affecting pharmacokinetic characteristics
- Drug product quality criteria (e.g., sterility, purity, stability and drug release) appropriate for the intended marketed product

The schematic in Figure 3 shows how the elements of the QTPP may be translated to the CQAs of the product. While not all potential attributes are critical for every product, a formal and concerted risk assessment can determine which are important to evaluate to demonstrate their relative influence on safety and efficacy for the patient. Ultimately, the QTPP may serve as the basis for deriving the product specification.

What is not present in Figure 3 schematic are the business attributes for which considerations must also be addressed in the QTPP. Decisions regarding commercialization of a product must be balanced with technical and regulatory expectations. The QTPP should address the following criteria as well:

- Patient population differentiation, that is, pediatric versus geriatric, which often determines preferential dosage forms.
- Ethnic and religious proclivities, that is, use of components derived from bovine or porcine sources.
- Specific market needs, for example, preferential formulations to accommodate local pharmacy practices, for example, sachets, crushed tablets, or for example, multiple packaging configurations/quantities to accommodate bulk distribution versus individual administration.

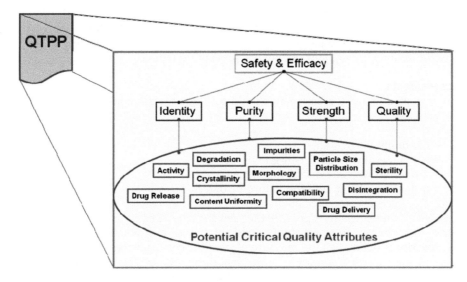

Figure 3   Derivation of critical quality attributes from the quality target product profile.

Finally, cost is, and should be, a consideration. While quality may be the primary driver for determining CQAs of the product, the cost of manufacturing will have an impact on the formulation options and manufacturing process selected to meet the commercial criteria in the QTPP.

QUALITY RISK MANAGEMENT

Quality risk management is seminal to QbD. While a robust pharmaceutical quality system serves as the foundation on which design space and control strategy are developed and managed, quality risk management provides the scaffold for creating meaningful design space and establishing an effective and robust control strategy. The inherent complexity of developing a drug product that behaves identically for each individual is a daunting challenge. As noted in ICH Q9, "the manufacture and use of the drug product necessarily entail some degree of risk" (10). Understanding risk and assessing which risks are important are at the core of QbD. In fact, it is the process of assessing, controlling, and reviewing risks throughout a product life cycle that instigates a systematic approach to developing process understanding and generates the development of design space and results in the establishment of a robust control strategy.

During the evolution of QbD, and particularly during the FDA pilot program and subsequent EMEA/EfPIA PAT Team sponsored workshops (30), the emphasis on quality risk management has engendered tremendous interest and engagement among industry and regulatory assessors and inspectors. Much of the focus has been on the quality of risk assessment justifications that lead to delineation of critical versus non-CQAs and process parameters. A robust quality risk management process typically requires the collaboration of a cross-functional team of experts from a variety of pharmaceutical science disciplines. Evaluating risk based on scientific knowledge that may reflect their collective prior experience or theoretical or conceptual analysis is extremely important to adequately address all of the potential sources of variability in a manufacturing process. Understanding what is known and recognizing and acknowledging uncertainty about what is not known is the beginning of the risk management process and can only be adequately addressed by adhering to the primary principles of quality risk management:

- The evaluation of the risk to quality should be based on scientific knowledge and ultimately link to the protection of the patient
- The level of effort, formality, and documentation of the quality risk management process should be commensurate with the level of risk (10)

Certainly an increase in the level of risk warrants concomitant and proportionate diligence in characterizing and evaluating and managing risk. In addition, transparency in describing and conveying the judgment basis of risk assessments, regardless of the level of risk, is useful for anticipating and potentially preventing failure.

Risk Assessments

Variables evaluated in a risk assessment should be judged relative to the following questions (10,31):

- What might go wrong?
- What is the likelihood it will go wrong?
- If it does go wrong, what is the impact?
- If it does go wrong, will the failure be detected?

Answers to these questions provide the relevant criteria by which risk is judged, namely the severity, uncertainty, and probability a risk may pose and whether or not a risk can be detected. Severity is defined as the measure of possible consequences of a hazard (10). Many companies in the industry employ a scale that differentiates catastrophic from negligible impact (32). Uncertainty is the unknown level of understanding for which the variability of a process parameter or quality attribute influences the severity and/or probability of risk to the safety, efficacy, and quality of the product. In many instances, experiments or studies can reduce the level of uncertainty. Probability is the likely occurrence of impact on the safety, efficacy, and quality of a product. Probability is generally characterized by an estimate of the degree of variability of a parameter or attribute to impact quality and may consider the combination of operational controls in place that reduces the level of occurrence. Detectability is the ability to discover or determine the existence, presence, or fact of a hazard (10). The ability to detect variability of a parameter or attribute and the relative sensitivity to variability can provide appropriate mitigation for a risk. The combination of these criteria are used together to assess the risk parameters and attributes may pose on the quality of the product.

Perhaps most significantly, quality risk management provides the basis for creating design space and establishing a robust control strategy as illustrated in Figure 4.

The QTPP yields a number of CQAs that for all intents and purposes represent the drug product specification. Process parameters functionally related to CQAs are evaluated to determine their relative risk to those CQAs. Risk assessments are judgments. They may rely on a combination of prior knowledge, experience, and/or experimentation. Figure 5 is illustrative

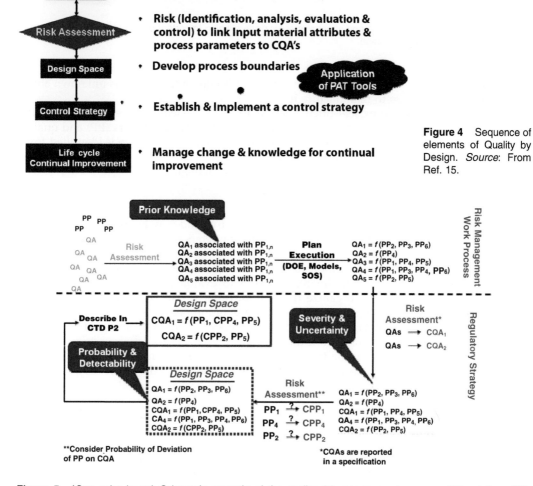

Figure 4 Sequence of elements of Quality by Design. *Source*: From Ref. 15.

Figure 5 (*See color insert*) Schematic example of the quality risk management process. *Abbreviations*: PP, process parameters; CQA, critical quality attribute; CTD, Common Technical Document. *Source*: From Refs. 22,23,26.

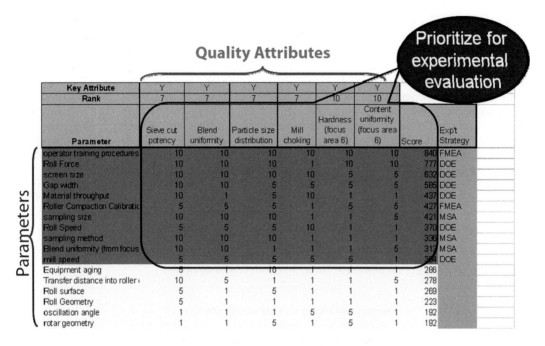

Figure 6   (*See color insert*) Cause-and-effect matrix for distinguishing important quality attributes and process parameters for subsequent evaluation.

of the approach that may be used to discriminate what is known from what is unknown about a product and the process to manufacture it.

A vast variable space exists—often referred to as "knowledge space" —that contains inputs and outputs that can be described and labeled as process parameters (PP) and quality attributes (QA), respectively. Through a risk assessment, what is not known is identified from what is known or judged to be understood. In this step of the process, the question of what can go wrong is addressed and a list of potential hazards is catalogued.

Risk Identification and Analysis
An estimation of the risk may be qualitative or quantitative and may be the result of ranking risk in a cause-and-effect matrix associating process parameters with their potential impact on quality attributes as shown in Figure 6.

The analysis of functional relationships can distinguish the level of risk and serves to prioritize relevant studies or experiments required to evaluate the risk. Another way to identify and analyze risks and organize them in an orderly fashion is to use an Ishikawa diagram. An example of a template for an Ishikawa diagram organizes potential causes into four categories as shown in Figure 7. Frequently, Ishikawa diagrams are used to identify potential causes of a specific problem. If the problem is *why isn't the telephone being answered on time*, the potential causes can be traced to specific sources. In much the same way, the risk of having production defects in a tablet can be traced to the potential sources of variability that create that risk. The Ishikawa diagram provides an alternative tool for identifying risks.

Other options for identifying and analyzing risk include, but are not limited to, the following tools:

- Quality function deployment—a qualitative and structured analysis that translates "customer" requirements into technical options.
- Influence matrix—quantitative measure of the effect a specific parameter has on a measurable product characteristic or attribute.

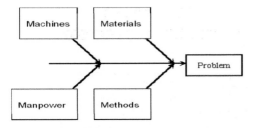

Reason Phone Not Answered

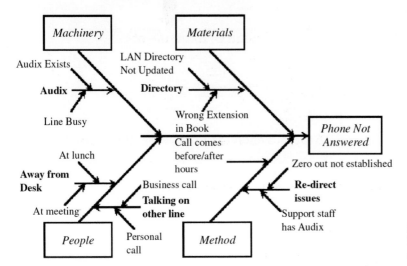

Figure 7 Example of Ishikawa diagrams. *Source*: From Ref. 33.

- Failure mode effects analysis (FMEA)—structured approach to identify, estimate, prioritize, and evaluate risk that aims at failure prevention and is primarily used to limit risk in change management.

Each of these tools alone or in combination with one another can provide a preliminary and systematic assessment of risk. However, it is subsequent evaluation of risk, where scientific experiments, models, and simulations can increase understanding of the risk and lead to design space to describe the area within which risk can be controlled. Subsequent assessments may distinguish acceptable risks from risks that require controls and/or methods for measuring control. The threshold of acceptable risk may ultimately be described in a design space and is fundamentally based on an evaluation of severity of impact, relative uncertainty, probability of occurrence, and an ability to detect variability.

Risk Evaluation
Identifying the sources of variability among process parameters that may pose risk to quality attributes allows for an analysis of the impact and probability of that risk causing harm. The importance or magnitude a risk poses often leads to the development of an experimental strategy to evaluate the level of risk. The functional relationships between process parameters and quality attributes within the focus areas of a manufacturing process provide the opportunity to evaluate the risk quantitatively and characterize boundaries of that risk through experimentation. Figure 8 provides an example of how a DoE may be derived from a set of focus areas containing several unit operations.

Design of Experiment (33–39) is a structured and statistical approach to evaluating the interactions of process parameters and their impact on quality attributes. Multiple parameters are studied simultaneously that allows estimation of interactions between factors. The designs

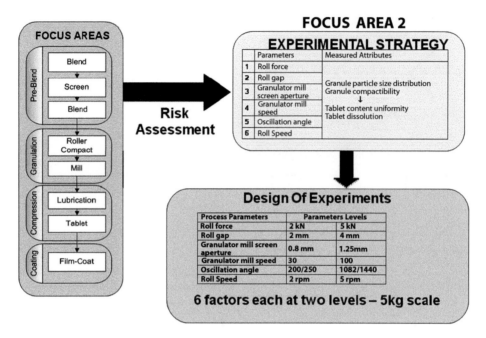

Figure 8   Translation of focus area variables into experimental strategy and plan.

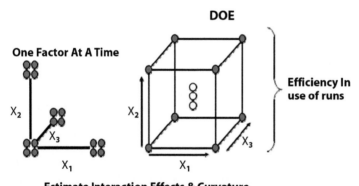

Figure 9   Generic example of design of experiment scheme.

can be structured to specific objectives, that is, factor screening or response surface exploration as well as identifying resource constraints. The use of DoE offers efficiency for estimating parameter effects and control over precision of response prediction in the case of response surface designs. They are comprehensive in nature, eliminate subjective assessments, and provide data with a wide inductive basis. Model building can be used to condense the raw data into systems of equations that describe relationships and thereby facilitate interpretation. Sequential experimentation provides for incremental understanding of the relationship between parameters and quality attributes by converging to conditions that produce the desired product.

Consider the case of three parameters evaluated over two levels. In Figure 9, the left-hand scheme represents the one-factor-at-a-time approach (OFAT). Four separate trials are run at 4 different points—16 total runs. In each instance, one factor is varied while the other two remain at the baseline level. This design allows estimation of each parameter effect (main effect) given that the other two parameters are at their baseline level.

The design scheme on the right requires only 11 runs and estimates the parameter effects with the same precision. In addition, interaction effects can be estimated. The replicated center points allow an estimate system curvature and pure experimental error. If any parameter does not alter the response, the design projects into a replicated design in the other parameters. Estimation of all effects, including interactions, provides a wider inductive basis for the experiment.

The statistical approach to DoE is useful in quantitatively characterizing the level of risk that any given parameter or attribute may pose in a multivariate expression. In addition, there are a variety of statistical designs, such as factorial, resolution of factorial, irregular fraction, D-optimal, Plackett–Burman, central composite, and mixture designs, etc., that may be employed to investigate specific effects or the extent to which a given parameter or attribute will impact quality attributes of a product. The variety of statistical approaches generate data that can be used to optimize the understanding of the boundaries of parameters and attributes in a design space and thereby improve the understanding their relative risk may have on the quality of the product.

Scientific- and risk-based assessments meet several fundamental objectives of QbD:

- Risk assessments are useful for characterizing and *ranking* attributes process parameters semiquantitatively relative to their impact on safety and efficacy.
- The risk assessment approach may be applied to drug substance synthesis and drug product manufacture.
- Use of formal risk assessment criteria to identify and differentiate critical from noncritical sources of variability and determine which variables are important to study and control.
- Design and performance of multivariate experiments to understand the interaction of variables with one another and their relative impact on quality attributes that affect patient safety and efficacy.
- Development of a well-characterized design space "or multidimensional combination and interaction of variables that demonstrates assurance of quality."
- Establishment of a coherent and concerted control strategy that may include the adoption of innovative technology, that is, PAT, to monitor or measure process variables directly.
- Sequential and iterative risk assessments → experimental plan → design space → control strategy.

The risk assessment process is iterative. As the life cycle of a product evolves from pharmaceutical development through technology transfer, during commercial manufacture and with the introduction of product enhancements and alternative formulations, the functional relationships between parameters and attributes and quality attributes of the product may change. Reassessing functional relationships, adjusting design space boundaries to accommodate changes in the manufacturing process, and establishing new design space increase process understanding and product knowledge and provide improved quality assurance of the product.

Risk Control

Decisions on what level of risk is acceptable have frequently centered on which parameters and attributes are "critical." Designating a level of criticality for attributes and parameters, that is, continuum of criticality, can be useful in delineating risk acceptance from risk reduction. Certainly, unacceptable risk requires mitigation or avoidance. However, there are risks where the severity of impact may be high, such as safety and efficacy to the patient, but the uncertainty and probability very low, for example, process parameters and attributes demonstrate that the risk is mitigated because the functional relationships are well understood and controlled within specified boundaries. For example, genotoxic impurities can be purged and controlled in the third step of a six-step synthesis of a drug substance. Stoichiometry, temperature, and pH of the reaction have been demonstrated to impact control of genotoxic impurities. Therefore, control of genotoxic impurities in the drug substance (a CQA of the product) is functionally related to the combination of these parameters.

$$CQA = f(PP_{stoichiometry}, PP_{temperature}, PP_{pH})$$

The design space described by the multivariate interactions of these parameters defines the boundaries within which control for this CQA is demonstrated. Subsequent demonstration that the process consistently operates within the design space reduces the risk.

Not all risk can be eliminated. In many instances, an appropriate risk management strategy will reduce the risk to an acceptable level where severity and probability may be mitigated by adherence to parameter and attribute boundaries, that is, control may be demonstrated by direct or indirect measurements of specific quality attributes. The acceptability of risk is often a decision that balances the presumed impact of the risk relative to appropriate controls to mitigate that impact. For example, the presence of genotoxic impurities produced during manufacture of a drug substance at residual levels that exceed the Threshold of Toxicological Concern (TTC) poses a safety risk to the patient. However, if the drug itself is mutagenic and is indicated for first-line therapy for breast cancer, the presence of these impurities should be balanced with the benefit of the drug and its duration of use. If reduction or elimination of genotoxic impurities is cost prohibitive or results in other quality issues, then acceptance of limits for these impurities that exceed the standard regulatory expectation may be justified.

Risk Communication and Risk Review

The other elements of quality risk management that support the scientific approach to decision-making are communication and review. Risks should be characterized by their respective and relevant relationship to quality attributes and process parameters and documented in a logical manner that shows the relationships between product quality and the attributes and parameters that influence quality. A general summary of the risk assessment approach and justifications for decisions regarding the attributes and parameters that warrant concern is helpful to regulatory authorities and should be transparent and reproducible.

In a regulatory submission, a description of the process used to evaluate and characterize risks should be provided. Regulators are keen to understand how a company distinguishes which attributes and parameters to study from those parameters and attributes that are noncritical (13,14,18–20). Summary examples of the evaluation tools and their respective results, for example, cause-and-effect matrix (Fig. 6), are useful ways to convey the outcomes from these risk assessments. In addition, descriptions of functional relationships between CQAs and the attributes and process parameters that may influence those CQAs provide context for describing the multivariate results from experiments and the design space created from those variable interactions.

DESIGN SPACE FOR DRUG SUBSTANCE

Design space can, and often is, the outcome of a robust quality risk management process. Table 2 and Figure 10 are a tabular summary and schematic representation of the outcomes from a comprehensive quality risk management process (40). Table 2 provides a summary and culmination of the results from risk assessment and evaluation including experimental results to establish design space for an example of drug substance manufacturing. Figure 10 is an example of the depiction of the design space created from these results. The drug substance manufacturing process can be separated into focus areas, that is, steps 1 and 2, steps 3 and 4, steps 5 and 6, etc., as shown in Figure 10, which may include one or more steps in the chemical synthesis. Once potential CQAs are defined, experiments can be performed to determine those process parameters that impact them, and subsequently identify acceptable operating ranges for which acceptable product is made. In some instances, edges of failure are identified for process parameters that lead to the production of an intermediate or drug substance that is not in compliance with the acceptance criteria for a CQA. However, most often edges of failure are not identified for the operating ranges investigated. The investigated range may be very large for the parameter being controlled, thereby providing more than enough "space" within which the

Table 2 Tabular Summary of Design Space Criteria

Quality attribute	Type	Process parameter or attributes impacting critical quality attribute	Type	Normal operating range	Design space range	Control strategy

Drug substance particle size	CQA	Temperature of API addition to acid
		Milling of API
			KPP
			KPP
				<60°C
				Screen: 12–20 mesh
				Impeller speed:
				  1500 ± 100 RPM
					Up to 60°C
					Screen: 12–20 mesh
					Impeller speed:
					  1000–2000 RPM
						Batch record, API
						  specification

Drug substance purity	CQA					Intermediate and API specifications

Quality of intermediate "B"	CQA	Catalyst filtration
		Quantity of reagant D
		Reagant "D addition rate
		pH of crystallization
			KPP
			KPP
			KPP
			KPP
				<20 ppm
				1.4–1.6 Meq
				60–90 min
				5.0–6.0
					<40 ppm
					1.0–1.9 Meq
					45–120 min
					3.0–8.0
						Batch record,
						  intermediate
						  specifications

Quality of intermediate "A"	CQA	Quantity of acid #1
		Quantity of acid #2
			KPP
			KPP
				2–3 Meq
				2–3 Meq
					1–4 Meq
					1–4 Meq
						Bach record,
						  intermediate
						  specifications

Abbreviations: API, active pharmaceutical ingredient; CQA, critical quality attribute; KPP, key process parameter.

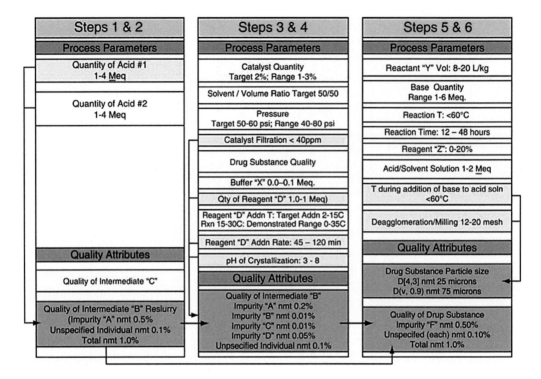

Figure 10   (*See color insert*) Schematic description of design space criteria.

process can effectively operate. In these cases, unless changes to the manufacturing process are anticipated, there is little business incentive to expand the investigated design space region any further. In other instances, attempting to expand the operating boundary ranges may exceed the capability of the equipment.

The pictorial representation of design space (Fig. 10) contains a series of columns, one for each focus area investigated. Each column is built from information contained in the summary table that summarized the knowledge obtained for each individual focus area. Cells that are shaded in pink highlight CQAs or process parameters, while those shaded in yellow contain key process parameters, those parameters that may influence CQAs, where risk assessment suggests probability and detectability warrant monitoring or further evaluation. Unshaded cells denote non-CQAs or process parameters. The focus areas are tied together by functional relationships that link the quality attributes and process parameters both within and across columns to other factors that they impact.

Biologicals

QbD principles are applicable to both small molecule drugs and large molecular biologics. However, the challenges of executing risk assessments are greater for a biological because the large size molecule is vastly more complex and the impact of attributes and process parameters on product quality attributes is generally more uncertain than for small molecules. In addition, the complicated nature of a biologics molecular generation from living organisms can lead to significant product heterogeneity. The inherent complexity of biological molecules can render the link between product attributes and clinical performance highly equivocal. The inability to associate quality attributes to safety and efficacy increases the level of uncertainty in assessing risk. Furthermore, the inherent difficulty to precisely characterize many biological molecules reduces the opportunities to develop concrete process understanding. However, examples and case studies describing the application of QbD principles and, in particular, quality risk management approaches have demonstrated limited success (41–45). In a similar fashion to Center for Drug Evaluation and Research (CDER), FDA's Center for Biologics Evaluation and Research (CBER) has initiated a pilot program for biological molecules that have been developed using QbD principles.

DESIGN SPACE FOR DRUG PRODUCT

Table 3 and Figure 11 illustrate the outcomes from the execution of a quality risk management process for a drug product (40). Focus areas that include one or more unit operations can be used to separate the drug product manufacturing process into manageable pieces. Statistically designed experiments and modeling can be used to identify CQAs and the process parameters that impact them. The use of univariate approaches to define the process may provide useful knowledge, but does not constitute a design space. Multivariate approaches (such as the use of factorial or central composite experimental designs or latent variable modeling) are encouraged as they detect interactions between multiple variables, which would otherwise likely go undetected. Following the completion of the development work for a given focus area, a table is prepared that summarizes the parameters investigated and the quality attributes that they impact. The table should also include columns describing the design space, the category that the quality attribute or process parameter falls into, and the control mechanism. Table 3 contains an example of a summary table of "knowledge space" i.e., all the formulation and process knowledge generated during product development for a dry granulation operation.

Table 3 Summary of Knowledge Space for a Dry Granulation Focus Area

Process parameters	Boundary results		Control	CPP/KPP
	Roller compactor "A"	Roller compactor "B"		
Roll force	X–Y kN	Y–Z kN	Batch record	KPP
Roll speed	X–Y RPM	Y–Z RPM	Batch record	No
Roll type	Pocketed Knurled	Serrated	Batch record	No
Gap width	X–Y	Y–Z mm	Batch record	KPP
Granulator screen size	X–Y mm	Y–Z mm	Batch record	KPP
Granulator speed	X RPM	Y–Z RPM	Batch record	No

Abbreviation: KPP, key process parameter.

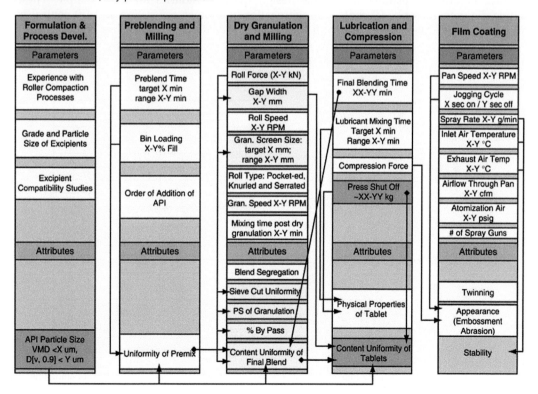

Figure 11 (*See color insert*) Design space for a drug product manufacturing process.

Figure 11 provides an example of the design space for the drug product. Much like the drug substance design space diagram, cells that are shaded in pink highlight CQAs or process parameters, while those shaded in yellow contain key process parameters, those parameters that may influence CQAs, where risk assessment suggests probability and detectability warrant monitoring or further evaluation. Unshaded cells denote non-CQAs or process parameters. Furthermore, links between drug substance and drug product attributes demonstrate the potential impact that drug substance quality attributes have on drug product quality attributes. For example, particle size distribution may be a CQA of the drug substance because it can influence drug product manufacturability. Drug substance particle size distribution translates to an important process parameter, or input, to the drug product CQA for content uniformity of the granulation blend or dosage form. Arrows show how the parameters and attributes are linked to each other, prospectively (feed-forward) and retrospectively (feed-backward), in the manufacturing process.

In addition to highlighting the parameters and attributes that define the design space (yellow and pink boxes), Figures 10 and 11 provide useful context to compare the relevant design space to noncritical parameters and attributes (white boxes) that constitute the knowledge space. Of course, these types of representations also reflect the outcomes from risk-based evaluation of the manufacturing process relative to CQAs of the drug product.

For a biological drug product, the formation of oxidative species and aggregation monomers can be reduced by introducing specific components in the formulation design to retard degradation. The effect of certain excipients, that is, surfactant and chelators, and pH on these CQAs can be built into the design of the formulation. On the basis of a risk assessment focused on limiting oxidation, a multifactorial DOE can be developed as described in the example presented in Figure 12.

On the basis of results from the DOE, a design space was created that demonstrates the optimal concentration of formulation process parameters, that is, excipient, chelators, and pH that will reduce generation of aggregate monomer and oxidative species and produce a stable product. Figure 13 illustrates the simple design space boundaries for this example. These simple diagrams describe (*i*) control of aggregation/monomer generation and (*ii*) control of oxidative species based on results from the multifactorial design described in Figure 12.

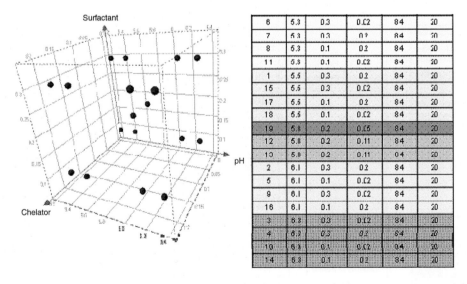

6	5.3	0.3	0.02	84	20
7	5.3	0.3	0.2	84	20
8	5.3	0.1	0.2	84	20
11	5.3	0.1	0.02	84	20
1	5.5	0.3	0.2	84	20
15	5.5	0.3	0.02	84	20
17	5.5	0.1	0.2	84	20
18	5.5	0.1	0.02	84	20
19	5.8	0.2	0.05	84	20
12	5.8	0.2	0.11	84	20
13	5.8	0.2	0.11	84	20
2	6.1	0.3	0.2	84	20
5	6.1	0.1	0.02	84	20
9	6.1	0.3	0.02	84	20
16	6.1	0.1	0.2	84	20
3	6.3	0.3	0.02	84	20
4	6.3	0.3	0.2	84	20
10	6.3	0.1	0.02	84	20
14	6.3	0.1	0.2	84	20

Figure 12 (*See color insert*) Example of multifactorial design to determine optimum concentrations of formulation parameters for a biologic.

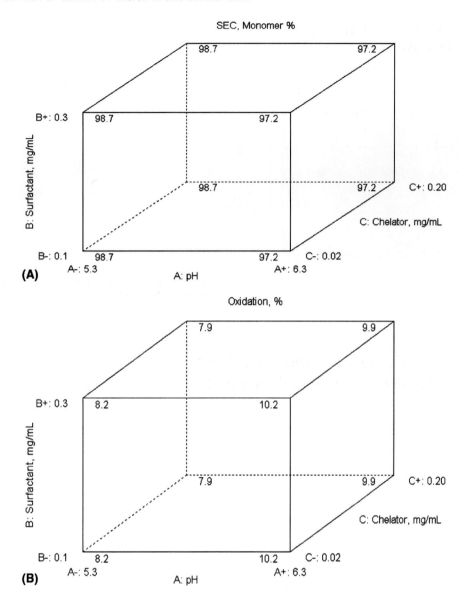

Figure 13 Design spaces for surfactant and chelator levels relative to pH that produce a stable product with respect to (**A**) aggregation/monomer content and (**B**) oxidation.

Finally, there are several ways to convey data from experiments where the multivariate interactions of attributes and parameters have been evaluated. Overlay or contour plots are frequently used to present design space and are useful for "seeing" where multivariate design space resides in contrast to areas where the risk of failure increases. Figure 14 is an example of a contour plot of design space for drug substance crystallization yield.

The area in the center of the plot (blue) defines the space within which the desired crystallized form of the drug substance is produced as a function of total solution volume, percentage of ethanol and yield. The area toward the edges reflects space that may lead to product with undesirable physical form. This contour plot shows the balance of process parameter boundaries required to deliver desirable physical form of the drug product.

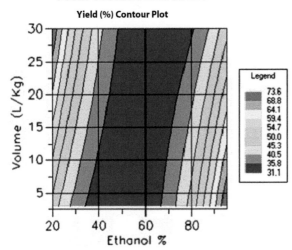

Figure 14 (*See color insert*) Example of a contour plot of design space for drug substance crystallization yield.

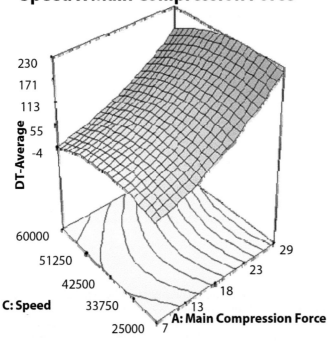

Figure 15 (*See color insert*) Contour plot describing design space for tablet disintegration time.

Similarly, a contour plot (Figure 15) describing the design space for tablet disintegration time relative to speed and compression force of the tablet press provides a profile for disintegration time that may influence tablet dissolution. The expression of data from experiments to describe a drug product attribute as a function of process parameters provides a reflection of how multiple variables interact to influence a drug product attribute.

CONTROL STRATEGY

The control strategy for a product is a comprehensive set of *planned* controls that reflect existing product knowledge and process understanding. From a holistic perspective, a control strategy includes reference to and demonstration of a robust pharmaceutical quality system and consists of appropriate qualitative confirmation and quantitative measurements that demonstrate risks to CQAs of the drug product are eliminated, reduced, or otherwise mitigated by a measure of control. A robust pharmaceutical quality system that effectively manages regulatory commitments adheres to and reinforces thorough and robust product release and is compliant with pharmaceutical quality standards is unequivocal. In particular, an effective and contemporary change management system and knowledge management process ensures continuity and consistency in the quality control of the product.

In adopting a QbD approach and applying the science and risk-based principles to assess quality attributes and process parameters, design space can be created to describe the boundaries within which unit operations of a manufacturing process may operate. In essence, design space can demonstrate control of variables that may impact a CQA, and a control strategy can be established to accommodate design space. In fact, a combination of well-defined design space boundaries and real-time release testing can effectively demonstrate and confirm control and serve as the basis for release of the product without the need for specific end-product testing. In fact, where the risk is understood and the severity and probability of impact are controllable, the demonstration of process control through the creation of design space could conceivably reduce the need to perform in-process testing as well. Continuous formal verification to demonstrate process capability in accordance with well-grounded design space criteria could serve as the basis for product release to a specification derived largely from CQAs.

CONTINUOUS IMPROVEMENT

QbD is by definition a mechanism to develop and improve process understanding and product knowledge. The approach and principles therefore are intended to be iterative.

The nature of quality risk management is and should be inherently iterative in that the development of product knowledge and process understanding stimulates regular reassessment to improve mechanistic understanding and potential control of variability. The characterization of the severity, uncertainty, probability, and detectability of risk through the life cycle also allows for accommodation of optimizations to support business objectives. The investment in QbD should therefore be construed as the appropriate cost of doing business, prospectively moving toward a paradigm of continual improvement rather than retrospectively reacting to unanticipated variability in the manufacture of products. Specific evaluations and studies are usually inserted/included into the development timeline or life cycle plan as a complement to or in concerted alignment with other business critical investments, such as, standard and in-use stability studies, impurity purge studies, formulation compatibility, packaging moisture vapor transmission, sterility evaluations, etc. However, QbD can, and perhaps should, be more than a collection of scientific exercises that incrementally improve understanding and may increase opportunities to improve a manufacturing process and reduce costs. In fact, several proponents of QbD have argued that the intrinsic value of QbD is the "full understanding of how product attributes and process variables relate to or influence product performance" (46).

The principles embodied in QbD provide valuable opportunities to increase understanding of how the quality of a pharmaceutical product contributes to patient safety and efficacy. Understanding properties and characteristics of raw materials and components; their relative combination and compatibility with one another; the influence of basic conditions of temperature, pressure, and time; and the operational criteria of manufacturing processes can collectively improve assurance of quality regardless of the product or process to which it is applied. While not all development timelines will permit a comprehensive execution of certain elements of QbD, that is, evaluation of all critical and important variables via complicated experiments, performing a preliminary risk assessment as part of process development is useful in delineating what may be important to control in a process and can provide the basis for subsequent systematic, mechanistic, and science-based studies retrospectively. The adoption and implementation of principles of QbD is a responsible and advantageous approach to managing the life cycle of pharmaceutical products.

ACKNOWLEDGMENTS

The authors recognize that the concept of Quality by Design can effectively be applied using several approaches. In fact, common conceptual threads that reflect how the principles of QbD have been prosecuted have progressed incrementally. The variety of tactics demonstrated through the application of actual practice has been useful substrate from which to draw general and definitive positions. In addition, frequent and exhaustive deliberations among many individuals within the pharmaceutical industry at multiple forums and conferences and among individual exchanges have yielded a tremendous wealth of meaningful scientific and regulatory experience and relevant opportunities to improve collective technical understanding of the merits of these concepts. We appreciate the contributions from many Pfizer colleagues, including Leslie Bloom, Tim Watson, John Groskoph, Graham Cook, Holly Bonsignore, T.G. Venkateshwaran, Ron Ogilvie, Zena Smith, and Alton Johnson. We also appreciate contributions from Chris Potter, John Berridge at ISPE, Chris Sinko (BMS), John Donaubauer (Abbott), Joanne Barrick (Lilly), Wim Oostra (Merck), Mette Kraemer-Hansen (Novonordisk), Tom Schultz (J & J), Chris Brook (GSK), and Steve Tyler (Abbott) for their valuable insights and experience. We are grateful for dynamic engagement with Moheb Nasr, FDA ONDQA, Joe Famulare and Rick Friedman, FDA OC, Yukio Hayama, MHLW Division of Drugs @ NIHS, and Jean-Louis Robert, Laboratoire National de Santé (LNS), Luxembourg for their concerted discourse and erudite perspectives. We welcome the collective enthusiasm to challenge prevailing dogma and traditional paradigms and encourage the pursuit of scientific- and risk-based approaches to pharmaceutical development and continuous improvement.

REFERENCES

1. Woodcock J. The concept of pharmaceutical quality. Am Pharm Rev 2004; 7(6):10–15.
2. FDA CDER Compliance Initiatives, Famulare J, DIA 42nd Annual Meeting, June 22, 2006, Philadelphia.
3. Pharmaceutical CGMPs for the 21st Century—A Risk-Based Approach, Final Report. U.S. Department of Health and Human Services, U.S. Food and Drug Administration, September 2004.
4. Guidance for Industry: PAT—A Framework for Innovative Pharmaceutical Development, Manufacturing, and Quality Assurance. U.S. Department of Health and Human Services, U.S. Food and Drug Administration, September 2004.
5. Draft Guidance for Industry: Comparability Protocols—Chemistry, Manufacturing, and Controls Information. U.S. Department of Health and Human Services, U.S. Food and Drug Administration, February 2003.
6. Guidance for Industry: Demonstration of Comparability of Human Biological Products Including Therapeutic Biotechnology-Derived Products—Protein Drug Products and Biological Products—Chemistry, Manufacturing, and Controls Information. U.S. Department of Health and Human Services, U.S. Food and Drug Administration, April 1996.
7. Draft Guidance for Industry: Powder Blends and Finished Dosage Units—Stratified In-Process Dosage Unit Sampling and Assessment. U.S. Department of Health and Human Services, U.S. Food and Drug Administration, October 2003.
8. Guidance for Industry: Quality Systems Approach to Pharmaceutical cGMP Regulations. U.S. Department of Health and Human Services, U.S. Food and Drug Administration, September, 2006.
9. ICH Q8(R2) Pharmaceutical Development, November 2005. Fed Regist 2006; 71(98). Annex incorporated December, 2008.
10. ICH Q9, Quality Risk Management, November 2005. Fed Regist 2006; 71(106):. 32105–32106.
11. ICH Q10 Pharmaceutical Development System, June 2008. Fed Regist 2009; 74(66):15990–15991.
12. Submission of Chemistry, Manufacturing, and Controls Information in a New Drug Application Under the New Pharmaceutical Quality Assessment System. (Docket No. 2005N-0262).
13. Chen C. (FDA) Implementation of Quality by Design, AAPS/ISPE/FDA, Pharmaceutical Quality Initiatives Workshop, February 28—March 2, 2007, Rockville, MD.
14. Moore C. (FDA) Update on Implementation of Quality by Design, ISPE Engineering Regulatory Compliance Conference, June 4, 2008, Arlington, VA.
15. Simmons S. (Wyeth) Application of QbD Principles to APi Stability Characterization, ISPE PQLI Conference, June 4–5, 2008, Arlington, VA.

16. Ahuja E. Merck Implementation of QbD: Case Study; Drug Product, ISPE PQLI Conference, June 4–5, 2008, Arlington, VA.
17. Stott P. (Aztra Zeneca) FDA CMC Pilot Program; Experiences and Lessons Learned from an Industry Perspective, DIA 44th Annual Meeting, 2008, Boston, MA.
18. Oliver T. (FDA) ONDQA CMC Pilot Program; FDA's Perspective Part 1, DIA 44th Annual Meeting, 2008, Boston, MA.
19. Ochetree T. (FDA) ONDQA CMC Pilot Program; FDA's Perspective Part 2, DIA 44th Annual Meeting, 2008, Boston, MA.
20. Vanstockem M. (Tibotec) FDA CMC Pilot Program; Prezista (darunivar) 600 mg Tablets; Lessons Learned, DIA 44th Annual Meeting, 2008, Boston, MA.
21. Smith Z. (Pfizer) Quality by Design, Leading by Examples, DIA, 2009, Bahrain.
22. Nosal R. (Pfizer) Industry Perspective of Risk-Based CMC Assessment Under Quality by Design, 2006 AAPS Annual Meeting and Exposition, San Antonio, TX, October 31, 2006.
23. Nosal R. (Pfizer) Integration of ICH Q8R, Q9 and Q10: Regulatory Evolution of Quality by Design during the FDA Pilot Program. 12th Annual GMP by the Sea Conference, Cambridge, MD, August 29, 2007.
24. The sequential representation of the QbD approach was originally presented by the EfPIA Working Group in 2006(?) but has subsequently been modified. This version was presented by Chris Potter, "PQLI Vision, Status and Next Step," ISPE India and ISPE Japan Annual Meetings, April 2009.
25. Several conversations and correspondence with Lawrence Yu and Frank Holcombe, FDA—OGD and representatives from GPhA, application of QbD principles among companies manufacturing generic pharmaceuticals has been limited.
26. Nosal R. (Pfizer) Practical Approaches to Implementation of ICH Q8R—Extending QbD to Continuous Processing, Real Time Release & Continuous Quality Verification 13th Annual GMP by the Sea Conference. Cambridge, MD, August 25, 2008. Representatives from excipient and generic pharmaceutical manufacturing companies acknowledge reluctance to invest in QbD and cite increase cost does not warrant investment for relatively small percentage of the pharmaceutical commercial market.
27. Rutten P, Fuhr T. (McKinsey & Company) The value of Quality by Design for Pharmacos, IPAC-RS Board Meeting, Canterbury, U.K., June 18, 2008.
28. Pfizer Inc reported cost benefits from QbD with one of the first NDAs approved as part of the FDA Pilot Program. As a direct result of increased process understanding, Pfizer was able to realize reduced regulatory expectations and expeditious implementation of multiple postapproval changes simultaneously. These changes in manufacturing site, scale, design space boundaries, etc., prevented an inventory stock-out prompted by rapid and unanticipated increase in sales volume for the product that prevented a projected loss of ∼$200 M in expected sales of the product. In addition, the waiver of regulatory criteria, stability data, etc., afforded a direct cost savings of ∼$500K. R. Nosal. Integration of ICH Q8R, Q9 & Q10—Regulatory Evolution of Quality by Design During the FDA Pilot Program, 12th Annual GMP by the Sea Conference, Cambridge, MD, August 29, 2007.
29. General correspondence with representatives from Merck, Wyeth, GSK, Abbott, Lilly, Schering-Plough, Organon, BMS confirm these companies have integrated relevant elements of QbD principles as a routine part of their respective development activities.
30. EMEA/EfPIA PAT Team workshops at pharmaceutical manufacturing sites in Ireland, April 2008; and QbD application Workshop in London, September 2009.
31. PhRMA Drug Product Technology Group, Risk Management in Pharmaceutical Product Development Process, White Paper, 2008.
32. ISPE PQLI CQA & CPP Task Team deliberations on the adoption of a severity scale to delineate the continuum of criticality revealed several options for characterizing degree of severity. Lilly, GSK, Pfizer, Novonordisk, Schering-Plough—Organon, provided examples of severity scales currently in use.
33. McCurdy V. (Pfizer) Pfizer internal training on performing risk assessments, 2004.
34. Anderson ML, Whitcomb P. Design of Experiments Simplified; Practical Tools for Effective Experimentation. 2nd ed. Portland, OR: Productivity, Inc., 2000.
35. Box GEP, Hunter SJ, Hunter WG. Statistics for Experimenters; Design Innovation and Discovery. 2nd ed. New York: Wiley Interscience, 2005.
36. Montgomery DC. Design and Analysis of Experiments. 6th ed. New York: Wiley Interscience, 2005.
37. Myers RH, Montgomery DC, Christine M, et al. Response Surface Methodology; Product and Process Optimization Using Designed Experiments. 3rd ed. New York: Wiley Interscience, 2009.
38. Cox DR. Planning of Experiments. New York: Wiley Interscience, 1958.
39. Cornell JA. Experiments with Mixtures (Design, Models and the Analysis of Mixture Data). 2nd ed. New York: Wiley Interscience, 2002.

40. Garcia T, Cooke G, Nosal R. PQLI key topics—criticality, design space and control strategy. J Pharm Innov 2008; 3(2):60–68.
41. Rathore AS, Winkle H. Quality by design for biopharmaceuticals. Nat Biotechnol 2009; 27(1):26–34.
42. EFPIA. Mock P2 for "Examplain" Hydrochloride. Available at: http://www.efpia.org.
43. Cook G. EFPIA. Mock S2 Project for Drug Substances. PDA Workshop on Quality by Design, Frankfurt, Germany, September 22, 2009.
44. CMC Biotech Working Group. A-Mab Case Study. Available at: http://www.ispe.org.
45. Rathore A, Mhatre R, eds. Quality by Design for Biopharmaceuticals: Principles and Case Studies. New York: John Wiley & Sons, Inc, 2009.
46. Nasr M. FDA, ISPE EU Conference. Paris, FR, April 18, 2007.

14 | Future of parenteral manufacturing

James Agalloco, James Akers, and Russell Madsen

We began the development of this chapter with an open mandate from the editors. We were asked to consider what parenteral manufacturing might be like in the future. As we had all separately and sometimes collectively developed papers and presentations on various elements of this subject, we believed we were up to the task (1,2). Our opening discussions revealed we had a substantial amount of work to do. We considered many more subjects for inclusion; the one's we have included met two important criteria. First, we were unanimous in our belief that they would be relevant in the next 10 to 15 years; and second, one or more of us felt sufficiently well versed in the technical area to make a meaningful effort as to what the future circumstance might be. The result is what you see here, a collection of brief essay's outlining what might be the state of the art in the not too distant future. We had one consolation in developing this, unlike a technical paper or even a commentary, references to what we are predicting would be beneficial, but their absence is a reflection of the future tense of this entire chapter. Undoubtedly, we will be wrong with respect to some portion of the following, but since we have made a rather substantial number of predictions, some of them might be completely accurate. Your difficulty as a reader will be to decide which is which, an effort that will be increasingly easier as the future becomes the present.

OVERALL CHANGES
Outsourcing

The pharmaceutical has expanded its use of contract manufacturing organization (CMO) substantially in recent years, and this is a trend likely to continue in the future. There had always been a segment of the parenteral industry that has utilized contract services. The predominant usage of CMOs for many years was largely in four areas:

- Filling of sterile penicillin and cephalosporin formulations
- Filling of prefilled syringes
- Filling of lyophilized formulation
- Filling of clinical trial materials

The motivation for the use of a contractor was customarily to avoid the added expense of separate or additional facilities for these unique formulations and presentations. These were perhaps the predominant use of contract manufacturing services until the advent of the biotechnology industry.

The first products developed by the biotechnology industry were large molecular weight proteins, many of which were lyophilized or required cold chain distribution. The simplest and most direct means for evaluation of these materials in a clinical setting required a parenteral dosage form, and in the majority of cases a freeze dried formulation was most appropriate. This led to an increase in the use of outside formulation and filling services. Initially, much of this outsourced production was accomplished by traditional parenteral manufacturers, although from the beginning CMOs sought to fill this increased demand. The biotechnology pioneers quite properly focused their attention to the unique parts of their processes, which was the fermentation and biochemical purification of these macromolecular entities. The seemingly scientifically simpler parenteral manufacturing steps that followed were important, but would be outsourced so that the many start-up biotech firms could rightly focus their technical and financial resources on the unique aspects of their technology.

The initial explosion in CMO usage came in the late 1980s and early 1990s as a result of two parallel but largely unrelated events. First, the initial success of biotechnology fostered a substantial increase in clinical (and later commercial) scale parenteral manufacture. By the early 1990s, the explosion in biotechnology product development resulted in approximately half of all Investigational New Drug (IND) applications being for these "large-molecule" products. Second, the restructuring of the world's major pharmaceutical firms in which mergers and a desire to "right size" led to a realization that there was substantial excess capacity in all areas of the industry. The result of this effort was a transfer of facilities to entrepreneurs, sometimes consisting of the prior employees of the site, and initial contracts for the production of prior products by the new CMO. In some instances, traditional pharmaceutical firms have supplemented their own operations at production sites by offering their services as CMOs to leverage their facility utilization.

These trends have continued and perhaps increased. Recent years have witnessed a new driver for CMO usage, the virtual pharmaceutical company mimicking a trend that originated in the consumer products and microelectronics industries. In its most extreme mode, the virtual company outsources all aspects of product life cycle from development, production, and marketing. The virtual firm by virtue of its small size can operate profitably at very modest volumes due to the absence of internal infrastructure. Slightly larger virtual firms operate similarly, through purchase of New Drug Applications (NDAs) and Abbreviated New Drug Applications (ANDAs) of older and lower volume products from larger firms, securing CROs (contract research organizations)/CMOs for production and using others for distribution and marketing.

Outsourcing in parenteral manufacturing has expanded in ways that have never been considered previously. These applications are less comprehensive than those of the fully functional CMO, but are nevertheless represent a clear trend across the industry.

- Suppliers of packaging components, primarily rubber closures, but occasionally of other items, have vertically integrated their offerings. They are offering for sale, at a premium of course, ready-to-sterilize or ready-to-use components. Many companies large and small find it economically attractive to buy components ready to use to avoid operating costs, capital equipment costs (and maintenance), as well as improving inventory turns on components.
- Contract laboratories offering a variety of chemical and microbiological testing to virtual firms or to CMOs that lack extensive internal laboratory capabilities.
- Contract development/research firms that assist with product formulation and process development activities.
- Contract service providers offer virtually every conceivable form of assistance necessary to firms large and small including consultation, regulatory submission, training services, and audit execution.

Some recent statements by executives at major pharmaceutical firms suggest that even greater usage of CMOs might be forthcoming. There was some quick retraction of that perspective, perhaps because of the potential backlash at internal production sites. That a large pharmaceutical firm might consider outsourcing all of its production related activities is certainly significant. There is a clear belief that manufacturing of pharmaceuticals, and perhaps most importantly, is a costly activity with substantial infrastructure and overhead that could perhaps be eliminated. Could a multinational pharmaceutical giant go "virtual"? The notion itself, suggests that the economics of virtual operation might be substantially better than those of more fully integrated firms. If that is the case, then the future will certainly see outsourcing play an increasingly important role in pharmaceutical manufacturing, and certainly in parenterals.

Harmonization of Regulations and Inspections

The pharmaceutical industry is highly regulated, primarily by agencies such as FDA (United States), EMA (Europe), and MHLW (Japan)[a]. These agencies seek to ensure the safety and efficacy of pharmaceuticals by means of registration requirements and through inspections

---

[a]Food and Drug Administration (FDA), European Medicines Agency, Ministry of Health, Labour and Welfare (MHLW).

designed to ensure that the manufactured products comply with the standards contained in their approved applications and that appropriate manufacturing controls required by the various Good Manufacturing Practice (GMP) regulations are in place. Organizations such as PIC/S also influence the direction of manufacturing practices for sterile products. The industry must also concern itself with pharmacopoeial standards contained in the USP, JP, and Ph. Eur.[b] Also, compliance with other regulations such as environmental and employee safety and health is a necessity in the industry.

As expected, there are differences in regulations in various regions and in the interpretation of those regulations by the agencies charged with ensuring compliance with them. This leads to variation in quality systems and manufacturing controls employed by multinational pharmaceutical companies, and it increases cost.

International Conference on Harmonization

The International Council on Harmonization (ICH), which was formed in 1990, has been effective in producing a series of guidance documents designed to provide a common framework for pharmaceutical regulatory compliance. ICH develops guidance through a Steering Committee that meets twice a year. Members include, EFPIA, MHLW, JPMA, FDA, and PhRMA[c], representing regulatory bodies and research-based pharmaceutical companies in Europe, Japan, and the United States. Nonvoting observers (WHO, EFTA[d], and Health Canada) provide a link between the ICH and non-ICH countries and regions.

These meetings have resulted in guidance documents that provide a framework of concepts and practices acceptable in the three regions. Examples include Q7 "Good Manufacturing Practice Guide for Active Pharmaceutical Ingredients," Q8 "Pharmaceutical Development," Q9 "Quality Risk Management," and Q10 "Pharmaceutical Quality System." Other important guidance documents have been produced covering the topics of Stability (Q1A–F), Analytical Validation (Q2), Impurities (Q3A–C), Pharmacopoeias (Q4, Q4A and B, with Appendices), Quality of Biotechnological Products (Q5A–E), and Specifications (Q6A–B) (3). Under ICH auspices, a Common Technical Document has been developed to facilitate a harmonized approach and format for regulatory filings relating to new drugs.

Pharmaceutical Inspection Convention/Scheme

Harmonization of regulatory inspections in terms of scope, focus, and interpretation of the applicable regulations and guidance documents is an important issue from the viewpoint of the regulated industry as well as the involved regulatory agencies. Consistency of inspectional approach leads to improved regulatory compliance since expectations are clearly defined and production and quality systems can be presented to the inspectors in ways that are understandable to all concerned parties.

PIC/S develops harmonized GMP standards and guidance documents, trains inspectors, assesses inspectorates, and promotes cooperation for competent authorities and international organizations. The participating authorities represent most of the countries of the European Union, as well as Australia, Canada, Singapore, and South Africa, to name a few. Partners and observers include the European Directorate for the Quality of Medicines and HealthCare (EDQM), and WHO.

The FDA is not, as of this writing, a member of PIC/S, but the agency has made application for membership. Inclusion of the FDA could result in progress toward the harmonization of international inspectional practices. As experience is gained, and with increased cooperation and interaction among inspectorates, harmonized inspections should become a reality. The hoped for outcome of harmonization at the inspection and enforcement

---

[b]United States Pharmacopeial Convention (USP), Japanese Pharmacopoeia (JP), and European Pharmacopoeia (Ph. Eur.).
[c]European Federation of Pharmaceutical Industries and Associations (EFPIA), Japan Pharmaceutical Manufacturers Association (JPMA), and Pharmaceutical Research and Manufacturers of America (PhRMA).
[d]European Free Trade Association (EFTA).

level would be the mutual recognition of inspections. This would obviate the need for the same firm operating under the jurisdiction of a competent authority that participated in the harmonization scheme from undergoing numerous inspections from other competent authorities. It is not uncommon for multinational firms to undergo 10 or more inspections per year, which typically cover the same technical and compliance subject matter. This duplication of effort is neither useful to the firms subjected to these redundant inspections nor does the end user accrue any benefit. Mutual recognition of inspections could significantly reduce costs for industry and regulatory agencies while providing necessary oversight to protect the end user.

Pharmacopoeias

Pharmacopoeias have been established in many countries, regions, and internationally in an effort to standardize the testing of active pharmaceutical ingredients, excipients, containers and closures, and pharmaceutical products and to ensure the efficacy of the medicines delivered to the consumer. There are many minor differences in the monographs and chapters of the various pharmacopoeias. These differences result in extra testing in the event pharmaceutical products are marketed in these countries and regions. Pharmacopoeial differences can be particularly troublesome for active pharmaceutical ingredients and excipients since these materials are often widely distributed. Pharmacopoeial harmonization can, therefore, result in better product uniformity, reduced testing, cost savings, and reduced regulatory burden. The Pharmacopoeial Discussion Group (PDG) was created to foster pharmacopoeial harmonization.

The PDG consists of members representing the EDQM, the USP, and the JP. The PDG usually meets in conjunction with ICH and provides the ICH Steering Committee with updates on pharmacopoeial harmonization issues. The PDG works to harmonize general chapters and excipient monographs in the three pharmacopoeias. A chapter or monograph is harmonized when "a substance or preparation tested by the harmonized procedure yields the same results and the same accept/reject decision is reached" according to the PDG. Full harmonization is not achieved until the text becomes official in all three pharmacopoeias.

In conclusion, there appears to be consensus among all parties involved, that is, regulators, the regulated industry, the pharmacopoeias, and organizations such as ICH and WHO to harmonize inspectional practices and regulations, resulting in improved compliance levels, patient safety, and decreased cost.

Globalization of Manufacturing

The majority of the world's production of pharmaceutical products was for many years the near exclusive province of multinational firms located in Japan, Western Europe, and the United States. While there certainly were plants located in other areas of the globe, distribution from those facilities was predominantly local. This began to change with the opening of mainland China to outside investment. The availability of extremely low cost labor for facility construction and operation along with financial support from government sources and less restrictive environmental regulation led to an influx of pharmaceutical manufacturing. India and Brazil offer similar opportunities. These countries offer added possibilities in the domestic and nearby markets.

The availability of low cost labor is perhaps the greatest single motivator in the placement of pharmaceutical facilities in these environments. Labor rates in these locales are a fraction of that paid in the traditional pharmaceutical manufacturing locations, and is clearly a major driver in the global transition to them. Parenteral manufacturing in these environments raises some concerns in the minds of many. Sterile products, especially those made using aseptic processing requires a proficient work force with unique skills. Training of operators to work in aseptic manufacturing is no mean task. In transitioning to these new production centers, firms must be prepared to make a substantial training commitment to bring their new workforce up to an appropriate level of competence. This is presumed easier in India, where the English language competency makes the task somewhat less daunting.

Recent experience in China as a production site has raised concerns relative to the integrity of its supply chain. The difficulties encountered in 2008 with contamination of heparin, baby food, and other materials have made it a difficult choice. These difficulties may be in the past; however, the negative experience is perhaps too recent for continued expansion to continue at the same pace. A commonly applied strategy has been to maintain tighter control over new high-value and high-profit products by continuing to manufacture them in the United States, Japan, or Europe while moving products to lower cost areas only near the end of their patent life. This enables firms to move older less profitable products out and newer and perhaps more technically challenging products in to take their place. This is also consistent with the philosophy of developing manufacturing sites within an organization that specialize in a particular technology. This core-competency approach has proven popular in an era of consolidation and rationalization of capacity, and in a world where redundancy of supply can be achieved easily through outsourcing rather than maintaining "back-up capacity" within the organization.

The drive toward globalized manufacturing is largely driven by economics, with the cost of labor in the developed world being one of the major factors. If the technology advances described elsewhere in this chapter reduce the labor content associated with parenteral manufacturing, then the labor cost driver is somewhat or even fully mitigated. A parallel element of this shift would be the need for those workers who do remain to be substantially better trained to enable them to operate and maintain the more sophisticated equipment that will be utilized. Also, if global environmental policies become more harmonized another major cost driver in moving manufacturing to a developing nation may no longer exist.

DEVELOPMENT
Elimination of Conventional Manual Filling
The production of sterile injectable products began with gowned personnel manually assembling containers. We could also envision that shortly after these first fills were performed that the innate hazards associated with intimate contact between personnel and sterile materials were recognized. Many years have passed since the origins of sterile product preparation by manned personnel, and with that passage our industry has witnessed substantial improvements in processing technology. The 1950s saw the introduction of HEPA filters and machine filling. The 1960s witnessed wholesale changes in sterilizing filtration with the adoption of the 0.2-μm filter. Validation as a requirement for sterilization and other processes came about during the 1970s. The next decade brought forth automated systems, parametric release, and isolation technology. The 1990s saw the first Restricted Access Barrier Systems (RABS) utilized for aseptic processing. In the 21st century, we have seen Process Analytical Technology (PAT), Quality by Design (QbD), and improved analytical technologies. Clearly, a lot has changed over the last 60 years, yet one surprising constant remains—hand filling of sterile products is still a common practice.

The extensive number of improvements adopted for parenteral manufacturing on a larger scale has undoubtedly improved the quality and safety of sterile products (4). That manual filling persists is likely due to the fact that it is somewhat hidden. Hand fills are the province of clinical supply production, orphan drugs, and very low volume operations, the very sort of operation unlikely to distinguish the firm using it. Would anyone tout the performance of their hand fill operations? It may be performed in a better environment, with superior gowning materials and even a modicum of automation, but in many ways it is little different from the practices of a long ago era. Gowned personnel still serve as the means for transfer of components from filler to stoppering and then to crimping. The operator's role in hand filling is largely unchanged from what it originally was, and that strikes us as an unacceptable compromise in today's far more capable and demanding environment.

The difficulty with manual filling lies in the required intimate involvement of personnel with sterile materials. Perhaps the simplest way to understand the increased contamination risk associated with hand fills is through the Agalloco-Akers Aseptic Risk Evaluation Methodology (5,6) One of the elements of this method is the Intervention Risk, which is the number of operator touches per container produced. In a manual filling process, this number cannot be less than one, and could approach three or four. When determined for a machine fill, the Intervention Risk can be anywhere from 100- to 10,000-fold lower. The continued use of an aseptic process that the "best case" is one hundred times more risky than the poorest machine supported process unnecessarily exposes the patient to potential microbial contamination, which could easily be avoided using other technologies.

It is clearly an anachronism, a relic of an earlier time that belongs in our past, not in our future. Closed isolators are a near ideal substitute for the production of small batches using largely the same practices used in hand filling in clean rooms. The isolator provides

- The ability to effectively decontaminate the processing environment in a more effective manner
- A near perfect separation of the operator from the sterile materials, approaching the goal of the "sterile field"
- Means for the safety introduction/removal of items without compromising the integrity of the isolator through the use of rapid transfer ports

There are isolators-based systems being developed that will take the entire hand fill process to an entirely new level. Vision equipped robots will be used to prepare sterile materials without operator involvement during the entire process. These offer opportunities for potent and hazardous materials where even the use of an isolator means that some operator risks are unavoidable. These systems also eliminate the fatigue factor common in a heavily manual process of any kind.

Given the availability of technologies that can dramatically enhance the certainty of the process, and thus raise the capability of hand filling to levels commensurate to commercial scale production (where isolators and RABS are becoming more and more common), the elimination of manual aseptic filling in clean rooms appears near certain. The elimination of manual filling could be hastened if inspections really did focus on end-user risk. There have been examples of inspections that focused on the enforcement of arcane local requirements for certain practices while overlooking hand-filling or manual aseptic interventions that would easily be either automated or separated by RABS or Isolator technology.

QbD/Design Controls

Ensuring the quality of pharmaceutical products and medical devices is a primary concern of manufacturers, regulatory agencies, and patients globally. QbD and associated design controls have been used for many years in the aerospace, automobile, and the electronics and computer-related industries to ensure the quality and reliability of manufactured products. The pharmaceutical industry has relied on pharmaceutical development, clinical studies, and conformance to the manufacturing processes and controls contained in the approved marketing applications to ensure product quality.

The FDA's GMP for the 21st century initiative effectively cleared the way for a different approach to drug development, licensing, and manufacturing (7). The initiative allowed manufacturers to continuously evaluate drug product quality during the development and marketing phases and make changes to improve those processes without undue regulatory burden. The ICH Q8 and Q9 guidance documents, mentioned in the previous section, provided manufacturers with a road map to improve the development and evaluate and manage risk.

QbD incorporates comprehensive prior knowledge about the product and process, some of which is derived from similar products and processes, scientific studies such as design of experiments, and quality risk management throughout the product's life cycle. This leads to improved understanding of the product and its manufacturing process, forming a basis for a more flexible regulatory approach. Regulatory flexibility is directly related to the level of scientific knowledge contained in the registration application. The application should contain information substantiating the scientific relevancy of the data submitted.

Important elements of a QbD-based pharmaceutical development program include accurately defining the product as it relates to dosage form, bioavailability, safety, efficacy, and stability; identifying critical quality attributes of the drug product so that they may be appropriately controlled; evaluating characteristics of the active ingredient(s) and excipients to ensure they possess the requisite attributes; selecting and defining the manufacturing process; identifying and implementing an appropriate control strategy; and providing a system of continuous improvement capable of integrating information and knowledge gained over the product's life cycle, which can be the basis for refining the development and manufacturing process.

QbD allows the use of a manufacturing process that is adjustable within the design space as contrasted with the one that is fixed. Instead of validation based on a few initial full-scale production batches, validation is carried out over the life of the product through continuous process verification and statistical process control. In such a system, process operations are tracked to support continuous improvements and postapproval. Instead of product specifications based on data available at the time of product registration, specifications become part of the control strategy based on product performance and process capability. In such a system, evaluation and control of product quality shifts further and further upstream (through "retroactive" development, API and excipient characteristics, process control) thereby eliminating problems before they occur. This allows for a risk-based control strategy rather than quality control, primarily through intermediate and finished product testing.

As QbD evolves, it is conceivable that the principles can be applied to custom products formulated for specific patients. These products are produced in small quantities, typically to meet a specific diagnostic or therapeutic requirement of a single patient. In therapeutic applications, the promise of these products includes improving safety with reduced risk from adverse reactions. Advanced QbD, design control strategies, and product development processes will provide the tools to ensure patient safety in this environment.

COMPOUNDING
Single-Use Technologies ("Disposables")
The production of parenteral formulations has typically involved the preparation of solutions and suspensions using vessels equipped with a means for agitation and a method for heating/ cooling of the contents. The scale of this activity varied from the laboratory bench to the large commercial scale equipment, but the principles were essentially identical. Cleaning of the equipment was always required between products, and customarily between lots of the same product; but it was not until the early 1990s that cleaning became a concern of substantially greater magnitude (8). Cleaning validation was certainly something that had been considered necessary for years by those responsible for validation at parenteral firms, but the paramount difficulty facing all was the selection of an appropriate (read that as acceptable to a regulatory agency) criteria for allowable residual. This was the inevitable stumbling block that no firm wanted to topple alone. The interest in cleaning validation was such that the PDA (Parenteral Drug Association) formed a task group that developed an industry guidance document that defined several means for the establishment of a suitable acceptance criterion (9). The logjam with respect to acceptance criterion removed, cleaning validation began in earnest across the industry. However, it soon became evident that the effort required in cleaning validation and subsequent in-process cleaning and testing was substantial and were not going to be reduced substantially by the availability of a limit for residues. Uncertainties with respect to selection of appropriate sampling locations, rinse versus swab testing, residual recovery uncertainty and confirmation of process parameters, and other issues have hampered progress on cleaning validation. With the growing frustration regarding the scope and difficulties associated, a Gordian knot–like solution was identified.

Disposables are perhaps the solution to cleaning validation by eliminating the issue at the very core. If an equipment item was not reused, it would not have to be cleaned. If it was not cleaned, there is no need for cleaning validation. In one simple stroke, the issues with cleaning validation could be swept away: both cleaning and cleaning validation could be entirely eliminated where a disposable item could replace a reusable one.

The earliest applications of disposables were rather surprising, it was not a simple hose connection or a modest size bag, but rather large containers utilized for buffer and media preparation in biologic manufacturing. The largest initial concern with respect to these applications was not their preparation or in their cleaning (they were all relatively simple solutions), but rather the desire to minimize initial capital expenditure. Also, disposables had the secondary positive effect of eliminating the need to validate production vessels including, in some cases, fermentors. This brought about additional cost savings and reduced facility construction and validation timelines. Using a single-use plastic container for the buffer made the overall process substantially easier and cheaper, at least initially. Extension of this disposable concept to single-use bioreactors was the next logical step. These vessels were somewhat more complex and the residues more diverse, so while the design of the disposable unit is more complex (and expensive) this is more than offset by the elimination of the difficult and lengthy cleaning process that must follow, as well as eliminating cleaning validation for this equipment.

Potential applications for disposable systems in parenteral manufacturing and filling are limited only by the imagination of the end user. Small-scale filling sets, process filtration assemblies, aseptic sampling systems, and other items are all available for use as disposable assemblies, which can be either a "standard" configuration or customized for a specific application. The increased use of disposable systems can be certain, given the operational advantages to their use, because the benefits extend beyond cleaning validation. The smaller disposable systems and assemblies offer substantial advantages, including increased certainty of assembly; reduced labor costs for system preparation, cleaning, and sterilization; reduced needs for internal sterilization time; enhanced confidence in sterilization; and greater system integrity.

The use of disposables does have some negatives to be considered. The polymeric materials utilized must not interact with the materials being processed in any way. The potential for extractables and leachables materials from the polymers to enter the product must be evaluated to assure that no adverse effects are present. Similarly, the adsorption/adherence of formulation components by the polymeric materials must also be considered. The issues here are comparable to those associated with the use of membrane filters; however, the exposure periods are likely to be longer and the conditions of use more variable. These issues can be overcome by careful selection and evaluation of the materials to be used. Environmental issues are of growing concern. Throwing away expensive and very large polymer containers and tubing after a single use runs counter to the current green ethos taking shape around the globe. It seems logical that sooner rather than later single-use technology vendors will consider means by which their products can be recovered and recycled. The use of disposable fluid systems offers enough advantages that increased use in the future is assured.

Filter Pore Size

Filtration has been used to remove microorganisms from parenteral and ophthalmic solutions for many years. During the first half of the last century, many sterilizing filters were made of asbestos or unglazed porcelain materials and achieved bacterial retention mainly through depth filtration and absorption. Microporous membrane filters came into widespread use for sterilization of pharmaceutical solutions during the 1950s. Use of these filters facilitated integrity tests, such as the bubble point, which could be directly correlated with the size of the largest pore(s) in the membrane, paving the way for validation studies that could be used to demonstrate the ability of a membrane to remove specific microorganisms from a solution.

The first microporous membranes used for sterilizing filtration had a pore size designation of 0.45 μm (10). When these filters were initially introduced it was thought that they retained particles and microorganisms exclusively through sieving. It was soon discovered, however, that other retention mechanisms were at work and microorganisms larger than the filters' pore-size rating were sometimes not retained. Dr Frances Bowman, an FDA scientist, discovered in 1960 that small numbers of certain microorganisms could penetrate 0.45 μm membrane filters and that this penetration was linked to the challenge level of those microorganisms, that is, penetration occurred when the challenge level was between 10^4 and 10^6 microorganisms per cm^2 of effective filtration area (11).

Further study revealed that this microorganism (at the time known as *Pseudomonas diminuta*) was retained by membrane filters rated at 0.2 μm. This led to the use of this microorganism to differentiate between 0.45- and 0.2-μm-rated filters. Another interesting fact emerged: *P. diminuta* could penetrate even 0.2-μm-rated filters if the challenge level was high enough (12). The concept of the absolute sterilizing-grade filter was abandoned.

The discovery that small bacteria could exist in pharmaceutical process streams under conditions where nutrient levels were extremely low led to the search for membrane filters of still smaller pore size. Currently, 0.1-μm-rated filters are used for process streams and products where these microorganisms may be present. If mycoplasma may be present, 0.1-μm-rated filters are generally employed. Use of these filters can restrict throughput and result in relatively long filtration times, and the filters tend to cost more than filters with larger pore-size ratings.

Membrane filters are classified according to pore size. Microfilters have nominal pore size ratings ranging from 1.2 μm to 0.04 μm. Ultrafilters are classified in kilodaltons, with pore-size ratings of about 0.05 μm to 0.0005 μm. Ultrafilters retain large molecules such as gamma globulin and serum albumin. Nanofilters have pore-size ratings of between about 0.005 μm and 0.0002 μm. Nanofilters are typically used for virus removal in the pharmaceutical and biopharmaceutical industries. Reverse osmosis membranes are rated with pore sizes between 0.005 μm and 0.00005 μm, and they are the tightest of the porous membranes. They find their use in the pharmaceutical industry primarily in water purification (13).

For all but the most unusual situations, 0.2- and 0.1-μm-rated membrane filters will continue to be used in pharmaceutical and biopharmaceutical sterilizing filtration applications, assuming viral removal is not an issue. These filters can be integrity tested and validated, and they have combinations of flow-rate and throughput that do not unduly impact their utility as process filters. It is unlikely that filters with smaller pore-size ratings will be used unless new technologies are discovered to increase throughput and to reduce expense.

FILLING
Plastic Containers

Glass vials, ampoules, and bottles have been used extensively to package parenteral products. Glass containers are clear, offer high chemical resistance, and can be securely closed either by fusion or with elastomeric closures, thereby preserving product sterility. These containers also withstand autoclaving and have low vapor and gas transmission characteristics.

Glass containers must be processed before they can be used. Washing and depyrogenation steps consume relatively large amounts of purified water and water for injection and electrical energy. Validation of washing and depyrogenation steps must be performed. Stoppers must also be washed to ensure cleanliness and freedom from endotoxins unless they are provided ready to use by the manufacturer. Failure to adequately prepare glass container closure systems can result in undesirable particulate matter in the finished product. Finally, glass containers are fragile. Breakage can be especially problematic if the pharmaceutical products are highly potent, toxic, or carcinogenic. Plastic containers solve many of these problems while offering some new ones of their own.

The first widespread use of plastic containers for injectable products was for large-volume parenterals (14). These infusion bags were easier and safer to handle in clinical practice than glass and did not require washing and depyrogenation. However, special autoclave cycles employing air overpressure had to be developed to prevent damage to the containers during postfilling terminal sterilization. Also, it was more difficult to inspect the finished product for the presence of particulate matter since the containers were not as clear as glass, although modern inspectional systems along with statistical sampling have provided means to ensure that appropriate control of foreign matter in infusion bags and other opaque or semiopaque products is achieved. Additionally, filters at the point of use can be employed with all products including infusion bags. Also, there are now infusion bags in which the drug and diluent are aseptically filled in advanced technologies such as automation and isolator technology.

Blow-fill-seal (BFS) and form-fill-seal technologies using both low- and high-density polyethylene and other polymers have been used for ophthalmic and for parenteral products (15). In this process, the container is formed from melted polymer pellets, filled, and sealed sequentially in a controlled environment designed to preserve the sterility of the components and the finished product. Container washing and depyrogenation are not required. Like infusion bags, the containers are translucent, which hampers visual inspection of the contents.

Plastic containers, however, have several disadvantages compared with glass. Leachable compounds from the polymers themselves, from plasticizers, and from inks and label adhesives can potentially find their way into the finished product. Careful formulation development and choice of polymers can effectively mitigate this problem. Plastic containers are in some cases relatively more permeable to atmospheric gases and water vapor than glass. Oxygen-sensitive will require careful consideration gas permeation properties in the development process; however, they can provide a vapor barrier that compares well with glass. Again, careful formulation development and packaging configurations utilizing vapor-resistant barrier overwraps can mitigate vapor and gas transfer issues. It is apparent that the percentage of plastic containers used for pharmaceutical packaging has been increasing. It is likely that the percentage will continue to increase, especially with the discovery of new polymers and packaging technologies that will reduce vapor transmission and leachables, improve container clarity, and provide increased safety for highly potent products. Industry has seen remarkable development in the properties of plastic containers over the last 15 to 20 years and we have reason to suspect that there will be an even more rapid rate of delivery system and container development over the next decade.

Closed Vial Technology

Glass vials and elastomeric closures have been used for decades to package parenteral products. Generally, the vials and closures must be washed and depyrogenated before use and in the case of aseptically filled products rendered sterile. These processes consume relatively large quantities of purified water and water for injection and the WFI production and glass depyrogenation processes are energy intensive. After filling, the closures are inserted and aluminum seals applied. Filling, stoppering, and sealing are performed in controlled environment areas, ISO Class 5 for aseptically filled products. The controlled environment must be monitored to ensure it remains within its designed operating parameters. Product contact surfaces of the vials and closures must remain sterile. This means that stopper hoppers, tracks, and vibratory bowls must be sterilized using validated processes and assembled so that they remain sterile throughout the filling and stoppering processes (16).

Closed vial filling offers an attractive alternative to conventional container closure systems. It eliminates the need for container and closure washing, depyrogenation, and sterilization, and it eliminates the need for the attendant systems such as purified water, water for injection, vial and stopper washers, depyrogenation/sterilization tunnels, and highly controlled environmental conditions related to container-closure preparation. Closed vial filling significantly decreases energy consumption, eliminates the costs of installation and operational qualification for these unneeded systems, and significantly reduces validation costs, especially those associated with environmental monitoring and media fills.

In the closed vial filling system, product is aseptically filled into presterilized closed containers inside the filling machine that maintains an ISO Class 5 environment, resulting in a high level of sterility assurance. Vials and stoppers are manufactured and assembled robotically in an ISO 5 clean room, resulting in very low levels of subvisible and no visible particulates. The assembled container-closure system is subsequently exposed to a γ-irradiation sterilization process at 25 kGy (minimum). The closed vial filling system incorporates e-beam irradiation for surface sterilization of the closure immediately prior to filling through the vial closure (stopper), laser resealing of the closure puncture, and application of the flip-away cap.

Other advantages include high levels of safety for operators, supply chain, and medical personnel when handling potent or cytotoxic products since the process design uses polymeric vials that eliminate breakage.

Separative Technologies
Separative technologies are defined as environmental control systems that fully separate the human operator from the aseptic production environment. Separative technologies by preventing direct human intervention proactively minimize the risk associated with human-borne and released contamination into the aseptic manufacturing environment. It has long been recognized that humans are the only significant source of contamination and hence risk in aseptic processing. Therefore, the implementation of separative technologies marks a very significant step forward in the effort to reduce microbiological risk to the end user. Two general categories of separative technologies are used in the aseptic processing industries: isolator technology and RABS.

Isolation Technology
Isolation technology can be expected to become the dominant environmental control system for the filling of aseptically filled products. The number of conventional human occupied clean rooms being constructed in our industry has been diminished in number over the last 10 to 15 years. Newly constructed facilities are most commonly isolator or RABS-based designs. The reasons for the introduction of these technologies are obvious; they offer an enhanced means to separate personnel from sterile materials. One of us was bold enough to predict that isolators would become the dominant means of production for aseptic production (17). It now seems certain that isolators will ultimately become the technology of choice for aseptic filling. RABS will play a role in the future of our industry, but it seems to us that that RABS cannot realize the performance and return on investment available with isolation technology and thus will play only a secondary role (see following text).

Isolation technology provides undeniable advantages over other environmental control systems for aseptic processing:

- Has the ability to decontaminate reproducibly and automatically
- Provides the best possible separation between operators and sterile materials
- Eliminates aseptic gowning for personnel
- Reduces external environmental conditions relative to other technologies
- Reduces operating costs
- Provides superior environmental conditions

For these reasons and others the trend toward increased use of isolators is likely to continue (18). We predict they will become the technology of choice for virtually all newly constructed aseptic processing facilities.

Although isolators have noteworthy advantages, they may not be the best choice for all product types or for all drug delivery system/drug product combinations. In 2010, when we think of isolator technology we immediately think of vapor-phase hydrogen peroxide (VPHP) decontamination of the system as well. However, protein and peptide products can, in some circumstances, be exquisitely sensitive to H_2O_2 residues. There have been residue targets as low as 10 to 50 ppb established, and this can be a practical difficulty depending on the need to decontaminate product contact materials in situ and also depending on the complexity of the processing equipment. It will be critical for research and development into alternative decontamination/sterilization methods to continue apace as we move forward, and it may also be necessary for industry and regulators to consider that decontamination as currently practiced may not be required in all cases.

Restricted Access Barrier Systems
The production of sterile products has been dominated by the conventional manned clean room since the 1960s. Around 1990, after they had been implemented for sterility testing at a number of firms, isolator-based filling systems were first introduced into the sterile products sector. Hailed as extraordinary breakthrough, the initial interest in their capabilities was overwhelmingly favorable. Regrettably, some of the initial enthusiasm was tempered by firms that had difficulty with their implementation. Whether this was the fault of the technology in

asking too much too soon or the result of overstated expectations on the part of some firms is unclear. What is certain was that there was a desire for a system with the performance capabilities of an isolator with the simplicity of a manned clean room. Thus, the RABS was born, as a less complex alternative to isolation technology. As this is written, the use of RABS is expanding, but evidence is starting to appear that the rate of growth may not match that of isolators. After something of a plateau in implementation rate in the late 1990s and early part of the current decade, it seems clear that isolator implementation is accelerating. This may be largely due to the fact that industry and the regulatory community have learned that many of the concerns associated with mouse holes, rapid transport systems, enclosure leakage, decontamination efficacy, mouse holes, and gloves were significantly exaggerated. Actually, none of these issues have proven to be significant contamination risks. Plus, the general understanding of these issues has improved and effective countermeasures have been implemented to mitigate the perceived risk associated with these devises.

The glove system is a wonderful example of industry responding in a logical commonsense manner to reports of higher than expected glove leak rates in the mid-1990s. A wholesale move to Hypalon gloves, which have proven more reliable than the Neoprene gloves originally used, and the implementation of effective glove leak testing, inspection, and management programs have resulted in a manifold reduction in risk. Similar commonsense approaches have reduced concern regarding the so-called RTP ring of uncertainty as well.

Perhaps most importantly though, the isolator has effectively matched the contamination control benefits advocates of this technology foresaw over two decades ago. The best argument in favor of isolators is that they have, when well designed and implemented, proven to be exceedingly successful. Isolator technology has further benefited from the redesign of processing equipment to make it more isolator friendly. Manufacturers have learned to design and engineer equipment that functions much better in the isolator environment and which require far fewer interventions using gloves. So effectively has equipment been designed to work in isolators that gloveless isolators which seemed a pipe dream in the near past are now quite possible.

It is evident to all that isolators have superior capabilities to RABS in some areas. There is a belief that the simpler RABS designs may offer advantages and indeed where VPHP decontamination is a persistent problem to the product, closed RABS systems would be a viable solution worth considering.

The long-term prognosis for aseptic processing technology is thus somewhat clouded. RABS appear to offer capability without the validation headaches sometimes associated with VPHP and thus have considerable appeal. It is certain that as a retrofit, or replacement of an existing filler system within an operating facility, RABS can perform better than the human scale clean room. Converting an existing parenteral operation over to isolation technology seems a near impossible task; the facility alterations are so significant that such a project seems fraught with pitfalls. RABS can be utilized as a suite-by-suite approach to upgrading an operating facility without the difficulties a conversion to isolation technology might entail. This is perhaps the greatest opportunity for their application, and given the longevity of parenteral facilities, it is in this mode that we might see the greatest application of RABS in the future.

The isolator by virtue of its superior decontamination, operator protection, and reduced operational cost will be the preferred approach for new construction. This has been the conclusion reached by many firms across the industry, and is likely to continue in the foreseeable future. Thus, some 20 years from now, we will likely see both RABS and isolators in everyday use across the industry.

Robotics

In 2004, the first robot designed to operate in a VPHP decontaminated environment was installed and validated in an isolator technology filling line in Japan. In this installation the robots were used for unloading stopper containers from the autoclave and also were used to charge the stopper bowl. Robots have also been used successfully in aseptic cell culture isolators and in radiopharmaceuticals applications. There are now at least two firms offering robots designed to operate in an isolator environment and therefore capable of routine, frequent exposure to VPHP.

If we consider that the real benefit of separative technology is the elimination of the direct human intervention, it becomes clear that the robot can be equally effective at mitigating contamination risk. Clearly, eliminating the human with automation can work just as effectively at controlling contamination as separating the human from the aseptic environment. Recently, at least one vendor has begun marketing robots that are both resistant to VPHP and capable of remote operation using hand controls. These hand controls can also be used to train the robot increasing flexibility and reducing the need for complicated programming when process change is required.

The reluctance toward the use of robotics in the pharmaceutical and biopharmaceutical industries is puzzling given the widespread implementation of robotics in nearly all manufacturing industries. Robots have proven to be economically advantageous not only in heavy industries but also in light manufacturing including electronic component assembly. As this chapter is written our industry continues to lag behind other manufacturing industries in robotic applications and in some other forms of machine automation as well. This is very likely to change over the next decade as our industry learns the myriad advantages of robots, and success breaks down the prevailing reluctance to use them.

Robots and automation will also reduce interest in finding lower cost manufacturing venues. In fact, logic tells us that as they continue to develop, countries like India, China, and others will see their standards of living and regulatory climates change such that they will not be able to compete on price advantage alone. Once, not so long ago, Japan was considered a relatively low-priced site of manufacture, but this is certainly no longer the case. There is no reason to believe that nations now moving up the development trajectory will not follow a similar course. Thus, at some point, shipping and short supply lines may be of more interest in low costs of labor, low taxation, and lax environmental compliance requirements, all of which are likely to go away as a country develops. Robots could make possible economic and low cost manufacturing closer to the target populations for a given group of patients.

Robots are also likely to play a greater role as more customized healthcare products are developed. These products will not require high-throughput operations but rather flexible manufacturing in ultralow risk aseptic environments, since in effect most of these products will be released aseptically and process validation in the manner we understand it now may not be fully possible. Gene therapy and regenerative medicine products promise to be game changers in the therapeutic world and they need to be made close to the point of use. Manual assembly or fill as already covered is not desirable. Therefore, the only logical solution will be robotics and quite often robotics in conjunction with separative technology of one kind or another.

BFS

BFS systems are quite simple in concept but extremely complex in terms of engineering, manufacture, and operation. Earlier generations of BFS systems required not infrequent interventions to clear solidified plastic from fill nozzles and other interior parts. These systems did not qualify as advanced aseptic processing systems as they did not effectively eliminate direct operator interventions.

Fortunately, in the current generation of BFS systems, intervention-free operation is not only possible, it is generally achieved. As previously stated, plastic resin is the starting material for the BFS container. These resins are fed from large holding containers to hoppers on the machine; the plastic resin generally in the form of beads is melted and blown into molds under relatively high temperature and sufficient pressure to form the container. Filling systems, generally of the piston-pump variety, dose product into the container, which is then heat sealed. All filing and sealing is accomplished under an air shower that provides unidirectional Class 5 HEPA filtered air to the aseptic critical zone of the equipment. The containers, which may range from single-dose ampoules of 1 mL volume or less to up to a liter or more, are effectively sterilized by the heat and pressure of molding. The filling systems are sterilized and in most cases cleaned in place, thus no aseptic connections are required. The result is a very low contamination risk aseptic production system.

There are only two risk modalities associated with the current generation of BFS equipment. The first is the maintenance of a sterile supply of drug to the filling system.

BFS lends itself to quite long campaigns that may reach or exceed seven days. Therefore, the ability of the compounding and filtration system to deliver sterile product over an extended period of time is vital. This requires very careful design and engineering to ensure that bioburden can be very well controlled through the campaign duration. Obviously, the higher the probability of the formulation supporting microbial growth and therefore being prone to the amplification of contaminants, the greater the inherent risk and the most careful the design, engineering, installation, validation, and process control requirements. The second contamination route that has been observed in BFS originates from the cooling system. Generally the molds are cooled by water, and the water circulation system is not sterile in any current design. Given the long running times these cooling water systems can be a source of contamination, should leaks occur. Antimicrobial agents could be used in the cooling water system, but this raises a risk of chemical contamination, should aerosols occur. Fortunately, advancements in the design and therefore safety of cooling water systems are evolving rapidly.

Blow molding in-line of bottles made of (polyethylene terephthalate) PET or other plastics is possible, and aseptic filling systems using such bottles are in use and validated. In most cases the bottles are subjected to a sterilization using a chemical or more recently e-beam sterilization. Prefilled syringes can also be blow-molded inline and sterilized en route to the filling process generally using e-beam.

In situ or in-line blow molding of plastic bottles is a technology that will continue to evolve and which can be applied to even rather complex dosing systems in the future. It seems logical that the closures compatible with hypodermic syringes can certainly be developed and implemented in the coming decade. In-line or in situ blow molding with or without instantaneous heat sealing is likely to be with us for many decades to come. Also, as is the case with robotics we may see blow molding in use with isolators or RABS systems. In fact, hybridization of different manufacturing, environmental control, and automation/robots systems in a single-production operation seems increasingly likely.

Aseptic Filling Systems

Dramatic changes in filling technology, particularly for aseptically processed containers are anticipated. The equipment would be highly automated and specifically designed to operate without human access and would include operating capabilities and features such as

- Provision for all routine interventions
- Elimination of corrective interventions
- Clean-in-place or sterilize-in-place capabilities for all product contact surfaces
- Weight verification or adjustment on all containers
- Container integrity control and confirmation on all containers
- Continuous monitoring of critical process variables
- The use of PAT where appropriate
- Automated in-feed and discharge of components without human intervention
- Automated environmental monitoring of isolator internal air and surfaces
- Automated setup and transition from clean-in-place or sterilize-in-place to aseptic filling
- Self-clearing filling systems (for jam-free operation)
- No-container, no-fill to eliminate spillage

The reader is encouraged to seek out the many detailed reference texts that cover the technologies introduced in this section in far more technical detail than this brief chapter allows. Suffice it to say that the variety of aseptic manufacturing systems in the near future is limited only by the imagination. The authors are convinced though that each of these systems will have one critical thing in common, which is that they will all fully eliminate the direct human intervention. The future of aseptic processing will not be just one technology for manufacturing or one technology for environmental control, rather many possibilities will exist. However, they will all completely eliminate human contamination risk and thus result in aseptically produced products that are so safe that whole new regulatory approaches will have to be required. A blind adherence to clean room–derived regulatory policies will slow development and implementation, and therefore result in both safety and economic harm to the end user.

STERILIZATION
ClO$_2$, H$_2$O$_2$ and O$_3$ Sterilization/Decontamination

Sterilization of materials is at the core of almost every parenteral manufacturing process. The processes utilized are dominated by heating processes, with steam and dry heat among the most common. Solutions of course have been sterilized by membrane filtration for many years, a situation that is unlikely to change given the unique nature of fluids. For materials that are susceptible to heat, radiation and ethylene oxide (ETO) are the predominant alternatives. Radiation sterilization is seeing increased use (see following text), but is not compatible with some materials. ETO sterilization is widely used in the sterilization of medical devices, as well as many plastic items utilized in the pharmaceutical industry including filters, wipes, and containers. Despite its sterilization efficacy and wide use, ETO is hardly a method of choice because of the extreme environmental, toxicity, and safety issues associated with its use. Much of the pharmaceutical industry moved away from in-house sterilization using ETO to avoid the extensive measures required to handle it safely. As a consequence, contract sterilization firms now provide the bulk of ETO sterilization capacity worldwide. Movement away from ETO mixtures with CFC-12 (dichlorodifluoromethane) has accelerated because of potential ozone layer depletion as a result of CFC emissions (19). As a consequence, the global healthcare industry has sought alternatives to ETO that could provide comparable sterilization effectiveness for heat-sensitive materials without its substantial negative consequences. Contract sterilization sites no longer utilize ETO/CFC mixtures, and many have converted to 100% ETO systems increasing the explosion hazards and worker safety concerns accordingly.

Chlorine dioxide, hydrogen peroxide and ozone have demonstrated a broad range of antimicrobial activity against both vegetative cells and spore-forming microorganisms (20,21). The broader application of these agents for sterilization has been in part limited by the extensive experience and installed base of ETO sterilization units. As long as ETO use is considered acceptable and contract sterilization is available, there is minimal incentive to pursue alternatives. Nevertheless, development of ClO$_2$, H$_2$O$_2$ and O$_3$ has been pursued by firms seeking a safer and environmentally friendly alternative.

Ozone decontamination of classified rooms was incorporated into the initial design of an aseptic processing facility for Novartis in Stein, Switzerland.[e] This facility also utilized tightly sealed double door airlocks with ozone for the sanitization of items being introduced into the aseptic core. Both installations have been proven effective for microbial control in the facility. TSO$_3$ of Quebec City, Canada, obtained 510K approval of an ozone sterilizer by the FDA in 2003.

The detection of *Bacillus anthracis* in the U.S. post office and government buildings required a means for removal of spores of this toxic microbe. Chlorine dioxide was one of the agents utilized for this treatment, and it demonstrated excellent efficacy with minimal complications. This experience has led to additional applications on facilities dealing with mold and other forms of microbial contamination. Chlorine dioxide also has been successfully utilized for decontamination of isolator environments (22).

These agents have both demonstrated excellent lethality against spore-forming microorganisms, and expanded usage of each agent can be expected in the future. The only drawback to their application is perhaps the substantial installed capacity of contract ETO units across the globe. Given the explosion hazard and worker safety issues associated with ETO, it can be anticipated that a slow shift to ClO$_2$, H$_2$O$_2$ and O$_3$ is possible. For firms seeking in-house gas sterilization capability, these agents may become more common as knowledge of their efficacy grows. They each offer the industry a sterilization alternative to current processes.

Sterilization by Radiation

Radiation is widely utilized in the medical device industry for the sterilization of a wide variety of items including bandages, implants, latex gloves, wipes, and countless other items using either γ rays or electron beams. The extensive use of radiation for medical devices has been supported through the development of consensus documents that aid the practitioner in defining, validating, and maintaining a consistent process (23,24). These documents formed

---

[e] J Agalloco, personal communication, 1995.

the basis for global standards for radiation sterilization with the device industry, but were poorly suited for applications in pharmaceuticals. The lethality assumptions inherent in the older radiation sterilization standards are based on bioburden assumptions from common materials utilized in medical devices. The methods relied heavily on microbial testing of large numbers of fractionally sterilized units, with the results utilized to establish the minimum amount of radiation required. The application of these methods for sterilizing pharmaceutical products was considered quite difficult, and as a consequence only a handful of pharmaceuticals were ever successfully sterilized using radiation.

The success with radiation sterilization processes within medical device industry led to increasing consideration for application in pharmaceutical processing. Perhaps the single greatest factor in the expanded use of radiation sterilization has been the emergence of the VD_{MAX} methods for establishing an effective radiation dose for sterilization of materials (25). In contrast to the other dose-setting methods utilized that are heavily device oriented, the VD_{MAX} method utilizes substantially smaller samples of materials to establish a sterilization process. This seemingly modest change is better suited to pharmaceutical development, where limited material availability and high cost make use of the other dose-setting methods impractical. Several finished dosage forms have been introduced using radiation sterilization in a terminal process, something that would have proved impractical and prohibitively expensive previously. Applications for postaseptic fill lethal treatment using adaptations of the VD_{MAX} dose-setting method are also possible (see following section).[f] The continued use of radiation within the medical device industry has aided other applications. Not only is there a growing body of knowledge regarding radiation sterilization, makers of plastic containers, elastomeric closures, and other plastic materials have developed formulations and polymers that are less susceptible to the damaging effect of radiation, expanding the possibilities for application in many settings. Another factor influencing radiation usage is the increase in isolator installations, where electron beam systems are well suited for use for continuous material in-feed of heat-sensitive materials.

Several new radiation sterilization technologies are in active development, with a range of applications as diverse as full pallet e-beam and X-ray systems, and small-scale systems that can be inserted into individual containers. As with any technology, once the initial resistance has been overcome, increasing usage follows. It seems clear that radiation sterilization will play an increasingly prominent role in future production methods for parenteral dosage forms, expanding upon the modest but very promising use now being experienced.

Postaseptic Fill Lethal Treatments

The preparation of sterile products has been dominated by two distinct approaches for many years. Products are either manufactured using aseptic processing or terminal sterilization. The distinction between the processes has always been rather sharp, the processes were considered distinct and separate. Although preference was always given to the use of a terminal approach because of its increased reliability and certainty, aseptic processing was utilized in the majority of formulation because the adverse impact of the expected terminal processes in use proved destructive of product properties in many cases.

Some years back, there was an exchange of missives among industry regarding the potential for a postaseptic treatment to provide a higher level of sterility assurance to the end product (26–29). The discussion went on for some time and while perhaps educational, none of those who participated seized on what was perhaps the salient point that lay just beneath the surface of that discussion on sterility assurance level (SAL). While we quibbled on terminology, we all glossed over what should have been the focus of our interaction— would a postaseptic fill treatment of some kind provide a safer product for the end user? We never quite reached that issue in our discussion. Reflecting back on that dialog, I suppose we all knew that it would, and yet somehow we never broached that question or its answer directly. Hindsight is of course 20-20, and the answer to that unasked question has to be a resounding—yes, of course it most certainly would.

---

[f]JB Kowalski and JP Agalloco, personal communication, 2008.

In recent years, the climate for regulation has changed dramatically, with FDAs proposal for Risk-Based Compliance dramatically influencing industry thinking (30). It seems obvious that, if a postaseptic lethal treatment is provided to what is already accepted as a sterile product then the likelihood of an isolated contaminant surviving in the container would be reduced. The FDA made the point during the revised Aseptic Processing guidance, that aseptically filled products were the cause of the vast majority of recalls for lack of sterility assurance (31). In discussions held with the FDA to review the first draft of that document, a brief discussion was held regarding the potential desirability of a supplemental treatment following aseptic processing, but as the Product Quality Research Institute (PQRI) group was pressed for time, we tabled the subject, and regrettably never returned to it subsequently (32). The European Community regulatory community approached the subject of process selection rather differently and unfortunately in an extremely rigorous manner (33). The decision tree sets forth strict requirements for minimal treatments that would be accepted as terminal sterilization. Any lethal process not attaining a minimum F_0 of eight minutes shall be produced by aseptic processing. Some years earlier, the FDA endeavored to mandate an explicit preference for terminal sterilization through a modification of the 211 regulations (34). There was substantial discussion of this proposal, and it was eventually tabled as something not easily implemented. One of the many discussion items at that time was a request by industry for a defined minimum F_0 threshold that FDA would accept (35). This would have resulted in a situation much like that resulting from the PIC/S decision tree, albeit somewhat earlier. The absence of a single minimum value has not hampered firms seeking to assure greater safety for their products, and we are aware of firms that have utilized a number of very flexible approaches and minimum F_0 targets well below eight minutes. Nevertheless, the conjunction "or" is far more commonly used than "and" when consideration of processes for sterile products involves both aseptic processing and terminal sterilization.

Given the desire to mitigate patient risk, and in full consideration of all that has occurred in the past, it seems obvious that subjecting an aseptically filled (supported by a process simulation program) to some form of lethal treatment afterward will become increasingly common in the not too distant future. Possible moist heat processes that could be utilized include an A_0 process in the range of 70 to 90°C (A_0 is a process for microbial control of vegetative cells in hospitals); intermediate temperature ($\sim$100°C) for destruction of non-thermophilic spore formers; or low F_0 processes (<8 minutes). Adaptations of the VD_{MAX} radiation process can be utilized for a comparable objective where modest radiation doses, below those considered minimal for sterilization, could provide comparable improvements in microbial control following an aseptic fill.

TESTING AND INSPECTION
Elimination of Sterility Testing for Terminally Sterilized Products
Sterility testing has been an accepted practice for the acceptance of sterile products since their inception. It has been a mandatory requirement since 1932 and first appeared in USP 11 and was official in 1936 (36). The test provides a laboratory test component that endeavors to establish that sterile products are not microbially contaminated. When first introduced into the pharmacopeia, the precepts of validation for sterilization and process simulation for aseptic processing were many years in the future. Under those circumstances, it was wholly appropriate to require that a test able to detect microbial contamination be added to monographs for sterile products. Batch sizes at the time were relatively small and production methods for sterile products were primitive by today's standards. Contamination levels in sterile products at that time are largely unknown, but presumably substantially higher than at present. Under those circumstances, testing of a sample for sterility might be a reasonable expectation. The first mention media fills defined a contamination rate for aseptic process in the 1970s of approximately 0.3%, a number that is at least an order of magnitude higher that what is considered attainable in the industry today (37).

With the passage of years since it became a compendial requirement, the sterility test monograph has gone through considerable change with respect to the methods, media, sample selection, response to positive results, and a number of other changes. These changes have adapted how the test is performed and the results interpreted, but throughout the years the basic objective of the test has remained unaltered. The sterility test has remained a mandatory part of sterile product release systems. Produced lots, whether manufactured by terminal sterilization or aseptic processing, must be subject to the requirements of the sterility test unless regulatory approval has been granted for parametric release.

The first efforts toward parametric release of terminally sterilized products by moist heat were developed within Baxter Healthcare, which received FDA approval for its submission in 1985 after a 4-year review period. After that approval, the FDA outlined its expectations, perhaps drawn from the Baxter submission in a 1987 FDA Compliance Policy Guide CPG 460.800 (38). That document outlined the regulatory expectations for submission to use parametric release in lieu of sterility testing. The next submission for pharmaceuticals was approved in the mid-1990s, and several additional filings have been submitted and approved subsequently. Parametric release submissions and approvals have also been granted for medical devices using ETO and radiation sterilization.

The PDA provided an initial consensus industry perspective on parametric release in which it endeavored to define expectations for application of parametric release (39). The document outlines the components of a validation effort that could support parametric release with the goal of defining practices prior to regulatory initiatives. An updated version of the document is currently in development dealing with technology changes and more recent regulatory pronouncements.

The European Medicines Agency for Evaluation of Medicinal Products (EMA) CPMP committee issued a parametric release position statement in 2001, which updated thinking and provided a broader regulatory perspective (40). This was followed closely by an Annex to the EU CGMPs on the same subject (41). These efforts addressed parametric release in the broader context of operations with pharmaceuticals; nevertheless they did specifically address application for sterile products. The perspectives voiced in these documents brought parametric release out of a guidance setting and into formal regulation. The essence of these documents relative to sterility testing can be summarized in a single sentence, "Elimination of the sterility test is only valid on the basis of successful demonstration that predetermined, validated sterilizing conditions have been achieved." The focus of parametric release clearly resides in the sterilization validation effort. The documents correctly speak to the severe statistical limitations inherent in the sterility test.

The FDA has provided long anticipated parametric release guidance in a 2008 document that included the following definition (42):

> *Parametric release* is defined as a sterility assurance release program where demonstrated control of the sterilization process enables a firm to use defined critical process controls, in lieu of the sterility test, to fulfill the intent of 21 CFR 211.165(a), and 211.167(a).

The basic principle behind the use of parametric release in lieu of sterility testing is a heavy reliance on rigorous production controls largely defined from sterilization validation. A firm using parametric release must establish criteria for evaluation of each sterilization cycle that are used to establish conformance to the validated sterilization process. To what extent does the validation effort required for parametric release differ from that for sterilization processes used for nonparametric application? The answer to this question is that all modern sterilization processes are validated in an identical manner. Sterilization processes for the preparation of components, filling parts, and product formulations where the end product is sterile but where the treatment is in-process are validated in an identical manner, but there are almost no in-process sterility tests performed on these materials. Thus, parametric release is widespread, but somewhat "sub rosa" in that these processes are not considered within the same context because they occur in-process rather than as a terminal treatment. For aseptically filled products, which comprise an estimated 85% of all sterile products, consideration is expected in release for a review of all records impacting the sterility of the finished product. Thus parametric release is in daily usage worldwide for virtually every in-process sterilization performed. Given that these processes are so widespread, without scrutiny, and followed by a far less certain aseptic process does not alter the inherent risk associated with this common practice.

Why then should terminal sterilization processes where there are no subsequent contamination introduction steps be held to a more rigorous standard of requiring formal approval prior to adoption of parametric release? In Annex 17, EMEA cautions that parametric release should not be considered for initial use, "It is unlikely that a completely new product would be considered as suitable for Parametric Release because a period of satisfactory sterility test results will form part of the acceptance criteria. There may be cases when a new product is only a minor variation, from the sterility assurance point of view, and existing sterility test data from other products could be considered as relevant." The FDA's more recent effort maintains consideration of prior history with sterility testing, "Experience with the proposed or similar product (and container closure system), the overall risks to sterility, and the steps you have taken to assess and control these risks." These cautions seem misplaced, historical performance with a statistically invalid test is really of no relevance. The proper perspective with respect to initiation of parametric release has perhaps been enunciated by Dr T. Sasaki of the Japanese Ministry of Health, "Everybody knows that sterility testing is meaningless for terminally sterilized products, but sterility testing is still carried out on terminally sterilized pharmaceutical products in many countries. In Japan, we have investigated the introduction of PR for terminally sterilized pharmaceutical products from a scientific viewpoint for the past two years " (43). The conclusion to that investigation is certainly somewhat different from what might be expected from a major regulatory body; the JP has included the following in its general notices (44).

> When a high level of sterility assurance is maintained consistently, based on the records derived from validation studies of the manufacturing process and the in-process controls, the sterility test usually required for the release of the products may be omitted.

Dr Sasaki indicates in his article that the JP did not restrict parametric release merely to terminally sterilized products, and that with appropriate controls it could be considered for aseptically produced materials as well.[8] The JP has provided a three-tiered approach to terminal sterilization process validation including a process in which an aseptically filled product is acceptable for release if its subsequent heat treatment it receives an $F_0 > 2$ minutes as a minimum.

The JP initiative addresses the subject of parametric release from a purely scientific perspective, and takes the discussion in an entirely new direction level. Objective science would certainly preclude the imposition of the sterility test for terminally sterilized products, where it quite literally serves no beneficial purpose. The statistical limitations of the test are such that if it were submitted to the global compendia as a proposed new general test, it would likely be rejected as lacking any real utility. It is our considered opinion that the sterility test will be eliminated for any material subjected to a validated terminal sterilization process. Sterility testing will be acknowledged for the anachronism that it is for terminal processes; a means for release of sterile products that was conceived in an era of substantially less capability that has been largely transcended by present day validation capabilities. Its continuance in the global compendia is not justifiable considering its statistical limitations that could have been overlooked at the time of its adoption, but given the present state of the sterilization proficiency is a useless and arbitrary constraint on operations. We can also envision a time in the future, where this same statement could be made with respect to some future aseptic processing technology.

Visual and Automated Parenteral Inspection
Visual inspection of parenteral products is driven by the need to minimize the introduction of unintended particulate matter to patients during the delivery of injectable medications. Such inspection also offers the opportunity to reject nonintegral units, such as those with cracks or incomplete seals, which pose a risk to the sterility of the product. The desire to detect these defects at a very low frequency and the randomness of their occurrence has resulted in the current expectation for inspection of each finished unit (100% inspection).

---

[8]Dr Sasaki explained that in the case of aseptic processing, there is no means to demonstrate a SAL of $<10^{-6}$ in terms of aseptic processing parameters. So at this time, it is impossible to permit parametric release for aseptically filled products. In the future, it may be possible to accept parametric release for aseptic processing from some parameter(s) as yet undefined or currently undeterminable.

Human-based visual inspection historically has been done against a white and a black background using 100-W incandescent lighting. Such inspection is labor intensive and the results are subjective, depending on the skill and training of the inspectors. The results can be influenced by the inspection technique vis-à-vis manipulation of the inspected containers to optimize the location of any particulate matter that may be present and interpretation of what constitutes a defect, for example, bubble in the wall of the container, tightness of aluminum seal, etc.

Automated inspection equipment has been designed to improve the efficiency and accuracy of the inspection process, eliminating the subjectivity of human inspection. Automated inspection equipment is designed to detect particles in the product being inspected and to detect container and closure defects that may compromise the product, that is, cracks, pinholes, leaks, and loose seals. While human inspectors can inspect for both types of defects, machine systems are usually designed to inspect for one or the other, although some automated inspection systems perform both tasks. Automated inspection equipment utilizes light extinction, light scattering, lasers, light-emitting diodes, high-intensity lighting, video cameras, polarization, moving and stationary inspection technologies coupled with micro-processors to detect and categorize the various types of defects.

Automated inspection should be validated to ensure the results are at least as good as those obtained by a visual inspection performed by a well trained and qualified human inspector performing the inspection according to pharmacopoeial standards.

New technologies will certainly be developed to improve the speed and accuracy of automated inspection. One important factor will be manipulation of the container to focus the location of visible particulate matter to improve its detection capability.

PAT for On-line Release of Lots

PAT was introduced to the pharmaceutical industry by FDA's Office of Pharmaceutical Science, a branch of CDER, in 1996 as the Process Analytical Technology Initiative. Part of the initiative was a document titled "Guidance for Industry PAT—A Framework for Innovative Pharmaceutical Development, Manufacturing, and Quality Assurance." According to the guidance, an important FDA goal was "to tailor the Agency's usual regulatory scrutiny to meet the needs of PAT-based innovations that (1) improve the scientific basis for establishing regulatory specifications, (2) promote continuous improvement, and (3) improve manufacturing while maintaining or improving the current level of product quality" (45). More than 10 years later, pharmaceutical companies are still wrestling with ways to implement PAT principles and technologies in their manufacturing operations.

Historically, and as required in the GMP regulations, manufacturers had produced pharmaceutical products and tested them at various intervals in the production process to ascertain whether they meet predetermined quality standards. The manufacturing process and testing requirements were predicated on the contents of the approved application, which could in general only be changed through resubmission and approval of a supplement. The FDA initiative titled "Pharmaceutical CGMPs for the 21st Century—A Risk-Based Approach," introduced in 2002, effectively paved the way for implementation of PAT. The initiative provided for the following:

- Encourage the early adoption of new technological advances by the pharmaceutical industry.
- Facilitate industry application of modern quality management techniques, including implementation of quality systems approaches, to all aspects of pharmaceutical production and quality assurance.
- Encourage implementation of risk-based approaches that focus both industry and agency attention on critical areas.
- Ensure that regulatory review, compliance, and inspection policies are based on state-of-the-art pharmaceutical science.
- Enhance the consistency and coordination of FDA's drug quality regulatory programs, in part, by further integrating enhanced quality systems approaches into the agency's business processes and regulatory policies concerning review and inspection activities.

PAT, coupled with robust and effective product development, risk analysis, and risk management practices, can improve manufacturing efficiencies and the quality of products reaching the consumer. Minor and unavoidable differences in active ingredients, excipients, and packaging components can be accommodated through the use of on-line, at-line sampling and testing coupled with automated process control. As sensor technology improves and automated control systems are implemented, PAT will replace conventional manufacture, hold, sample, test, and release quality control practices.

A PAT system used for tablet production provides an example of the efficiencies and process control advantages of such a system. The process starts with the automated addition and in-line blending of the active ingredient and excipients, which are discharged into a process hopper. Rates of addition and mixing are automatically controlled. Sensors in the hopper, coupled with mass-balance sensors in the feed system, ensure the correct concentration of the active ingredient and monitor blend uniformity.

The hopper discharges into a series of high-speed tablet presses equipped with hardness monitoring and weight control sensors. The produced tablets are also examined on-line with optical sensors to detect physical defects that may be present. Automated filling machines package and label the tablets as they are discharged from the compression step. Cap torque and correct labeling, control number, and expiration date are automatically verified. The packaged tablets are packed into shippers, the labeling of which is verified on-line, and the shippers are transferred by an automated conveyor system to a quarantine area in the warehouse, awaiting final batch record review and release. The batch records are generated automatically with input from the on-line process control systems and are reviewed by the quality control unit. Once production is complete, the process train, with the exception of the tablet presses, is cleaned in place. PAT systems are used in clean-in-place systems, monitoring and controlling those processes to ensure systems are clean and free of objectionable levels of residues. The process is estimated to reduce time between the start of production and batch release by 80% compared with the conventional production system it replaced.

PAT is likely in the next 10 years to become the norm for many types of products, stimulated by improved levels of process understanding and control and production efficiencies. The PAT and GMP for the 21st century initiatives, coupled with ICH Q8, 9, and 10, will make this a reality.

CONCLUSION

The materials included in this chapter likely represent only a portion of those likely to impact the manufacture of sterile products. Those who might read this chapter 20 or more years in the future will be either surprised by our insight, or humored by our misconceptions. In either case, we have little doubt, that by that time the preparation of sterile products will have been altered by one or more of the potential influences described above. More than 2000 years ago, the Greek historian, opined that "the only constant is change" (46). We wholeheartedly agree.

REFERENCES

1. Agalloco J, Akers J. The future of aseptic processing. Pharm Technol Aseptic Process 2005:S16–S23.
2. Madsen R. The future of aseptic processing. Pharm Technol Aseptic Process 2003:S41–S42.
3. ICH Secretariat, c/o IFPMA, 15 ch. Louis-Dunant, P.O. Box 195, 1211 Geneva 20, Switzerland.
4. Agalloco J, Madsen R, Akers J. Aseptic processing—a review of current industry practice. Pharm Technol 2004; 28(10):126–150.
5. Agalloco J, Akers J. Risk analysis for aseptic processing: the Akers-Agalloco Method. Pharm Technol 2005; 29(11):74–88.
6. Agalloco J, Akers J. Simplified risk analysis for aseptic processing: the Akers-Agalloco Method. Pharm Technol 2006; 30(7):60–76.
7. FDA. Pharmaceutical CGMPs for the 21st century—A risk-based approach: Final report, September, 2004.
8. Agalloco J. Points to consider in the validation of equipment cleaning procedures. J Parenter Sci Technol 1992; 46(5):163–168.
9. Madsen R, Agalloco J, Brame W, et al. Points to consider for cleaning validation, PDA Technical Report #29. PDA J Pharm Sci Technol 1999; 53(1 suppl).
10. Meltzer TH. Filtration in the Pharmaceutical Industry. New York, NY: Marcel Dekker, Inc, 1987.

11. Bowman FW, Holdowsky S. Antibio Chemother 1960; 8:508.
12. PDA, Technical Report No. 26. Sterilizing Filtration of Liquids. Bethesda, MD: Parenteral Drug Association, Inc, 1998.
13. Jornitz MW, Meltzer TH. Filtration Handbook: Liquids. Bethesda, MD: PDA, 2004.
14. Uoitla JA, Santasalo NT. New concepts in the manufacturing and sterilization of LVP's in plastic bottles. J Pharm Sci Technol 1981; 35(4):170–175.
15. Ljungqvist B, Reinmuller B, Lofgren A, et al. Current practice in the operation and validation of aseptic blow-fill-seal processes. PDA J Pharm Sci Technol 2006; 60(4):254–258.
16. Verjans B, Thilly J, Vandercasserie C. A new concept in aseptic filling: closed-vial technology. Pharm Technol Aseptic Process Suppl 2005: s24–s29.
17. Agalloco J. Opportunities and obstacles in the implementation of barrier technology. PDA J Pharm Sci Technol 1995; 49(5):244–248.
18. Agalloco JP. Thinking inside the box—application of isolation technology for aseptic processing. Pharm Technol Aseptic Process Suppl 2006; 30(5):S8–S11.
19. United Nations Environment Program. Montreal Protocol on Substances that Deplete the Ozone Layer, 1987.
20. Jeng DK, Woodworth AG. Chlorine Dioxide Gas Sterilization under Square Wave Conditions. Appl Environ Microbiol 1990; 56(2):514–519.
21. Foegeding PM. Ozone inactivation of *Bacillus* and *Clostridium* spores and the importance of spore coat to resistance. Food Microbiol 1985; 2:123–134.
22. Elyath A. Successful sterilization using chlorine dioxide gas: Part 1 Sanitizing an aseptic fill isolator. Bio Process Int 2003; 1(7):52–56.
23. AAMI. Process Control Guidelines for the Radiation Sterilization of Medical Devices, No. AAMI RS-P 10/82. 1982.
24. AAMI/ISO 11137-2:2006 Sterilization of health care products—Radiation. Part 2: Establishing the sterilization dose.
25. Kowalski JB, Tallentire A. Substantiation of 25 kGy as a sterilization dose—a rational approach to establishing verification dose. Radiat Phys Chem 1999; 54:55–64.
26. Enzinger RM. Sterility assurance from post-filling heat treatment. J Parenter Sci Technol 1990; 44(6):294–295.
27. Akers J, Agalloco J. Comments on the calculation of sterility assurance levels. J Parenter Sci Technol 1990; 44(6):293.
28. Sharp J. Apples, oranges and additive assurance. J Parenter Sci Technol 1991; 45(3):122–123.
29. Agalloco J, Akers J. Letter to the editor—re: apples, oranges and additive assurance. J Parenter Sci Technol 1992; 46(1):2–3.
30. FDA. Pharmaceutical CGMPs for the twenty-first century—a risk-based approach, 2004.
31. FDA. Guideline on Sterile Drug Products Produced by Aseptic Processing, 2004.
32. PQRI. Points to consider for aseptic processing. PDA J Pharm Sci Technol 2003; 57(2 suppl)1–72.
33. PIC/S. Decision Trees for the Selection of Sterilization Methods (CPMP/QWP/155/96), 1999.
34. FDA. Proposed Current Good Manufacturing Practices in the Manufacture, Processing, Packing or Holding of Large Volume Parenterals, Federal Register 22202-22219, June 1, 1976, Rescinded—December 31, 1993.
35. PDA. PDA's response to FDA's proposed rule: use of aseptic processing and terminal sterilization in the preparation of sterile pharmaceuticals for human and veterinary use. J Parenter Sci Technol 1992; 46(3):65–68.
36. Akers M, Larrimore D, Guazzo DM, eds. Parenteral Quality Control, Sterility, Pyrogen, Particulate and Package Integrity Testing. 3rd ed. New York: Informa USA, 2002.
37. WHO. Annex 4, General Requirements for the Sterility of Biological Substances. Part A, Section 2, 1973.
38. FDA. Sec. 460.800 Parametric Release—Terminally Heat Sterilized Drug Products (CPG 7132a.13), 1987.
39. PDA. Technical Report No. 30. Parametric release of pharmaceuticals terminally sterilized by moist heat. PDA J Pharm Sci Technol 1999; 53(4):217–222.
40. The European Agency for the Evaluation of Medicinal Product (EMEA); Committee for Proprietary Medicinal Products: Note for Guidance on Parametric Release, CPMP/QWP/3015/99, 2001.
41. EMEA. Annex 17, Parametric Release, 2001.
42. FDA. Guidance for Industry Submission of Documentation in Applications for Parametric Release of Human and Veterinary Drug Products Terminally Sterilized by Moist Heat Processes, 2008.
43. Sasaki T. Parametric release for moist heated pharmaceutical products in Japan. PDA J GMP Validation Jpn 2002; 4(1):7–10.
44. Japanese Pharmacopeia. Article 6, 2001.
45. FDA. Guidance for Industry PAT—A Framework for Innovative Pharmaceutical Development, Manufacturing, and Quality Assurance, 2004.

Index